Quantitative Methods for Conservation Biology

Quantitative Methods for Conservation Biology

Springer
New York
Berlin
Heidelberg
Barcelona
Hong Kong
London
Milan
Paris
Singapore
Tokyo

Scott Ferson   Mark Burgman

Editors

Quantitative Methods for Conservation Biology

With 73 Figures

Springer

Scott Ferson
Applied Biomathematics
100 North Country Road
Setauket, NY 11733
USA
scott@ramas.com

Mark Burgman
School of Botany
University of Melbourne
Victoria 3010
Australia
m.burgman@botany.unimelb.edu.au

Cover illustration: See Figure 7.1, page 98.

Library of Congress Cataloging-in-Publication Data
Quantitative methods for conservation biology / edited by Scott
    Ferson, Mark Burgman.
        p. cm.
    Includes bibliographical references.
    ISBN 0-387-95486-4 (softcover : alk. paper)
        1. Conservation biology—Statistical methods. 2. Conservation biology
    —Mathematical models. I. Ferson, S. II. Burgman, Mark A.
    QH75.Q36 2000
    333.95′16′0151—dc21   99-36218

ISBN 0-387-95486-4   Printed on acid-free paper.

First hardcover printing, 2000.

9 8 7 6 5 4 3 2 1   SPIN 10874639

www.springer-ny.com

Springer-Verlag New York Berlin Heidelberg
A member of BertelsmannSpringer Science+Business Media GmbH

Preface

Quantitative methods are needed in conservation biology more than ever as an increasing number of threatened species find their way onto international and national "red lists." Objective evaluation of population decline and extinction probability are required for sound decision making. Yet, as our colleague Selina Heppell points out, population viability analysis and other forms of formal risk assessment are underused in policy formation because of data uncertainty and a lack of standardized methodologies and unambiguous criteria (i.e., "rules of thumb"). Models used in conservation biology range from those that are purely heuristic to some that are highly predictive. Model selection should be dependent on the questions being asked and the data that are available. We need to develop a toolbox of quantitative methods that can help scientists and managers with a wide range of systems and that are subject to varying levels of data uncertainty and environmental variability.

The methods outlined in the following chapters represent many of the tools needed to fill that toolbox. When used in conjunction with adaptive management, they should provide information for improved monitoring, risk assessment, and evaluation of management alternatives.

The first two chapters describe the application of methods for detecting trends and extinctions from sighting data. Presence/absence data are used in general linear and additive models in Chapters 3 and 4 to predict the extinction proneness of birds and to build habitat models for plants. Chapters 5 and 6 explore the application of probabilistic models for decision support in wildlife management. Chapter 7 examines the use of population abundance estimates in making conservation decisions. Chapter 8 describes some approaches to model development to regulate harvesting, with a particular focus on whale population management. Chapter 9 provides a new way of synthesizing qualitative and quantitative information in a form that is designed to support wildlife manager decisions. Chapter 10 describes matrix model approaches to population modeling. Chapters 11 to 13 outline different approaches to building stochastic population models, including frequency-based models, individual-based models, and branching processes. Chapters 14 to 16 provide a review of genetic techniques that have applications in conservation biology and suggestions for the development of laboratory-based protocols to support conservation management. Chapter 17 outlines the strategies that might be used to design a conservation reserve network.

This book is intended for graduate students, working conservation biologists, and others interested in broadening their knowledge of the opportunities available for using numerical and mathematical tools to solve conservation problems. The methods described here are oriented to solving problems for the management and conservation of single species, rather than multiple-species systems or eco-systems. Some are more or less well established (e.g., matrix populations models, structured stochastic simulation models, and regression-based habitat models). In these cases, the authors outline the strengths and limitations of the methods, bringing to the task a broad range of experience and suggestions for applications and improvement. Other methods are perhaps not as widespread. In these, the authors illustrate the utility and potential applications with descriptions of details and example applications. The broadest of these chapters describes methods for designing conservation reserves that use the presence of individual species as the currency for decision making.

The methods cover a broad range from inferences based on simple statistical models, through individual-based dynamic models, to laboratory-based procedures for evaluating genetic considerations. However, the list of methods is not comprehensive. The book does not cover, for instance, the incidence function approach to modeling metapopulation dynamics, management for multiple species, landscape process models, predator prey or host-parasitoid dynamics, physiological models, geographical information systems applications, gap analysis, formal decision theory, or other quantitative methods that have useful applications in conservation biology. Many of these deserve (and have) entire books devoted to them.

We hope this book is of some use in bringing together a relatively broad set of possibilities for the management of individual species, from which a conservation biologist might select the right tool to solve a problem.

We wish to thank Claire Drill for proofreading the document and making the index, Heather Anderson for creating and editing the figures, Anne Findlay for additional proofreading, and Jane Elith for her editorial work. We are grateful to Robin Smith from Springer-Verlag New York, Inc., for seeing this book through to publication.

Setauket, New York, USA   *Scott Ferson*
Victoria, Australia   *Mark Burgman*

Contents

Contributors

Sandy Andelman, National Center for Ecological Analysis and Synthesis, 735 State Street, Suite 300, Santa Barbara, CA 93101, USA

David Andrewartha, School of Forestry, University of Melbourne, Creswick 3363, Australia

Ian Ball, Department of Applied and Molecular Ecology, Waite Campus, The University of Adelaide, PB 1, Glen Osmond, South Australia 5064, Australia

Chris Boek, Department of Computer Science, University of Melbourne, Victoria 3010, Australia

Mark S. Boyce, Department of Biological Sciences, University of Alberta, Edmonton AB T6G 2E9, Canada

Mark Burgman, School of Botany, University of Melbourne, Victoria 3010, Australia

Deborah T. Crouse, Center for Marine Conservation, 1725 DeSales Street NW, Washington, DC 20036, USA

Larry B. Crowder, School of the Environment, Duke University, 135 Duke Marine Lab Road, Beaufort, NC 28516-9721, USA

Donald L. DeAngelis, Oak Ridge National Laboratory, P.O. Box 2008, Oak Ridge, TN 37831-6038, USA

Jane Elith, School of Botany, University of Melbourne, Victoria 3010, Australia

Scott Ferson, Applied Biomathematics, 100 North Country Road, Setauket, NY 11733, USA

Michael J. Firko, Scientific Services, Plant Protection and Quarantine, Animal and Plant Health Inspection Service, U.S. Department of Agriculture, 4700 River Road, Unit 133, Riverdale, MD 20737-1236, USA

Richard. Frankham, Department of Biological Sciences, Macquarie University, New South Wales 2109, Australia

Lev Ginzburg, Department of Ecology and Evolution, State University of New York, Stony Brook, NY 11794, USA

Lloyd Goldwasser, Southwest Fisheries Science Center, NMFS, 3150 Paradise Drive, Tiburon, CA 94920, USA

Frédéric Gosselin, Centre d'Ecologie Fonctionnelle et Evolutive, CNRS 1919, route de Mende 34 293, Montpellier cedex 5, France

Thomas Helser, National Marine Fisheries Service, Northeast Fisheries Science Center, Woods Hole, MA 02543, USA

Selina S. Heppell, School of the Environment, Duke University, 135 Duke Marine Lab Road, Beaufort, NC 28516-9721, USA

Paul T. Jacobson, Langhei Ecology, LLC, 14820 View Way Court, Glenelg, MD, 21737, USA

Marie R. Keatley, School of Forestry, University of Melbourne, Creswick 3363, Australia

Jean-Dominique Lebreton, Centre d'Ecologie Fonctionnelle et Evolutive, CNRS 1919, route de Mende 34 293, Montpellier cedex 5, France

Danny C. Lee, USDA Forest Service, Pacific Southwest Research Station, 801 I Street, Room 419, Sacramento, CA 95814, USA

Sabine S. Loew, Behavior, Ecology, Evolution and Systematics Section, Department of Biological Sciences, Illinois State University, Normal, IL 61790-4120, USA

Bruce R. Maslin, Department of Conservation and Land Management, Locked Bag 104, Bentley Delivery Centre, Western Australia 6983, Australia

Yiannis G. Matsinos, Graduate Program in Ecology, University of Tennessee, Knoxville, TN 37996-1300, USA

Michael McCarthy, Centre for Resource and Environmental Studies, Australian National University, Canberra 0200, Australia

Brendan Moyle, Massey University, Auckland, New Zealand

Edward V. Podleckis, Scientific Services, Plant Protection and Quarantine, Animal and Plant Health Inspection Service, U.S. Department of Agriculture, 4700 River Road, Unit 133, Riverdale, MD 20737-1236, USA

Hugh Possingham, Department of Applied and Molecular Ecology, Waite Campus, The University of Adelaide, PB 1, Glen Osmond, South Australia 5064, Australia

Andrew Solow, Woods Hole Oceanographic Institution, Woods Hole, MA 02543, USA

Yoshinari Tanaka, Laboratory of Theoretical Ecology, Institute of Environmental Science and Technology, Yokohama National University, Tokiwadai 79-7, Hodogaya, Yokohama 240-8501, Japan

Barbara L. Taylor, Southwest Fisheries Science Center, P.O. Box 271, La Jolla, CA 92038-0271, USA

Paul R. Wade, National Marine Mammal Laboratory, 7600 Sand Point Way N.E., Seattle, WA 98115, USA

Wilfried F. Wolff, Institut für Biotechnologie 3, Forschüngszentrum Jülich, Jülich D-52425, Germany

1
Detecting Extinction in Sighting Data

Andrew Solow and Thomas Helser

Introduction

Extinction is probably the most dramatic, if not necessarily the most important, form of change in a biological community. There is currently great concern that human activities are directly or indirectly contributing to increased extinction risks for a wide variety of plant and animal species (Diamond 1989). Extinctions are rarely observed directly, and there is considerable uncertainty not only about the overall rate of extinction (Smith et al. 1993) but also about the extinction of individual species whose existence is known only through chance sightings.

This chapter summarizes some recent work on statistical methods for detecting extinction from sighting data (Solow 1993a,b). The basic idea underlying these methods is that it should be possible to base inference about the extinction of a species on the time since the most recent sighting. Loosely speaking, the question addressed in this work is, how long must a species go unsighted before it is reasonable to conclude that it is extinct? The answer to this question depends on a number of factors. One important factor is the way in which the observation effort varies through time. For example, even a short period since the most recent sighting may be significant evidence of extinction if the recent observation effort has been high. Conversely, even a long period since the most recent sighting may not provide significant evidence of extinction if the observation effort has been nil or inefficient. Although it is possible to incorporate varying observation effort into the methods discussed in this chapter, we assume that it is approximately constant over the observation period. This would be the case, for example, for a fixed observation program replicated over time. It would also be the case if sightings were purely accidental, such as opportunistic collections in museums and herbariums.

A second important factor influencing the significance of the time since the most recent sighting is variation in the population size of the species. We consider two cases. In the first, the pre-extinction population is assumed to remain approximately constant for the period during which it was observed. This model would be appropriate for a chronically small population subject to relatively rapid extinction due, for example, to the loss of critical habitat. Incidentally, the same model would be appropriate if there were no dependence of the sighting rate on population size. In the second case, population size is assumed to decline exponentially prior to extinction. This model would be appropriate for a species experiencing a negative net rate of population growth (e.g., due to overharvesting).

Testing for Extinction in a Stable Population

Suppose that during the period of observation $(0, T)$ sightings occur at the set of ordered times $t_1 < t_2 \ldots < t_n$. A word is in order about the definition of the observation period. If the observation program has a clearly defined beginning and a clearly defined end, then these will define the beginning and end of the observation period. In many cases, it is difficult to define the beginning of the observation program because historical data usually only record the presence of a species and neglect to record absences. Indeed, it may be difficult to define the observation program itself. In that case, the beginning of the observation period should be taken to coincide with the initial sighting (in which case, the initial sighting is dropped from the sighting record) and the end of the observation period should be taken to coincide with the present time.

In statistical terminology, the sighting record is said to arise from a point process (Cox and Lewis 1978). A point process is characterized by a rate function (i.e., the interval between any pair of consecutive events follows a common probability distribution), which gives the expected number of events (in this case, sightings) in a unit time interval. In the stationary case, the rate function does not depend on time. The first model that we consider is that the sighting record follows a stationary Poisson process with rate function

$$\lambda(t) = \lambda \qquad 0 \leq t \leq T_E$$
$$0 \qquad t > T_E \tag{1.1}$$

where both the pre-extinction sighting rate λ and the extinction time T_E are unknown. Properties of the Poisson process are described in Taylor and Karlin (1984). As noted, the stationary model is appropriate for cases in which pre-extinction population size is stable.

Under this formulation, interest centers on testing the null hypothesis that extinction has not occurred—that is, $H_0: T_E = T$ (or, equivalently, $T_E \geq T$)—against the alternative hypothesis that it has—that is, $H_1: T_E < T$. Let the random variable T_i be time of the ith sighting (i.e., the random variable of which t_i is a realization). A natural statistic for testing H_0 against H_1 is the time of the most recent sighting, T_n, with H_0 being rejected if T_n is small (or, equivalently, if the time since the most recent sighting, $T - T_n$, is large). Formally, H_0 is rejected at significance level α if $T_n < c(\alpha)$ where the critical value $c(\alpha)$ is chosen to satisfy

$$\text{prob}_0[T_n < c(\alpha)] = \alpha \tag{1.2}$$

with the subscript 0 denoting that the probability is calculated under H_0.

Under H_0, the sighting record is a realization of a stationary Poisson process on the interval $(0, T)$. To choose the critical value $c(\alpha)$, it is useful to exploit the following property of the stationary Poisson process. Conditional on their number n, the sightings are independently, uniformly distributed over the interval $(0, T)$. Thus, under H_0, the time T_n of the most recent sighting has the same distribution as the maximum of n independent random variables uniformly distributed over the interval $(0, T)$. It follows from the distribution theory of the sample maximum (David 1981) that the significance level of the observed value of T_n (or p value) is

$$\text{prob}_0(T_n \le t_n) = (t_n/T)^n \tag{1.3}$$

Alternatively, the critical value for testing at significance level α is

$$c(\alpha) = \alpha^{1/n} T \tag{1.4}$$

The same argument can be used to find the power of this test. Under the alternative hypothesis, the n sightings are independently, uniformly distributed over the interval $(0, T_E)$ with $T_E < T$. The power of the α-level test is given by

$$\text{prob}_1[T_n < c(\alpha)] = 1 \qquad 0 \le T_E \le \alpha^{1/n} T$$
$$\alpha (T / T_E)^n \qquad \alpha^{1/n} T < T_E \tag{1.5}$$

with the subscript 1 denoting that the probability is calculated under the alternative hypothesis H_1. For example, if $n = 10$ and $\alpha = 0.05$, H_0 is sure to be rejected if $T_E/T \le 0.74$ (if extinction occurs earlier than 74% of the way through the observation period).

Testing for Extinction in a Declining Population

The test described in the previous section assumes that the pre-extinction sighting rate is approximately constant. If the pre-extinction population size is declining, then to the extent that sighting rate depends on population size, this test will tend to give spuriously significant results. The reason for this is that, even under the null hypothesis that extinction has not occurred, the sightings will tend to be concentrated in the earlier part of the record. In particular, the time of the most recent sighting will not represent the maximum of n independent random variables uniformly distributed over $(0, T)$.

In this section, we describe a test for extinction that can be used when the pre-extinction sighting rate declines. Specifically, suppose that the sighting record follows a nonstationary Poisson process with rate function

$$\lambda(t) = \exp(\beta_0 + \beta_1 t) \qquad 0 \le t \le T_E$$
$$0 \qquad t > T_E \tag{1.6}$$

with $\beta_1 \le 0$. As before, the sighting rate parameters β_0 and β_1 and the extinction time T_E are unknown, and interest centers on testing the null hypothesis $H_0: T_E = T$ (or, equivalently, $T_E \ge T$) against the alternative hypothesis $H_1: T_E < T$.

The particular choice of the form of $\lambda(t)$ in Equation (1.6) is to some extent arbitrary. It is consistent, however, with a model in which population size follows Brownian motion with drift parameter β_1 (Lande and Orzack 1988) and the sighting rate is proportional to population size.

The following theoretical development is necessarily a little complicated, although the test itself is easy to apply. It can be shown that, conditional on n, the sighting times represent an ordered sample from the exponential distribution with mean β_1^{-1} truncated on the right at T_E (Cox and Lewis 1978). The probability density function of this distribution, which does not depend on β_0, is

$$f(t) = \beta_1 \exp(-\beta_1\, t)/[1 - \exp(-\beta_1\, T_E)] \qquad 0 \le t \le T_E \qquad (1.7)$$

Let $S = \Sigma\, T_1$ with realized value s. The uniformly most powerful unbiased test of H_0 against H_1 at significance level α is to reject H_0 if $T_n < c'(\alpha)$ where the critical value $c'(\alpha)$ is chosen to satisfy

$$\mathrm{prob}_0[T_n < c'(\alpha) \mid S = s] = \alpha \qquad (1.8)$$

(Beg 1982). The statistic S is sufficient for β_1 under H_0. It follows that the conditional distribution of T_n given $S = s$ is the same as the largest gap in $n - 1$ points independently uniformly distributed over the interval $(0, s)$ such that the largest gap does not exceed T. Finally, from the results of Fisher (1929) on the distribution of the largest gap, it follows that

$$\mathrm{prob}_0(T_n \le t_n \mid S = s) = F_s(t_n)/F_s(T) \qquad (1.9)$$

where F_s is defined by Solow (1993b). The significance level of the observed value of T_n is given by Equation (1.9), and H_0 is rejected at significance level α if the p value is less than α.

The conditional power of this test can be found by noting that, under H_1, the conditional distribution of T_n given $S = s$ is the same as that of the largest gap in $n - 1$ points independently uniformly distributed over the interval $(0, s)$ such that the largest gap does not exceed T_E. Thus, the test has conditional power 1 if $0 \le T_E \le c'(\alpha)$ and $\alpha F_s(T)/F_s(T_E)$ if $c'(\alpha) < T_E \le s$. Solow (1993b) presented some simulation results indicating the conditions under which the unconditional power of this test is reasonably high.

Mark Burgman (personal communication) pointed out that, for t_n fixed, the significance level in Equation (1.9) does not fall to zero as T approaches ∞ but is equal to $F_s(t_n)$ for $T \ge s$. However, this is the conditional significance level given $S = s$. Because the distribution of S under H_0 depends on T (specifically, S tends to be larger for larger T), the unconditional probability that $T_n \le t_n$—which is the unconditional significance level—does fall to zero as T approaches ∞.

Illustrative Examples

In this section, the tests outlined above are illustrated by using sighting data for two northwest Atlantic fish species. These data were taken from the annual autumn bottom trawl survey of Georges Bank conducted by the Northeast Fisheries Science Center (Clark 1979). These examples are for illustrative purposes only. It is important to note that, because of the limited nature of the survey, in these applications extinction refers to *local* species loss (strictly, extirpation during the autumn season).

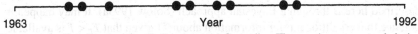

1963   Year   1992

FIGURE 1.1. Sighting data for the Long-nosed Grenadier. The species was recorded nine times between 1967 and 1984 (see text).

The first application is to the Long-nosed Grenadier (*Coelorhynchus carminatus*). In this case, the sighting record began in 1963 and terminated in 1992, so the observation period was taken to be (0, 29). During the observation period, this species was sighted in 1967, 1968, 1970, 1975, 1976, 1978, 1979, 1983, and 1984, so that $n = 9$. This sighting record is illustrated in Figure 1.1. There is no evidence of a declining sighting rate, so the test outlined above was applied. In terms of the rescaled observation period, the value of t_n is 21. The corresponding p value is about $(21/29)^9 = 0.055$, so that there is moderately strong evidence against the null hypothesis.

The second application is to the Snakeblenny (*Lumpenus lumpretaeformis*). In this case, the sighting record began in 1969 and again ended in 1992, so the observation period was taken to be (0, 23). During this period, this species was sighted in 1970, 1971, 1972, 1973, 1974, 1975, 1977, 1978, 1980, 1981, and 1986, so that $n = 11$. This sighting record is illustrated in Figure 1.2. If the test for extinction in a stable population outlined above is again applied, the p value is about 0.036, which again represents moderately strong evidence against the null hypothesis. Note, however, that there is some indication of a declining sighting rate (e.g., the intervals between sightings appear to be increasing on average). To account for this possibility, the test for extinction in a declining population outlined above was also applied to this sighting record. In this case, the p value is about 0.348, which constitutes no real evidence against the null hypothesis.

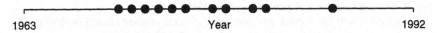

1963   Year   1992

FIGURE 1.2. Sighting data for the Snakeblenny. The species was recorded 11 times between 1970 and 1986 (see text).

Discussion

The methods outlined above for formal inference about extinction from sighting data make minimal use of biological information. If such information is available, then its use may strengthen inference about extinction. For some species, however, little is known (at least about population dynamics) beyond what is available in the sighting record. In such cases, it may be necessary to fall back on empirical methods such as those described here. Of course, these methods are based on certain assumptions (which, to some extent, can be checked). Substantial violations of these assumptions may lead to problems in inference.

One way in which auxiliary information can be incorporated into the methods described here is through a Bayesian approach (Solow 1993a). It may happen in practice that conditional prior information about T_E given that $T_E < T$ is available. For example, critical habitat may be disturbed or lost as a result of a hurricane. Alternatively, more effective fishing technology may be introduced. Information of this kind can be converted into a conditional prior distribution for T_E and used in a Bayesian test of H_0 against H_1.

In applying the methods described here, it is important to think carefully about the meaning of extinction. For example, in the applications described in the previous section, due to the limited nature of the observation program, conclusions are limited both spatially and seasonally. Moreover, in this example, extinction may be related to temporary environmental changes (e.g., related to ocean circulation), and the species may return in the future.

There are several possible extensions of this work. For example, Solow (1996) considered the problem of testing for a common extinction time for two or more species that are known to be extinct. This would be of interest in determining whether the extinctions had a common cause.

Acknowledgments. The helpful comments of two anonymous reviewers are gratefully acknowledged.

Literature Cited

Beg MA (1982) Optimal tests and estimators for truncated exponential families. Metrika 29:103–113

Clark SH (1979) Application of bottom trawl survey data to fish stock assessment. Fisheries 4:9–15

Cox DR, Lewis PAW (1978) The statistical analysis of series of events. Chapman and Hall, London

David HA (1981) Order statistics. Wiley, New York

Diamond JM (1989) The present, past and future of human-caused extinctions. Philosophical Transactions of the Royal Society of London B 325:469–477

Fisher RA (1929) Tests of significance in harmonic analysis. Proceedings of the Royal Society Series A 125:54–59

Lande R, Orzack SH (1988) Extinction dynamics of age-structured populations in a fluctuating environment. Proceedings of the National Academy of Sciences 85:7418–7421

Smith FDM, May RM, Pellew R, Johnson TH, Walter KR (1993) How much do we know about the current extinction rate? Trends in Ecology and Evolution 8:375–378

Solow AR (1993a) Inferring extinction from sighting data. Ecology 74:962–964

Solow AR (1993b) Inferring extinction in a declining population. Journal of Mathematical Biology 32:79–82

Solow AR (1996) A test for a common upper endpoint in fossil taxa. Paleobiology 22:406–410

Taylor HM, Karlin S (1984) An introduction to stochastic modeling. Academic Press, London

2

Inferring Threat from Scientific Collections: Power Tests and an Application to Western Australian *Acacia* Species

Mark Burgman, Bruce R. Maslin, David Andrewartha,
Marie R. Keatley, Chris Boek, and Michael McCarthy

Introduction

The classification of threat by the World Conservation Union (IUCN 1994) is used widely in conservation biology, particularly to assist in developing priorities for management. Originally, the classification was based on subjective estimates and qualitative classes of threat. Mace and Lande (1991) specified quantitative thresholds for risks and time horizons for categories of threat. These criteria were extended in the IUCN (1994) revised criteria. They may be static (e.g., population size fewer than 250 mature individuals), retrospective (e.g., an observed decline of 50% in the past 10 years), or prospective (e.g., probability of extinction of 20% within the next 20 years), and may include the absolute population size and distribution, the size of the adult population, or changes in population size or range. Quantitative criteria such as trends in range or abundance may be costly to estimate because they involve collection of field information at repeated intervals over time, parameter estimation, and predictive modeling.

When dealing with rare species, it may be necessary to make inferences about the decline or extinction of a species on the basis of a handful of collections or opportunistic field observations (Solow and Helser, this volume). If the abundance of a species declines or its range contracts through time, then this may be represented by less frequent collections and relatively long periods during which the species is not observed or collected. Often, these data will be the only information available to establish conservation priorities.

The first developments of formulas for inferring extinction from observation data were proposed by Solow (1993; see Solow and Helser, this volume). Burgman et al. (1995) explored the use of some more general tests in these circumstances. McCarthy (1998) extended Solow's test to include variation in the intensity of the observation process. Deviations from the assumptions of the tests will occur if species ranges or abundances decline through time. The tests are designed to detect such changes. However, it is difficult to know how sensitive the equations will be to changes in the underlying abundance of a species, even if all the assumptions of the methods are met. The use of these statistical formulas in various circumstances may be measured by their statistical power (Peterman 1990).

Western Australia harbors approximately 600 species of *Acacia* and approximately 10,000 species of vascular plants (these estimates include formally described species and an estimate of the numbers of undescribed species). Many of these species are endemic to the state, particularly the southwest region. Perhaps as many as 20% may be threatened by a number of processes including agricultural land clearance, changed fire regimes, and soil-water–borne pathogens (particularly *Phytophthora cinnamomi*).

Western Australia has legislation that affords special protection to threatened flora (and fauna). The state agency responsible for administering the Wildlife Conservation Act, the Department of Conservation and Land Management, maintains a register of formally gazetted threatened flora, declared rare flora, together with an informal list of priority flora taxa. Declared rare flora and priority flora taxa are referred to here collectively as "conservation taxa." The priority flora list includes taxa for which there may be some reason for conservation concern; many of these taxa are poorly known and require further study to assess accurately their conservation status and, thus, consideration for being afforded the special protection of a declared rare flora taxon. The priority flora list therefore provides a means of setting conservation and research priorities in Western Australia for taxa being considered for declaration as rare flora. It comprises four categories:

- Category P1: taxa that are known from one or a few (generally fewer than five) populations that are under threat
- Category P2: taxa that are known from one or a few populations, at least some of which are not believed to be under immediate threat
- Category P3: taxa that are known from several populations, at least some of which are not believed to be under immediate threat
- Category P4: taxa that are considered to have been adequately surveyed and that, although being rare in Australia, are not currently threatened by any identifiable factors (K. Atkins, personal communication)

The purposes of the present study are to undertake power tests of formulas used to infer threat from data of the kind found in herbaria and museums and to identify patterns that may indicate trends in population size or range by testing collections of 192 *Acacia* conservation taxa from Western Australia. The results of these tests will help to clarify the capabilities and limitations of these tools.

Methods

Description of the Equations

The equation by Solow (1993) was designed to estimate the probability of extinction from sighting data. Burgman et al. (1995) wrote the equation in an analogous, discrete-time form (assuming a Poisson process with a fixed chance of observation) including the number (frequency) of sightings. In this form, Solow's test is

$$p = \left(\frac{C_e}{C_T}\right)^n \tag{2.1}$$

and it gives the likelihood that a species has not become extinct since the last observation.

The time period $(0, T)$ is partitioned into C_T equally sized units of time (months or years). C_e is the number of time intervals between the start of observations and the last collection. The total number of observations for the species is n.

Grimson et al. (1992) developed a test for the longest run of empty cells in a time series composed of counts. It is a generalization of Bradley's (1968) runs test, which calculates the probability of the longest run of empty cells (time cells with no observations) in the sequence of cells, C_T. Grimson's test assumes that the occurrences of observations (counts) are independent. The probability of a run of empty cells as long or longer than the longest run of observed consecutive absences is

$$p = C_T^{-n} \sum_{\substack{i \geq 1 \\ k \geq 1}} (-1)^{k+1} \binom{j+1}{k} (C_T - rk)_i S(n, j) \tag{2.2}$$

where $(\ )_j$ is a falling factorial (for an integer, a, the falling factorial $(a)_j = a(a-1)$ ... $(a-j+1)$, and $S(\ )$ is a Stirling number of the second kind for which Grimson et al. (1992) provide a table of values for small values of n and j. The term r is the maximum number of consecutive empty units of time within the sequence of C_T units of time, and n is the total number of cases arising in the C_T units.

Detecting systematic changes in natural populations is a type of trend analysis. McCarthy (1998) suggested that changes in abundance (represented by the number of collections of a species per unit time) through time may be detected more efficiently by incorporating an index of collection effort such as the total number of records of, say, all plant species in the relevant database. He rewrote Solow's (1993) equation as

$$p = \left(\frac{\sum_{i=1}^{C_e} e_i}{\sum_{i=1}^{C_T} e_i}\right)^n \tag{2.3}$$

where e_i is the index of collection effort each time step, C_e, and C_T and n are defined above. When the collection effort is constant, the equation reduces to Solow's (1993) equation, and McCarthy (1998) called it the partial Solow equation, because collection effort is treated as a covariate. Solow's equation determines the probability of the run of empty cells occurring at the end of a sequence, giving the probability that a taxon still exists. The equation is a special case of Grimson's equation, with the run of empty cells restricted to the end of the sequence. Solow's equation results in relatively large p values (i.e., small probabilities of extinction) if a single observation of a species was made relatively recently, irrespective of any evidence of population decline through the observation period. Grimson's equation is sensitive to patterns of observations other than the time since the last observation.

Simulations

Power tests for these equations involve setting up a scenario in which the "true" circumstances concerning a species' decline or loss are known. Then, the scenario is sampled repeatedly, and the equation to be tested is applied to the resulting data. Power is measured by the probability that an equation results in a significant test whenever there is a real difference.

There is some probability of observing an extant species in each time step. The applications described above assume implicitly that this probability is a function of the underlying population size of the species. If the underlying scenario involves no real change in population size, the equations should produce, on average, a significant result fewer than once in 20 replications (i.e., the type I error rate is less than 5%). If the underlying scenario involves a real change in the population size, the best method will be the one that is most likely to produce a significant result. The type II error rate is the probability of concluding the patterns are random, when, in fact, there is an underlying change. A more powerful method is more likely to detect a true change. In these examples, we assumed that survey intensity remained constant throughout the period, so that McCarthy's (1998) equation reduced to Solow's (1993) equation, and the results from the two equations were identical.

Three scenarios were developed. In the first, population size did not change for the duration of the observation period. The mean number of observations per time step was set at one of five levels (i.e., 0.05, 0.1, 0.2, 0.5, and 1.0 observations per unit time). The frequency of observations in each cell (time step) was sampled from a Poisson distribution, and each level of the mean number of observations per time step was replicated 1,000 times. This first scenario was used as a control to test the α level (the type I error rate) of the formulas.

In the second scenario, the underlying population size declined linearly. The rate of decline was set at one of four levels: 0.5, 1, 2, and 5%. The mean number of observations per time step declined from a maximum of one in each case (Fig. 2.1). The third scenario involved a stepped decline at the midpoint of the time period, the 50th time cell. The magnitude of the decline was set at one of five levels: 20, 50, 70, 90, and 100%. As in the first two scenarios, the mean number of observations per time step declined from a maximum of one per time step in each case (Fig. 2.2). Each combination of decline and frequency of observation for each of the scenarios was replicated 1,000 times.

Analyses of the Acacia Data

The information associated with collections of all Western Australian conservation Acacia taxa toward the end of 1995 was extracted from the specimen database

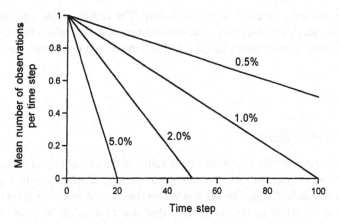

FIGURE 2.1. Four rates of decline of a population over a period of 100 time steps and the associated mean number of observations per period.

of the Western Australian Herbarium. From this, the date of collection, collector, and determination of each specimen were recorded. The subset of the specimen data containing date and collector information was scanned to locate records of the same taxon, made by the same collector within the same month. These records were considered not to be independent of one another and in each instance only one of each set was retained for the analyses.

Probabilities were calculated by using Solow's and Grimson's tests for each taxon based on collection information and their current taxonomic status. Partial correlations were also calculated between the time of collection and the number of collections per year for each species, using the number of conservation *Acacia*

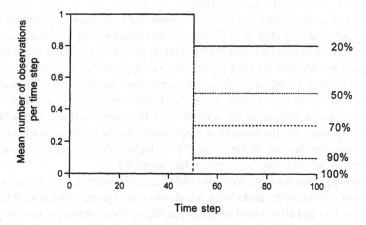

FIGURE 2.2. Five different magnitudes of a stepped decline in underlying population size at the halfway point of a period of 100 time steps.

species collected in each year as a covariate. The results of the analyses were compared with current listings of conservation taxa and were interpreted by one of us (BM) familiar with their abundance, distribution, and taxonomic status.

Results

Type I Error Rates

The type I error rate for a formula (the α value) is the probability of rejecting the null hypothesis that the observations are randomly distributed when the null hypothesis is, in fact, true. Convention has set the acceptance criterion for the type I error rate at 0.05, or the probability that one case in 20 or fewer will be considered significant when there is no true significance. The formulas are well specified for the convention of $\alpha = 0.05$. For the scenario in which there was no change in the underlying population, the equations returned a type I error rate that was equal to or less than 0.05, for all levels of the mean number of observations per time period.

Statistical Power

The power of a statistic is the probability of detecting a change when there is a change present (i.e., the probability of rejecting the null hypothesis when it is false). Statistical power equals $(1 - \beta)$, where β is the type II error rate, the probability of accepting the null hypothesis when it is false (i.e., the probability of failing to detect a change when a change has occurred). The statistical power of a formula in this study was equivalent to the proportion of replications detected as being significantly different from random in each of the scenarios that involve a continuous or a stepped decline. Power will generally increase with increasing sample size (Siegal and Castellan 1988).

Neither of the equations was effective at detecting continuous declines for rates of 0.5% per time step (Fig. 2.3). There was no corresponding increase in power with an increase in mean number of observations per time step. When the underlying population declined by 1% per time step, the equations were likely to detect a change if the mean number of observations per time step was greater than about 0.25 for the period of 100 time steps. At decline rates of 1% and higher, an increase in mean number of observations per time step corresponded with an increase in power. The likelihood that a true change would be detected approached 100% when the rate of decline was 2% or higher and the mean number of observations per time step was greater than about 0.5.

Solow's equation was the most powerful in circumstances in which there was a continuous decline in the underlying population size. This conclusion held for all rates of decline and all levels of the mean number of observations per time step. In most cases, most of the data generated by random sampling resulted in single observations within time cells. The relative power of Grimson's equation may

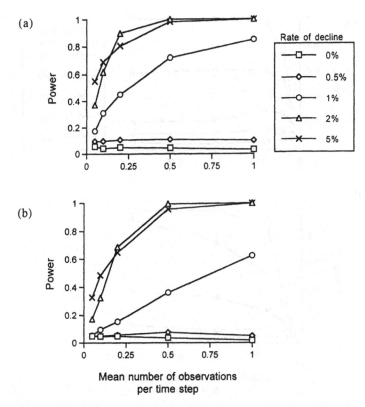

FIGURE 2.3. Power curves for three equations for a scenario in which the underlying population declined linearly at 0.5, 1, 2, and 5% per time step. (a) Solow's equation; (b) Grimson's equation.

improve in circumstances in which cells with more than a single observation are relatively common.

The general shape of the stepped-decline function curves was similar to those generated by the continuous-decline scenario (Fig. 2.4). The equations performed better when the mean number of observations increased and when the change in the status of the underlying population was more marked. A 100% reduction in population size halfway through a 100-time-step sequence was much easier to detect than was a 20% reduction.

Neither of the equations was effective at detecting stepped declines of less than about 50%. There was little corresponding increase in power with an increase in mean number of observations per time step at these levels of decline. When the underlying population declined by more than 50% at the halfway point, the equations were likely to detect a change if the mean number of observations per time step was greater than about 0.25. The likelihood that a true change would be detected approached 100% when the magnitude of the decline approached 100% and the mean number of observations per time step was greater than about 0.25.

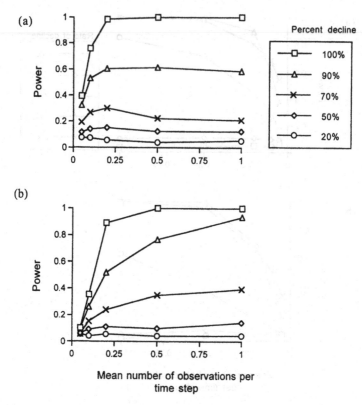

FIGURE 2.4. Power curves for three equations for a scenario in which the underlying population declined by 20, 50, 70, 90, or 100% halfway through the 100-time-step sequence. (a) Solow's equation; (b) Grimson's equation.

Solow's equation was the most powerful in circumstances in which there were relatively few observations per time step. However, the power of these tests was reversed when observations were relatively frequent. Grimson's test was more powerful when there were more than about 0.5 observations per time step and when the magnitude of a stepped decline was greater than about 50%.

Acacia *Analyses*

A steady increase in collection frequency is readily apparent in data on Western Australian conservation taxa of *Acacia* (Fig. 2.5). The total number of collections in each year is an index of collection effort in that year, and it was very closely correlated with other statistics such as the number of different species collected per year (although these correlations are not shown here). The numbers of records of conservation taxa of *Acacia* in each year (Fig. 2.5) were used as an index of collection effort when assessing the pattern of collection for each *Acacia* species.

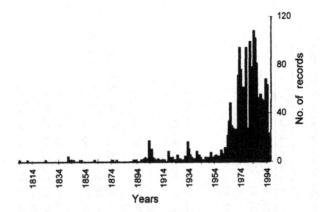

FIGURE 2.5. Number of collections made each year of all *Acacia* species held by the Western Australian Herbarium listed on the declared rare flora list and the priority species list.

The 41 taxa in Table 2.1 represent those for which Solow's test had a value of less than 0.01, Grimson's tests had a value of less than 0.01 and the run of absences was in the latter half of the time series for the species, or the partial correlation test (McCarthy 1998) had a value of less than 0.05. The choice of these probability levels was somewhat arbitrary. These 41 taxa represent more than 20% of the conservation taxa of *Acacia* from Western Australia. Collection patterns for 23 of these taxa considered to be at greatest risk were examined in detail by one of us (BM; see Appendix). In nine cases, after reexamination of the collections and other relevant information, it was concluded that the taxon was threatened or that additional surveys were warranted. They included *A. aprica, A. auratiflora, A. depressa, A. incaena* subsp. *conformis, A. kingiana, A. manipularis, A. megacephala, A. microneura,* and *A. prismifolia.*

If collections of Western Australian conservation taxa of *Acacia* were purely opportunistic and were a simple random sample of taxa, they would reflect the actual distribution and abundance of the species from which they were sampled. Changes in distribution and abundance would be reflected in changes in the frequency of collection. However, there will be some apparently unusual patterns of collections that are due to chance alone, even if the underlying abundance and distribution are stable. Thus, by chance, in circumstances in which there were no deterministic changes in the distribution and abundance of any species, we could expect about ten species of the 192 conservation taxa to have probabilities of less than 5%, and we could expect the distribution of probabilities to be approximately uniform. Interestingly, there was a preponderance of values less than 0.1 for all the statistics (Fig. 2.6).

One of the most outstanding features of these results is that when we examined the pre-1970 taxonomy of the 192 *Acacia* conservation taxa in this study, we found that 44 of them were neither formally described nor even informally

TABLE 2.1. Collection data for those species for which the frequencies of collection are significantly nonrandom on the basis of the equations applied here[a].

Species	Solow	Grimson	Partial	Class[b]	Collection dates[c]
A. ancistrophylla var. perarcuata	.483	.011	.003	P3	37, 39, 68, 70, 71, 71, 73, 76, 79, 89
A. aprica	.059	.458	.009	P3	57, 62, 67, 70, 73, 73, 76, 82
A. ataxiphylla ssp. ataxiphylla	.256	.072	.044	P3	27, 31, 71, 75, 77
A. auricoma	.010	.085	.052	P3	72, 73, 74, 78, 80, 83, 83, 83
A. auratiflora	.001	.007	.018		68, 72, 73, 74, 74
A. benthamii	.341	.973	.026	P2	19, 41, 65, 75
A. bifaria	.001	.008	.002	P3	62, 64, 70, 71, 71, 74, 75, 76, 79, 80, 80, 80
A. botrydion	.331	.000	.037	P1	03, 73, 73, 74, 75, 76, 79, 80, 80, 80, 85
A. aff. eremophila	.028	.160	.038	P3	74, 75, 79, 79, 80, 84
A. depressa	.065	.451	.005	R	63, 64, 64, 66, 70, 76, 83
A. desertorum var. nudipes	.839	.009	.038	P1	31, 62, 64, 70, 71, 79, 79, 81, 82, 82, 92
A. dissona var. indoloria	.255	.791	.003	P3	32, 33, 47, 64, 79, 79, 82
A. drummondii ssp. affinis	.218	.049	.002	P3	57, 58, 59, 70, 70, 71, 71, 71, 72, 73, 73, 75, 76, 76, 79, 79, 79, 80, 90
A. errabunda	.045	.398	.045	P3	64, 64, 71, 71, 73, 75, 76, 80, 84, 85
A. euthyphylla	1.00	.003	.459	P3	82, 83, 83, 84, 84, 84, 84, 85, 92, 93
A. excentrica	.141	.398	.046	P3	78, 83
A. gemina	.646	.009	.013	P2	38, 64, 70, 71, 80, 80, 82, 82, 82, 87, 91
A. glaucissima	.013	.149	.044	P3	64, 64, 70, 71, 76, 79, 83, 83, 83, 83, 84, 85
A. grisea	.89	.002	.038	P4	16, 64, 70, 70, 70, 75, 76, 92, 92
A. incaena ssp. conformis	.159	.004	.007	P1	16, 31, 65, 71, 73, 76, 79, 82, 82, 82, 84, 84, 84
A. insolita ssp. recurva	.031	.136	.016	P2	84, 86, 86, 86, 88

A. kerryana	.002	.008	.013	P2	80, 80, 81, 81, 83
A. kingiana	.014	.028	.01	E	23
A. lirellata ssp. compressa	.384	.249	.031	P2	32, 33, 45, 68, 71, 79, 84, 86
A. manipularis	.04	.114	.041	P1	59, 65
A. megacephala	.006	.009	.009	P1	32, 56, 61, 63, 64, 64, 67, 67, 70, 70, 70, 71, 71, 72, 76
A. microneura	.554	.426	.003	P1	1841, 01, 01, 72
A. mutabilis ssp. rhynchophylla	.358	.088	.003	P3	28, 32, 64, 70, 71, 72, 84
A. newbeyi	.012	.12	.009	P3	69, 71, 71, 73, 73, 74, 76, 82, 82, 85
A. nigripilosa ssp. latifolia	.348	.164	.01	P1	31, 64, 70, 76, 81
A. obesa	.092	.685	.025	P3	64, 65, 66, 68, 70, 78, 79, 82, 86
A. oncinophylla ssp. oncinophylla	.703	.003	.028	P3	1839, 1899, 01, 01, 16, 16, 19, 23, 39, 42, 61, 68, 76, 77, 79, 84, 84, 90
A. phaeocalyx	.002	.022	.000	P3	62, 66, 68, 68, 69, 70, 70, 71, 71, 73, 75, 76, 76, 79, 82, 83
A. prismifolia	.126	.374	.013	E	01, 33
A. pygmaea	.001	.003	.028	R	77, 80, 80, 80, 80
A. rhamphophylla	.002	.007	.115	P1	80, 82, 82
A. richardsii	.044	.431	.036	P3	74, 74, 76, 78, 79, 80, 80, 81, 81, 82, 82, 85, 89, 89
A. semicircinalis	.326	0	.046	R	03, 62, 70, 71, 74, 75, 76, 79, 79, 80, 80, 82, 82, 86
A. subflexuosa ssp. capillata	.007	.014	.172	P1	82, 82
A. subrigida	.019	.063	.04	P2	79, 81, 82
A. subsessilis	.549	.062	.026	P2	31, 66, 71, 76, 82, 87

[a] Partial correlations were calculated by using the total number of collections of *Acacia* conservation taxa as a covariate.

[b] The classes refer to CALM priority list classes and classification on the declared rare flora list of Western Australia (see text).

[c] Years are from the 20th century, unless otherwise indicated.

(a)

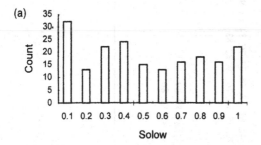

(b)

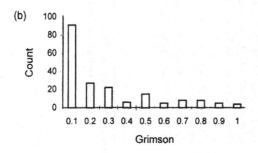

(c)

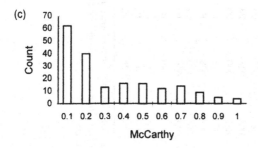

FIGURE 2.6. Frequencies of probabilities generated by (a) Solow's test, (b) Grimson's test, and (c) partial correlations (McCarthy 1998) for the 192 conservation *Acacia* taxa.

recognized. Thus, both science and the relevant management agencies were unaware of their existence until relatively recently.

Discussion

It is not obvious from inspection of the figures and tables, but the different equations do not always identify the same sequences as being significantly different from random. It is therefore possible to obtain a higher detection rate (i.e., better statistical power) by using the different formulas together. The use of several tests in concert is discussed by Grimson et al. (1992; Grimson 1993). It involves the use of complementary tests, so that there is less chance of missing significant changes simply because one test happened to fare poorly on a given data set. This is particularly relevant in the analysis of small data sets, conventionally with fewer than 30 observations. Levels of accuracy may be unsatisfactory or unknown for asymptotic methods applied to sample sizes larger than 30. Combinatorial methods provide "exact" rather than approximate or asymptotic results, and their applications require relatively mild conditions or assumptions about the data (Grimson 1993).

All the methods were poor at detecting relatively small changes in the underlying population. Changes such as a 20% step decline and a 0.5% continuous decline were difficult to detect because of the low probability of there being empty cells in the sequence. Grimson et al. (1992) anticipated this result, suggesting that Grimson's equation should have good power mostly in situations in which there is a rationale for zero occurrences in certain periods of time.

Solow (1993) conducted power tests for Equation (2.1) and found that the test had low power when the extinction time was near the end of the sequence, and when prior to extinction, the population had declined to a level at which the sighting rate was close to zero. The results here suggest that the formulas had low power when the rate of decline was 1% or less or when the magnitude of a step decline was less than about 50%.

Perhaps counter-intuitively, Solow's equation was more likely to detect a change in sequences with a 2% continuous decline than in sequences with a 5% continuous decline, when the mean number of observations per time step was less than 0.2 (Fig. 2.3). This is due to a relationship between the length of a run of zeros at the end of the sequence and the number of observations made before the population became extinct. The scenario involving a 2% continuous decline has a minimum of 50 empty cells at the end of the sequence, and the 5% continuous decline has a minimum of 80 empty cells at the end of the sequence. The number of time cells from the first observation to the last observation is divided by the total sequence length (100 time cells) and is then raised to the power of the number of observations (Equation [2.1]). As it takes longer for the population experiencing a 2% decline to reach extinction than it does for the population experiencing a 5% decline, the value of n will be higher in the scenario involving a 2% decline. The effect of the increased sample size results in it being easier to detect situations with more observations and a shorter run of empty cells.

The power of the formulas to detect the step declines (Fig. 2.4) was slightly higher than their ability to detect the corresponding continuous declines (Fig. 2.3). For example, a stepped decline of 100% may be considered equivalent to a continuous decline of 2% because both result in the loss of the population in the 50th time step. The stepped decline scenario results in more observations before the loss of the population. As a result, all equations are more likely to detect a change in a population that experiences a step decline than in a population that experiences a continuous decline, even if the populations are lost at the same time.

Relevance of Taxonomy

The observation that 44 conservation taxa were not recognized prior to 1970 serves to make the somewhat obvious point that conservation research, and the setting of conservation priorities depends on sound taxonomy, for without it, conservation has nothing to work with. Clearly defined taxa are the result of taxonomic research, and the names applied to the biological entities are the means whereby information about them is exchanged. Thus, in circumstances in which the conservation of species is a priority, the first priority for taxonomy should be straightforward species circumscription and description (alpha taxonomy).

Of the taxa that were described before 1970, 43 were found subsequently to be confounded with at least one other taxon. In these cases, subsequent revision has separated these closely related or similar species and circumscribed the new taxa in their own right. Again somewhat obviously, such changes lead to substantial changes in the evaluation of the conservation status of the taxa. For example, *A. bifaria* was segregated from *A. glaucoptera* in 1995. Before then, specimens of the two taxa were not differentiated. *A. bifaria* was not afforded any protection because *A. glaucoptera* is common and widespread. Similarly, *A. browniana* was composed of taxa currently recognized as *A. browniana, A. grisea, A. lateriticola, A. luteola, A. newbeyi,* and *A. subracemosa,* all of which are listed on the priority flora list. Few of the species currently recognized in the declared rare flora list or the priority flora list would have been guaranteed protection in Western Australia if the taxonomy of the *Acacias* had not progressed beyond that known in 1970.

Efficacy of the Methods

The methods applied here are intended to enhance the ability of managers to set conservation priorities, not to replace existing procedures. The objective of these methods is to highlight unusual patterns in collection frequency and find reasons for them. The relative magnitudes of the probabilities for the taxa are important, not the absolute values. Biological explanations such as range contraction or population decline should be considered only after more mundane explanations such as changes in taxonomic interest (whether or not a taxon is the subject of current taxonomic research) have been discounted. The results of the equations provide a mechanism to stimulate conservation and land management authorities to consider why there are unusual collection patterns for different taxa. If there are no simple explanations in terms of other factors, then managers may have to consider specific surveys to establish the current conservation status of the taxa. For example, of the 192 Western Australian *Acacia* taxa recognized as declared rare and priority flora list taxa, 41 were detected by one or other of the equations applied here as having unusual collection patterns. Of these, 23 were scrutinized carefully, and of these, nine species were suggested to warrant further survey. This does not mean that the nine taxa are on the brink of extinction. Rather, the information resulting from our analyses has been used together with other relevant information to improve the assessment of their conservation status.

The discrepancies between taxa identified by our results and those identified by other means do not indicate a failure of the respective methods. Range size, population size, number of populations, and the rates of change of these parameters are central in setting conservation priorities (IUCN 1994). Population size, the number of populations, and geographic range are static properties of a taxon and are not considered by the equations applied here. Instead, these equations provide some insight into dynamic population attributes in that they integrate changes through time in population size and geographic range. On the basis of district management plans, the Department of Conservation and Land Management (CALM) gave priority for survey to approximately 20 taxa in the Katanning district that were part of the analyses conducted here. CALM's priorities were set on the basis of information from regional and district surveys and the conventional wisdom of ecologists, taxonomists, and collectors. Only one of the 20 species, *A. auratiflora*, was detected in our analyses. Similarly, CALM is systematically surveying all P1 and P2 taxa, a total of 16 *Acacia* taxa, within their south coast region. However, only two of these (*A. microneura* and *A. prismifolia*) were highlighted in our analyses. In total, only four species were highlighted in these results and were recommended independently for detailed survey for consideration as declared rare flora: *A. aprica*, *A. microneura*, *A. prismifolia*, and *A. kingiana*.

The different equations produce different results. None of the equations provides better or worse information than the others. Rather, each is sensitive to different patterns of change and therefore to different patterns of deviation from a simple random process (Burgman et al. 1995). All have a useful role to play in the exploration of collection information.

Collection Patterns

The preponderance of small probabilities resulting from the equations (Fig. 2.6) may suggest that there exists in Western Australia a statewide process (or processes) that is affecting the distribution and abundance of many species. It may also suggest that there are systematic factors at play such as collection protocols and herbarium curatorial procedures that affect the collection patterns of many species in the same way.

Collections are made for many different reasons, including scientific research, biological survey, and curiosity. For example, a taxonomist working on any group is also likely to collect relevant material at the time of working on the taxon. However, once the taxonomist is able to describe adequately the nature of the morphological variation, the taxon is unlikely to be a target of further sustained collection. Future collections by the specialist are likely to include only geographic range extensions, and thus collections may diminish. Plants that are rarely collected are more prone to the biases that result from variations in taxonomic interest. For example, *A. pygmaea* and *A. semicircinalis* are both endemic to the Wongan Hills, and neither has declined in abundance or distribution within the past one or two decades. However, *A. pygmaea* is more rarely collected, and the probabilities resulting from the equations are smaller, reflecting the greater influence of variations in collection intensity due to taxonomic interest. Similarly, if collections are neither taxonomically nor geographically informative, they may be destroyed to save the cost of processing and housing specimens. It is also possible that after a taxon has been formally circumscribed and published, and thus brought to the attention of the wider community, collections of it may increase. Some of the constraints affecting collection patterns include time, money, seasonal conditions, and land tenure changes.

The collections of the Western Australian Herbarium may not be entirely representative of collections elsewhere because CALM has a structured process for monitoring taxa of conservation concern. The process includes legislation to declare and then protect plant taxa, state government processes to maintain a conservation list of both declared and undeclared taxa, and formal survey and monitoring protocols. The interplay of these factors may affect collection patterns and confound the application of the methods described above. For example, once a plant species is declared rare, it is protected by legislation and almost all casual, opportunistic, and scientific collections will cease unless special authority is given by the state. However, immediately prior to being protected, the same taxon is likely to be relatively heavily collected because departmental protocols require that the range and abundance of the declared rare flora taxon be established through extensive surveys. A rash of collections followed by a dearth of collections could be expected for declared rare flora, a pattern independent of any change in conservation status. Other factors affect collecting patterns in an unpredictable way. The observations in Appendix I reveal that the following processes play a part in determining unusual collection patterns in Western Australia's *Acacia* conservation taxa:

- geographic location
- taxonomic uncertainty
- taxonomic interest
- legislative controls
- establishment of monitoring programs
- survey attention (priority list taxa are more likely to be surveyed than unlisted taxa)
- timing of district surveys
- particular interests of individual collectors
- herbarium curatorial procedures (e.g., "uninformative" specimens such as collections from a single location made at different times may be excluded)

Conclusions

Type II errors are made in setting conservation priorities when a conclusion is reached that a taxon is not threatened, when, in fact, it is. The frequency with which type II errors are made depends in part on the quality of the data. Any tool that makes better use of existing information will improve the quality of decision making in setting conservation priorities. The methods applied here are one such tool that provide better tests of the dynamic properties of populations than are currently available.

The best strategy is to use many different methods together. If one uses a set of complementary tests, there is less chance of missing significant changes simply because one test happened to fare poorly on a given data set (Grimson et al. 1992; Grimson 1993). This is particularly relevant in the analysis of small data sets, conventionally with fewer than 30 observations, in which levels of accuracy may be unsatisfactory or unknown for asymptotic methods that are usually applied to sample sizes larger than 30. Paradoxically, it is easier to detect decline and extinction in species that are well represented in a database than in species for which there are few records (Solow 1993).

Other simple rules may be applied to identify taxa that may need protection or management. For example, one may flag all species known from fewer than three collections, or all species not collected for more than 20 years, or all species for which collections may be enclosed by a convex polygon of less than 10 km^2 (IUCN 1994).

Extinctions are difficult to observe and they are a crude measure of impact (Burgman et al. 1995). Conservation biology is more concerned with threatened species, not those already lost. Observational data, particularly those in herbaria and museums, hold valuable information that may be used to make inferences about the status of species. There are two important points to consider in applying the methods described above. The first is that they do not assume that the process of collection has been fixed. Rather, they assume a stationary random process. The second is that they do not replace conventional wisdom or the application of more usual devices to establish conservation status. If they are informative or raise questions in the minds of those who establish conservation priorities, then they will have served their purpose. The application outlined here was successful because several taxa were identified for survey that would not otherwise have attracted priority attention. Explanations were found for most of the very unusual patterns. However, for several taxa there was no plausible explanation, and further surveys may be indicated. In this way, the methods may assist in the setting of conservation priorities. The lesson is that the results of analysis from the equations should not be interpreted in isolation. Rather, they should be used in conjunction with a sound understanding of the day-to-day practices that affect collection rates.

Acknowledgments. The authors gratefully acknowledge Kirsten Howlett, Ken Atkins, David Coates, and Terry Walshe for their thoughtful and valued comments on this project. This work was supported by a small grant from the Australian Research Council and by funding from the Australian Nature Conservation Agency (Environment Australia) administered by Andrew Taplin.

Literature Cited

Bradley JN (1968) Distribution-free statistical tests. Prentice Hall, Englewood Cliffs, New Jersey

Burgman MA, Grimson RC, Ferson S (1995) Inferring threat from scientific collections. Conservation Biology 9:923–928

Grimson RC (1993) Disease clusters, exact distributions of maxima, and *P*-values. Statistics in Medicine 12:1773–1794

Grimson RC, Aldrich TE, Wanzer Drane J (1992) Clustering in sparse data and an analysis of Rhabdomyosarcoma incidence. Statistics in Medicine 11:761–768

IUCN (1994) International Union for the Conservation of Nature, draft red list categories. World Conservation Union, Gland, Switzerland

Mace GM, Lande R (1991) Assessing extinction threats: toward a reevaluation of IUCN threatened species categories. Conservation Biology 5:148–157

McCarthy MA (1998) Identifying declining and threatened species with museum data. Biological Conservation 83:9–17.

Peterman, RM (1990) The importance of reporting statistical power: the forest decline and acidic deposition example. Ecology 71:2024–2027

Siegal S, Castellan NJ Jr (1988) Nonparametric statistics for the behavioral sciences. McGraw-Hill, New York

Solow AR (1993) Inferring extinction from sighting data. Ecology 74:962–964

Appendix: Interpretation of Collection Information

The paragraphs in this appendix provide details of interpretations of collection patterns on 23 conservation taxa highlighted by the analyses.

Acacia ancistrophylla var. *perarcuata* is a relatively unspectacular taxon that has only recently been recognized formally. Most collections were probably random. The variety is unlikely to be at risk, despite the paucity of collections since 1979. Variation in taxonomic interest is the most likely explanation.

A. aprica was known from only two populations, one on a road verge and one in a small reserve, at the time of these analyses. It was the subject of active taxonomic study in the period up to 1982. The absence of collections since 1982 may be because it was not included in the priority flora list until recently and because of the death of an active collector in the region where it occurs. Subsequent to our analysis, an independent survey located three new populations, all in close proximity to the original populations and all on degraded road verges. The continued degradation of road verges in the area where this species occurs raises concerns for the long-term survival of this species (D. Papenfus, personal communication).

A. auratiflora was collected five times between 1968 and 1974. The absence of records since 1974 lends support to the need for further survey.

A. auricoma has distinctive facies, and it is likely to be collected if encountered. The reason for the lack of collections since 1983 may relate to the fact that it has a restricted distribution in the remote ranges of central Australia. Collections may be linked to the timing of expeditions.

A. bifaria is superficially similar to, and only recently segregated from, *A. glaucoptera,* a common and widespread species that is readily recognized by most field workers. There have been no opportunistic collections of *A. bifaria* since 1980 when the taxon was the subject of active taxonomic study, and the absence of collections since 1980 is probably due to changes in taxonomic interest.

A. aff. *eremophila* is attractive and floriferous and is therefore likely to be collected opportunistically if encountered. The lack of collections since 1984 may relate, at least in part, to the fact that the taxon has a patchy distribution in remote areas of the southwestern arid zone. However, most collections appear to have been made opportunistically by a variety of people. Collection frequency does not coincide with regional surveys. Since these analyses were completed, a single specimen of *A.* aff. *eremophila* collected in 1991 and misclassified as *A. eremophila* has been added. The taxon is unlikely to be at risk.

A. depressa is distinctive and well known. Its distribution is very restricted and coincides with farmland in Western Australia. It was gazetted as a declared rare flora taxon in 1980 and is protected by a law that prohibits opportunistic collection, perhaps explaining the absence of collections since a formal survey in 1983. However, the absence of any collections suggests that a survey may be in order to establish its current status.

A. dissona var. *indoloria* is a relatively unspectacular taxon that has only recently been recognized formally. Most previous collections were probably random. It is unlikely that the variety is at risk, despite the paucity of collections since 1982.

A. errabunda is uncommon and is not particularly distinctive; however, it is unlikely to be threatened. At least some of the unusual collection pattern may have been caused by the death of a particularly energetic collector who lived close to the range of the species.

A. incaena ssp. *conformis* is relatively nondescript and is superficially similar to a number of its close relatives. It has only recently been recognized formally as a distinct entity, and most previous collections were probably opportunistic. It was included recently in a district vegetation survey and was flagged as being in need of urgent survey. Our data lend support to that conclusion.

A. kerryana was described in 1982 and last collected in 1983. It occurs in a few scattered populations between Coolgardie and Norseman. It occurs in a remote area that is nevertheless subjected to substantial mining activity. Mining activities are required to account for the species before they proceed. A collection was made in 1994 but was not processed in time for inclusion in these analyses. The taxon probably is not at risk, and the patterns of collection may be explained by its remote location and by variation in taxonomic interest.

A. kingiana is known from a single collection made in 1923 and is presumed extinct.

A. manipularis grows in the remote rugged Kimberley region of northern Western Australia. It has been collected from only two localities and not since 1965. Its current status is uncertain, and these results lend support to further survey.

A. megacephala resembles the widespread and common *A. pulchella*. The species was described in 1972, and it is possible that local wildlife officers are familiar with it and have not bothered to collect it. Several collections were made in 1994 but were not processed in time to be included in these analyses. Nevertheless, it occurs in a degraded roadside population and may be at risk.

A. microneura has been collected only four times since 1841, the most recent in 1972. It was already identified as needing urgent further survey, to establish its current conservation status.

A. mutabilis ssp. *rhynchophylla* was last collected in 1984. Although uncommon, it is not rare and probably is not threatened.

A. newbeyi habitat is restricted to one kind of landscape (laterite hills), and its distribution is restricted. It is relatively nondescript, and the absence of collections in 1985 is probably due mainly to the death of an active collector in the region.

A. phaeocalyx was described in 1978 and was last collected in 1983. It is distinctive (albeit small and spiny) and is not uncommon within its range. It is unlikely to be at risk.

A. prismifolia is known from three collections (two in 1901 and one in 1933), and it is presumed to be extinct.

A. pygmaea is rare and is confined to a single small population in the Wongan Hills of Western Australia. There have been no collections since 1980, perhaps because the population is well known and experienced collectors have avoided affecting it by taking further collections. In addition, any more collections may be discarded because they come from the same population and may therefore be considered geographically and taxonomically uninformative.

A. richardsii is a medium-sized *Acacia* species with a scattered distribution across a moderate geographic range and was collected frequently up to 1989 and not since. The species is probably not in need of urgent attention. It occurs in a remote area of the Kimberleys. Variation in collection frequency is most likely related to the timing of regional surveys.

A. semicircinalis is endemic to the Wongan Hills, but unlike *A. pygmaea*, it is common within its range. It is an insubstantial plant with a prostrate wiry habit and few flowers. Most collections originate from the period in the 1970s and 1980s during which it was described and during which a botanical survey of the hills was undertaken. These activities are likely to be responsible for the non-random collection pattern. It is unlikely to be in need of urgent attention. Variation in collection frequency is likely to be due to variation in taxonomic interest, the timing of regional surveys, and the fact that subsequent collections may have been discarded as uninformative.

A. subrigida has only been collected three times, between 1979 and 1982. Its range is remote and poorly collected. The collections were all made by specialists. The species is unlikely to be in need of urgent attention.

3

Identifying the Ecological Correlates of Extinction-Prone Species: A Case Study of New Zealand Birds

Brendan Moyle

Introduction

Risk of extinction is a variable used in a number of conservation management arenas. For instance, the U.S. Endangered Species Act classes species according to risks of extinction and mandates differing levels of federal protection based on these classifications. The international agreement on trade in endangered species (CITES) uses a similar system in which differing levels of trade restrictions are applied to species for which risk of extinction is an important criterion (Lyster 1985). The IUCN (World Conservation Union) categories attempt a more precise definition of extinction risks but without any associated policy prescriptions (Mace and Lande 1991). Nonetheless, these extinction risks are intended to provide species managers with information to aid in decision making.

Although extinction risk is widely recognized as being important and as having applications for conservation policy, there remains a gap between theory and application. One problem is that methods for determining extinction risk can be demanding. For instance, field studies may be necessary to determine whether a species is in decline. Population viability analysis (PVA) may be used, but this technique usually requires detailed population data; consequently, relatively few species have actually been the target of such analysis (Mace and Lande 1991; Boyce 1992; Beissinger and Westphal 1998).

The impact of the requirements for additional empirical information and PVA has several adverse effects. If the conservation organization avoids such exercises (and their associated costs), it may bias conservation policy toward those species that are already well known. Typically, these are the charismatic fauna (Metrick and Weitzman 1996). Mammals and birds tend to be popular recipients of recovery effort relative to the less well-known invertebrates or lower plants.

In addition to the substantial costs associated with field studies and data collection exercises for PVA, response lags may occur. There may be significant delays between the point at which a species crosses some extinction risk threshold and the point at which this is observed or recognized, compromising recovery efforts. For example, recovery work on the Little Spotted Kiwi (*Apteryx owenii*) was delayed initially by the erroneous belief that a large population persisted on the west coast of New Zealand's South Island (Williams 1986).

In the context of species recovery work, it is desirable to have management tools that are objective and transparent. A failure to generate such tools may perpetuate a bias toward well-studied or charismatic groups (Moyle 1998). However, the managers may simply lack the resources or time to engage in comprehensive analysis of extinction risk. In this setting, decision heuristics (rules of thumb) may be adopted that are believed to yield satisfactory results.

Decision heuristics that have been adopted include rules based on genetic theory and island biogeographic theory (Diamond 1975) and point-scoring systems used to establish priorities for species conservation activities (e.g., Molloy and Davis 1992). Applications of criteria derived from genetic theory in the 1980s used population size to assess population viability, with a consequence that some populations were judged to be nonviable populations and were excluded from consideration in recovery work (Hartley 1997). Applications based on island biogeographic theory relied on island (and habitat patch) sizes and arrangements to provide some guidance for target areas and management strategies for threatened species. Heuristics based on genetic theory and island biogeography ignored a host of ecological considerations and the impact of various management decisions.

More recently, decision rules (IUCN 1994) and point-scoring systems (e.g., Millsap et al. 1990) were developed in attempts to provide efficient means of estimating extinction risks. For instance, the species priority ranking system in New Zealand (Molloy and Davis 1992) incorporated a variety of management and biological factors into a point-scoring system. These included the security of existing ranges from disturbance, population size, threats, and success of management tools such as captive breeding or cultivation. Some factors were not directly related to extinction risk, but the system was dominated by attributes expected to be closely associated with it (Davis et al. 1992). Each factor was ranked ordinally from 1 to 5, in which risk was inversely related to rank. The ranks for each attribute were then summed to produce a cardinal number. However, summation is not an appropriate way to deal with ordinal numbers, and the cardinal sum was, unfortunately, nonsensical. Unweighted summation is sensitive to the number of factors that are included in the system and may not reflect the threats that make the greatest contribution to a species' vulnerability to extinction.

Although a point-scoring system is more demanding of data than the genetic or island biogeographic decision heuristics, it can still be attempted with rather coarse data. The requirement to collect ordinal data means that some of the costs of obtaining more accurate data can be avoided. If a conservation agency lacks the time or resources to conduct comprehensive surveys, these types of decision heuristics may improve the quality of the decisions made by managers. In this regard, they may still be a useful model even if their grounding in ecological realism is untested. To obtain better models with improved ecological realism may simply be unachievable, given available resources.

If comprehensive surveys, detailed ecological studies, and the development of explicit population models are not feasible, then agencies may consider other methods to generate extinction risks or estimates of vulnerability to assist in setting management priorities. An alternative is to develop simple models with the data at hand. Although population data are difficult to come by, morphological data are often easier to obtain or infer. Insofar as morphological traits are correlated to ecological factors, morphology may provide proxy measures of unobserved ecological factors.

Data on the distribution of the threatened species frequently are coarse and unreliable. For instance, data may be composed of a mixture of presence records and presence-absence records from a small sample of locations (see Elith, this volume). The purpose of this work is to describe approaches to estimating extinction risks that use taxonomic, morphological, and spatial information to develop models that explain extinction risks, using maximum likelihood estimation. The methods are intended to make better use of information that is commonly available to biologists and to enhance the ability of managers to make decisions that are sensitive to the relative risks faced by different populations and species.

Maximum Likelihood Estimation

Generalized linear models (GLM) provide an effective means of developing models for categorical response variables. A distribution is assumed for the response variable, commonly binomial for binary data and Poisson for count data. When using nonnormal response data, the expectations of the response have to be constrained (e.g., when modeling a binary event such as extinct-extant, estimates have to be constrained to be between zero and one). A link function, with a corresponding probability density function, achieves this. For binary response data (e.g., extinct-extant or present-absent), logit and probit links are common. In the probit case, the normal distribution is used as the probability density function. In the logit case, the logistic function is used as the probability density function. The resulting expectations are expressed as a probability (of success, if 1 = success).

In GLMs, likelihood functions are used to estimate parameters for the explanatory variables (Amemiya 1981). The values given to the parameters are those that maximize the probability of obtaining the observed set of data. The equations that determine the maximum likelihood estimations are nonlinear, and with GLMs, iterative algorithms are used for solving them. For mathematical reasons, it is actually the log-likelihood that is maximized. The use of a probability criterion for categorical response variables, generally nonlinear in form, differs from least-squares techniques. Least-squares models apply a linear distance criterion that can be represented in the Cartesian space formed by the explanatory and dependent variables. This forms the basis of goodness-of-fit measures in least squares. Classical regression models deal with response variables that have normally distributed errors. However, the probabilistic nonlinear relation between the explanatory and dependent variables means that least-squares measures are inappropriate estimators of goodness of fit.

In this study, the logistic and probit models (based on logit and probit link functions) are used to relate extinction data to morphological and other attributes, in an effort to predict the attributes of species that predispose them to extinction risk.

Correlates of Extinction Risk: The Case of New Zealand Terrestial Birds

The New Zealand avifauna has suffered important losses. Human settlement triggered a wave of deforestation and introductions of exotic mammalian pests. This was associated with the extinction of many bird species (Gill 1991). For many species, extinction in the period before European settlement means that there is little reliable information on population densities and distribution (East and Williams 1984; Atkinson and Millener 1991).

These extinctions are amenable to analysis with logistic regression. The dependent variable is the current status of the species: either extinct or extant. Survey data and the details of the ecology of most species are scant. The majority of the information about these species is morphological.

Determining the vulnerability to extinction of these various species may be of interest for a number of reasons. From a strictly ecological perspective, understanding why one species rather than another is extinct may be useful in explaining the pattern of observed extinctions. From a policy perspective, an understanding of what constitutes the risk factors facing New Zealand birds may aid conservation planning. Such a model may also complement field studies or anticipate the decline of other species.

In this study, only terrestrial birds were included (Mills and Williams 1984). Aquatic and marine birds are likely to be less vulnerable to some of the landscape changes associated with human settlement. For instance, deforestation may be presumed to have a greater impact on the terrestrial birds rather than the marine or aquatic. Two terrestrial species (the Kea [*Nestor notabilis*] and the Rock Wren [*Xenicus gilviventris*]) that had an alpine rather than lowland distribution were also excluded from the set. Again, the rationale was to focus on a narrow landscape. Hence the study is really concerned with those species that were distributed over forests and their terrestrial margins. This generated a sample of 80 taxa.

Model Specification

The dependent variable was the current status of each species: extinct or extant. In this study, species that were recognized in 1984 as endangered (Mills and Williams 1984) were classified as extinct because the persistence of these species was attributable to human efforts at preservation rather than to any resilience of the species to extinction pressure. In other words, if major efforts had not been made to save these species, they would have been either extinct or functionally extinct by the mid- or late 1980s.

The explanatory variables included the species' body mass, trophic level, flying capability, and taxonomic status. In addition, species endemic to the Chatham Islands were labeled in the analysis. The Chatham Islands have a peculiar representation of birds, and several common mainland species are absent (Atkinson and Millener 1991). Explicit recognition was necessary in the model to offset the peculiarity of this distribution. This was the only instance in which subspecies were included in the data set.

The selection of these variables was based either on hypotheses from the literature on correlates of extinction risk or from past empirical studies. For instance, body mass is correlated with reproductive rates, range requirements, and possibly population density. All these factors have been shown to be correlated with extinction risk (Pimm et al. 1988). The level of endemism was proposed by Mills and Williams (1984) as a trait linked to extinction risk. Small islands frequently have a higher turnover of species (Pimm et al. 1988), implying that this type of isolation may be a risk factor. Trophic level has been linked to range requirements, and as many reserves are believed to be too small to maintain viable populations of high trophic-level carnivores, trophic level may also be an extinction risk correlate. Flightlessness and poor flying ability have also been proposed as extinction correlates for New Zealand birds (Gill 1991; Mills and Williams 1984). Hence in the first instance, variable selection was determined by biological realism.

Model specification generates a range of problems. The first is that some explanatory variables may manifest multicollinearity. This may place restrictions on how a particular biological factor is represented. For instance, the first iterations of the model used several different measures of endemism. These variables were highly correlated amoung themselves and with other variables. New Zealand had many flightless birds, so some measures of endemism were correlated with "flight ability." The variable representing endemism that was least correlated with the others was used.

The particular approach to parameter selection in this example was based on the Hendry method (Gilbert 1986). The starting point for the regression is a model with as many parameters as possible that are likely to explain the phenomenon of interest. Then, the variable with the least significant coefficient is eliminated. Another regression is performed, and the process is repeated until the simplest possible model is found that includes all important explanatory variables. In this example, variables that were not significant at a 95% level of confidence for at least one of the regressions (i.e., whose coefficient was not significantly different from zero) were eliminated.

The advantage of this approach is that usually it results in a model with good explanatory power, without retaining variables that do not contribute to explaining the response variable. Multicollinearity can sometimes be detected informally with this approach, as regression coefficients are usually sensitive to specification in its presence. Large swings in the value of a coefficient in a new model tend to indicate multicollinearity. Nonetheless, if forecasting is the goal of the model, the Hendry method may not be appropriate. Although the Hendry method generates models that have high explanatory power, the models do not necessarily have good predictive power. Other approaches to the reduction in the number of explanatory variables and lowering of the influence of multicollinearity may be better at predicting but usually will be worse at explaining the data at hand.

In most applications, it is useful to contrast several different models, especially those that have similar explanatory power. For instance, Moyle (1997) used a variation of the model presented below. In developing the model below, it was suspected that body mass might have a nonlinear relationship with extinction risk because it ranges over five orders of magnitude. Formal testing for nonlinearity showed that the linear assumption for body mass was valid. Nonetheless, transformations to smooth the relationship were performed and the model reestimated. Not surprisingly, the fact that a raw linear relationship was valid meant the model was robust to either the raw body mass or the log-transformed body mass. Different models reveal whether the inferences of the model are robust as well as adding to the knowledge of the conservation manager. As with all statistical models, diagnostic tests on the explanatory variables are required, and ideally, some cross-validation information on the values of the parameters should be generated (see Elith, this volume).

The Model

Body mass was measured in kilograms, and the data were obtained mostly from the work of Atkinson and Millener (1991). When body mass is not recorded in this chapter, Dunning's data (1993) were used instead. This created the potential for a measurement bias as the two sources did not always agree on the body mass values. Complicating this issue is the fact the body mass estimates for many birds are based on museum samples or show great dependence on well-studied populations. Hence estimates of body mass have not been generated by representative samples of these species.

Species and subspecies endemic to the Chatham Islands (CHATHAM) were recorded as a dummy variable (i.e., represented as a binary variable). The working hypothesis was that species that are endemic to a small island have higher extinction risks (Pimm et al. 1988). Trophic level might also be related to extinction risk. Species at the top of the food chain might require a larger range to maintain a viable population. Herbivorous bird species in New Zealand may be less able to defend themselves against mammalian predators than more aggressive birds in higher trophic levels. The guilds used in the study of Atkinson and Millener (1991) were classed into four trophic levels and tested for statistical significance. For reasons of parsimony, only the top trophic level (TROPH1) and the bottom trophic level (TROPH4) were retained in this model.

Another factor that may be relevant for New Zealand species is flying ability. As many birds that are unable to fly are also relatively heavy, the variable most

TABLE 3.1. Logit regression.

Variable name	Coefficient	Standard error	t ratio
WEAK	5.5424	1.7177	3.2267
CHATHAM	6.4516	2.3182	2.7830
KG	1.3129	0.60606	2.1662
TROPH1	4.8906	1.9156	2.5531
TROPH4	2.6639	1.5320	1.7389
ENDEMIC	2.3008	1.0831	2.1243
Constant	−6.5554	2.2707	−2.8870
Maddala R^2	55.93%		
Cragg-Uhler R^2	74.966%		
Chow R^2	68.223%		
Prediction success rate	91.25%		

closely associated with vulnerability to mammalian predators included both size and flying ability. In the model below, all birds that were either weak fliers or flightless and that weighed less than 2 kg were classed as WEAK. A more appropriate measure of the ability to fly would be wing morphometrics and their relation to body size. However, these data were not available. The restriction to small birds was necessary to avoid multicollinearity problems with body mass and flight ability.

Finally, species endemic to restricted areas in New Zealand (ENDEMIC) may be more vulnerable to the impacts of human settlement than indigenous species whose natural ranges cover a variety of landscapes and threats.

In all but one instance, a strong statistical relationship was observed between the independent variables and the dependent variable (current extinction status; Tables 3.1 and 3.2). Aside from TROPH4 in the probit model, all variables were statistically significant at a 95% level of confidence. Hence this particular approach has at least successfully identified some of the correlates of extinction risk in New Zealand birds. Given that the maximum likelihood estimation has a

TABLE 3.2. Probit regression.

Variable name	Coefficient	Standard error	t ratio
WEAK	2.8575	0.74952	3.8125
CHATHAM	2.8741	0.79795	3.6018
KG	0.65481	0.30414	2.1530
TROPH1	2.2702	0.78822	2.8801
TROPH4	0.99562	0.60489	1.6460
ENDEMIC	1.0092	0.48865	2.0652
Constant	−3.0417	0.82094	−3.7052
Maddala R^2	55.33%		
Cragg-Uhler R^2	74.166%		
Chow R^2	64.351%		
Prediction success rate	87.5%		

nonlinear objective function, traditional measures of goodness of fit based on linear least-squares procedures are inappropriate. Several alternative measures of goodness of fit have been proposed and some reported here (Tables 3.1 and 3.2). The model diagnostics suggest that the models can successfully explain about 90% of the events relating to the persistence or extinction of these 80 New Zealand birds (prediction success is represented by a predicted probability greater than 0.5, corresponding to an event). Such results may be expressed as false-positive and false-negative rates, sensitivities, and specificities (see Elith, this volume, for some more comprehensive summary techniques).

The regression results are presented for the models, in which the predicted probability for each species is used as an index of extinction proneness. These

TABLE 3.3. Vulnerability indices for extinct species.

Species	Logit	Probit
Little Bush Moa (*Anomalopteryx didformis*)	1.00	1.00
Chatham Is. Bellbird (*Anthornis melanura melanocephala*)	0.993	0.969
Adzebill (*Aptornis defossor*)	1.00	1.00
Adzebill (*Aptornis otidiforms*)	1.00	1.00
Chatham Is. Fernbird (*Bowdleria rufescens*)	1.00	1.00
Snipe-rail (*Capellirallus karamu*)	0.839	0.843
Eyles' Harrier (*Circus eylesi*)	0.875	0.848
Chatham Is. Snipe (*Coenocorypha chathamica*)	0.508	0.461
New Zealand Quail (*Cortunix novaezelandiae*)	0.857	0.810
Giant Chatham Is.Rail (*Diaphorapteryx hawkinsi*)	1.00	1.00
Large Bush Moa (*Dinornis novaezealandiae*)	1.00	1.00
Slender Bush Moa (*Dinornis struthoides*)	1.00	1.00
Giant Moa (*Dinornis giganteus*)	1.00	1.00
Eastern Moa (*Emeus crassus*)	1.00	1.00
Coastal Moa (*Euryapteryx curtus*)	1.00	1.00
Stout-legged Moa (*Euryapteryx geranoides*)	1.00	1.00
Chatham Is. Coot (*Fulica chathamensis*)	1.00	1.00
Hodgen's Rail (*Gallinula hodgeni*)	0.904	0.866
Haast's Eagle (*Harpagornis moorei*)	1.00	1.00
Huia (*Heteralocha acutirostris*)	0.999	0.999
Owlet-Nightjar (*Megaegotheles novaezealandiae*)	0.999	0.999
Upland Moa (*Megalapteryx didinus*)	1.00	1.00
Chatham Is. Kaka (*Nestor meridionalis "chatham"*)	0.431	0.351
Crested Moa (*Pachyornis australis*)	1.00	1.00
Mappin's Moa (*Pachyornis mappini*)	1.00	1.00
Heavy-footed Moa (*Pachyornis elephantopus*)	1.00	1.00
Grant-Mackie's Wren (*Pachyplichas jagmi*)	0.793	0.803
Yaldwyn's Wren (*Pachyplichas yaldwyni*)	0.795	0.804
New Zealand Crow (*Palaeocorax moriorum*)	0.868	0.805
Dieffenbach's Rail (*Rallus dieffenbachi*)	0.594	0.530
Chatham Is. Rail (*Rallus modestus*)	0.496	0.450
Laughing Owl (*Sceloglaux albifacies*)	0.785	0.714
Stephen's Is. Wren (*Traversia lyalli*)	0.789	0.799
Piopio (*Turnagra capensis*)	0.683	0.619
Bush Wren (*Xenicus longpipes*)	0.787	0.798

results indicate the predicted probability that the species will become extinct under each of the models, given the values for the independent variables for the species (e.g., body weight, trophic level, endemic status). These values may be used as an index of extinction proneness for the extant species in setting priorities for conservation efforts. In this instance, vulnerability is expressed as an index in which 1 is the most vulnerable to extinction and 0 is the least. Vulnerability indices are presented for the set of extinct species (Table 3.3), the set of endangered species (Table 3.4), and the set of nonthreatened endemic breeders (Table 3.5).

Discussion

The models are successful in explaining past patterns of extinction. All but three extinct species had vulnerability scores greater than 0.5. The Chatham Island Rail, the Chatham Island Kaka, and the Chatham Island Snipe in the probit model were the only extinct species with vulnerability scores less than 0.5. That is, a few species modeled as not very vulnerable are already extinct. Given the stochastic nature of extinction, the number of events modeled, the simplicity of these models, and the complex nature of extinction processes, the occurrence of some extinctions among species with lower vulnerabilities (between 0.3 and 0.5) may

TABLE 3.4. Vulnerability indices of endangered species.

Species	Logit	Probit
North Is. Brown Kiwi (*Apteryx australis*)	0.274	0.346
Great Spotted Kiwi (*Apteryx haastii*)	0.374	0.434
Little Spotted Kiwi (*Apteryx owenii*)	0.948	0.949
Fernbird (*Bowdleria punctata*)	0.791	0.802
Kokako (*Callaeas cinerea*)	0.986	0.976
Chatham Is. Snipe (*Coenocorypha pusilla*)	0.503	0.456
Yellow-crowned Parakeet (*Cyanoramphus auriceps*)	0.021	0.022
Forbes' Parakeet (*Cyanoramphus a. forbesi*)	0.993	0.984
Orange-crowned Parakeet (*Cyanoramphus malherbi*)	0.022	0.023
Antipodes Is. Parakeet (*Cyanoramphus unicolor*)	0.022	0.023
New Zealand Falcon (*Falco novaeseelandiae*)	0.242	0.305
Weka (*Galliralus australis*)	0.518	0.639
Wood Pigeon (*Hemiphaga novaeseelandiae*)	0.324	0.270
Chatham Is. Wood Pigeon (*Hemiphaga n. chathamensis*)	0.997	0.991
Yellowhead (*Mohoua ochrocephala*)	0.014	0.014
Kaka (*Nestor meridionalis*)	0.283	0.240
Stitchbird (*Notiomystis cincta*)	0.176	0.155
Chatham Is. Tit (*Petroica macrocephala chathamensis*)	0.479	0.438
Black Robin (*Petroica traversi*)	0.481	0.439
Saddleback (*Philesturnus carunculatus*)	0.676	0.613
Takahe (*Porphyrio mantelli*)	0.512	0.467
Chatham Is. Tui (*Prosthemadera n. chathamensis*)	0.993	0.972
Kakapo (*Strigops habroptilus*)	0.998	0.998

TABLE 3.5. Vulnerability indices for nonthreatened endemic breeders.

Species	Logit	Probit
Rifleman (*Acanthisitta chloris*)	0.014	0.021
Bellbird (*Anthornis melanura*)	0.174	0.154
Pipit (*Anthus novaeseelandiae*)	0.001	0.001
Chatham Is. Pipit (*Anthus n. chathamensis*)	0.490	0.442
Red-crowned Parakeet (*Cyanoramphus novaezelandiae*)	0.022	0.023
Chatham Is. Grey Warbler (*Gerygone albofronta*)	0.477	0.436
Grey Warbler (*Gerygone igata*)	0.001	0.001
Brown Creeper (*Mohoua novaeseelandiae*)	0.014	0.021
Whitehead (*Mohoua albicilla*)	0.014	0.022
Morepork (*Ninox novaeseelandiae*)	0.192	0.255
New Zealand Robin (*Petroica australis*)	0.001	0.001
Tit (*Petroica macrocephala*)	0.001	0.001
Tui (*Prosthemadera novaeseelandiae*)	0.190	0.167

not be surprising. The model demonstrates that the past pattern of extinction has been associated with species that have particular traits: they tend to be large bodied, poor flyers, herbivorous, and endemic.

No endemic breeder currently classified as nonthreatened had a vulnerability measure in excess of 0.5. For the endangered species, vulnerability measures were more variable. The model had much lower predictive power for the set of endangered species. Some of the most endangered species (Little Spotted Kiwi, Kokako, Forbes' Parakeet, Chatham Island Wood Pigeon, and Kakapo) had vulnerability measures greater than 0.9. Many other species had intermediate vulnerabilities, whereas some of the smaller birds that were also agile fliers had relatively low vulnerability measures. These include the various parakeet species from the genus *Cyanoramphus* and the Yellowhead. Given their endangered status, these results appear anomalous.

The presence of anomalies is partly a consequence of the modeling technique. Models are by their nature simplifications of real and complex phenomena. As a result, they will not fit the data perfectly. In this case, anomalies can be expected as a result of the coarse and sometimes biased data and the modeling protocols. Further, the results also show that for some species in the intermediate range, the choice of the probability distribution function (i.e., the type of link function) can affect the value of the vulnerability index.

Anomalies may also appear if a species is subject to a novel extinction threat that is not correlated with the independent variables used in the model. For instance, disease has been suggested as a source of extinction risk for some of the smaller forest birds. The introduction of the possum (*Trichosorus vulpeca*) from Australia is a novel threat and may be only weakly related to the variables considered here that rely on past data. Possums compete with some native birds and have been observed to consume eggs from some threatened species.

Species may also have particular traits that make them more or less vulnerable than these estimates. For example, a predatory bird that also consumes carrion may have smaller range requirements and therefore lower vulnerability to patch reductions. Disturbance to landscapes through clearing forests or the introduction of new food sources or irrigation may enhance survival of some species. The presence of a trait that increases extinction risk but is not included in the model will lead to an underestimation of vulnerability.

GLMs can generate useful extinction models, even in circumstances in which the data quality is low. As the example shows, such an approach can inform managers about traits that are correlated with extinction risk and hence offer clues as to which species face elevated extinction risks. Evaluating the ecological correlates of extinction proneness has been a useful strategy in other situations (e.g., Laurance 1991; Pimm 1993; Angermeier 1995). In this application, the model confirmed some recent field work on the Weka that suggested it is more vulnerable to extinction than was supposed previously. As in all modeling situations, there is no technique that can fully compensate for poor data, and the quality of the model will be improved if better data are gathered. Maximum likelihood modeling within the context of GLMs provides one tool that may allow a modeler to make good use of the limited data that are available.

Literature Cited

Amemiya T (1981) Qualitative response models: a survey. Journal of Economic Literature 19:1483–1536

Angermeier PL (1995) Ecological attributes of extinction prone species: loss of freshwater fishes of Virginia. Conservation Biology 9:143–158

Atkinson IAE, Millener PR (1991) An ornithological glimpse into New Zealand's pre-human past. Acta XX Congressus Internationalis Ornithologici 1:129–192

Beissinger SR, Westphal MI (1998) On the use of demographic models of population viability in endangered species management. Journal of Wildlife Management 62:821–841

Boyce MS (1992) Population viability analysis. Annual Review of Ecology and Systematics 23:481–506

Davis A, Bellingham M, Molloy J (1992) Who goes into the ark: how to decide which species are the most threatened. Forest and Bird 23:38–41

Diamond JM (1975) The island dilemma: lessons of modern biogeographic studies for the design of natural reserves. Biological Conservation 7:129–146

Dunning JB (ed) (1993) CRC handbook of avian body masses. CRC Press, Boca Raton, FL

East R, Williams GR (1984) Island biogeography and the conservation of New Zealand's indigenous forest-dwelling avifauna. New Zealand Journal of Ecology 7:27–35

Gilbert CL (1986) Professor Hendry's econometric methodology. Oxford Bulletin of Economics and Statistics 48:288–307

Gill B (1991) New Zealand's extinct birds. Random Century, Auckland, New Zealand

Hartley P (1997) Conservation strategies for New Zealand. New Zealand Business Roundtable, Wellington, New Zealand

IUCN (1994) International Union for the Conservation of Nature, draft red list categories. World Conservation Union, Gland, Switzerland

Laurance WF (1991) Ecological correlates of extinction proneness in Australian tropical rainforest mammals. Conservation Biology 5:79–89

Lyster S (1985) International wildlife law. Grotius Publications Limited, Cambridge, UK

Mace GM, Lande R (1991) Assessing extinction threats: towards a reevaluation of IUCN threatened species categories. Conservation Biology 5:148–157

Metrick A, Weitzman ML (1996) Patterns of behaviour in endangered species preservation. Land Economics 72:1–16

Mills J, Williams G (1984) The status of New Zealand's endangered birds. In: Archer M, Clayton G (eds) Vertebrate zoogeography and evolution in Australasia. Hesperian Press, Carlisle, Western Australia, pp 1107–1120

Millsap BA, Gore JA, Rundle DE, Cerulean SI (1990) Setting priorities for the conservation of fish and wildlife species in Florida. Wildlife Monographs 111:1–57

Molloy J, Davis A (1992) Setting priorities for the conservation of New Zealand's threatened plants and animals. Department of Conservation, Wellington, New Zealand

Moyle B (1997) Generating extinction risks: the MLE approach. Society for Conservation Biology Annual Meeting, University of Victoria, British Columbia

Moyle B (1998) Species conservation and the principal-agent problem. Ecological Economics 26:313–320

Pimm SL (1993) Life on an intermittent edge. Trends in Ecology and Evolution 8:45–46

Pimm SL, Jones HL, Diamond J (1988) On the risk of extinction. The American Naturalist 132:757–785

Williams M (1986) Native bird management. Forest and Bird 17:7–9

4

Quantitative Methods for Modeling Species Habitat: Comparative Performance and an Application to Australian Plants

Jane Elith

Introduction

Habitat evaluation procedures are used to quantify the value of land as habitat for a species. They result in maps of species habitats that are useful for applications involving estimation of the impacts of management alternatives; ecological assessment; conservation planning; and the identification of steps that may be taken to avoid habitat losses or to compensate unavoidable habitat losses due to a proposed action (Van Horne and Wiens 1991; Gray et al. 1996; Rand and Newman 1998). For example, species that have narrow geographic distributions but are abundant in at least one place would benefit from a protected area. Species that are widely distributed within specialized habitat may benefit from targeted management prescriptions within a multiple-use landscape. Trade-offs may be identified so that biodiversity objectives may be met efficiently. The effectiveness of such management decisions depends critically on reliable knowledge of the current and potential distribution of species.

Frequently, management decisions are based on "known locations." These are geographic coordinates that accompany field observations or specimens in museums and herbariums. Such records are generally used as the basis for the construction of distribution maps, which are composed of polygons that enclose all known locations, conditioned by available ecological knowledge and subjective extrapolations based on topography or the spatial distribution of environmental variables. These maps are rarely validated, and their reliability is rarely reported.

The hard boundaries represented in most maps of species distributions are abstractions (Gaston 1994). In reality, habitat suitability is determined by the response of species to a set of spatially distributed, random variables. Hence the definition of habitat requires an understanding of the way in which environmental variables affect the likelihood that a species will be found at a given location (e.g., Austin 1985; Burgman 1989; Prober and Austin 1990; McIntyre and Lavorel 1994; Coates 1996). Species distributions are strongly influenced by processes such as successional dynamics and disturbance (Main 1984; Attiwill 1994).

Mapped distributions may be influenced by the time of year and the inclusion or exclusion of nonreproducing individuals and vagrants (Gaston 1994). More useful measures of the area of occupancy of species reflect habitat suitability and probability of occurrence on a continuous scale.

Different places support different densities of individuals of a species, depending on a range of historical, ecological, and demographic parameters. Potential habitat does not necessarily coincide with a species current or past distribution, even if these are known without error. Van Horne (1983) pointed out that areas in which a species is found relatively frequently do not necessarily represent the areas in a landscape associated with greatest survival or reproductive success. Individuals of a species may not congregate in the most suitable locations because of behavior, intraspecific competitive exclusion, interspecific interactions, or dispersal dynamics. As a result, population sinks may have high population densities but be of relatively limited value in contributing to the likelihood of persistence of a species.

Some of these procedures are based on qualitative information, subjective estimates, and expert judgment, e.g., habitat suitability indices (USFWS 1980; Crance 1987; Rand and Newman 1998). This chapter describes several quantitative modeling methods that can be used to map predicted species distibutions from various combinations of species and environmental data. Five of them were applied to data from several plant species in a forested landscape in Australia. The resulting predictions were assessed in the context of the ability of the models to provide results that could be useful in the management of land for conservation.

Modeling Methods

A broad range of computer modeling methods has been applied to the problem of predicting species distributions, and some of the more common approaches are summarized with key references and examples in Table 4.1. Often, the choice of method for prediction is related to availability of software and local expertise. Few studies involve comparison of modeling methods. Exceptions include comparisons of generalized linear models (GLM), generalized additive models (GAM), decision trees, and genetic algorithms by Austin and Meyers (1995), and ANUCLIM, GLMs, GAMs, and decision trees by Ferrier and Watson (1996). In this section, the attributes and advantages of five of the methods are described and evaluated.

ANUCLIM (also known as BIOCLIM) is an example of a climate-mapping approach to modeling. It uses presence data as point locations together with elevation data and climatic surfaces developed from long-term rainfall, temperature, and radiation records to construct a climate profile for a species. With several climatic parameters, the aggregated profile forms a multidimensional rectangle known as a "bounding box" or "climatic envelope." Several sets of bounds can be defined within these profiles: they can be based on the range of all observed points, on pairs of percentiles, or on means and standard deviations. The mapping

TABLE 4.1. Modeling methods: key references and examples.

Method	Key references	Example of ecological applications
Bioclimatic envelopes	ANUCLIM: Nix (1986) and CRES (1999)	ANUCLIM: Kauri Pine in New Zealand (Mitchell 1992), Eucalypts in South Africa (Richardson and McMahon 1992), Leadbeater's Possum (Lindenmayer et al. 1991b), and Myrtle Beech (Busby 1986) in Australia Other: plant species in United States (Box et al. 1993)
Canonical correspondence analysis (CCA)	Jongman et al. (1995) and ter Braak (1986)	Rock outcrop vegetation, United States (Wiser et al. 1996), and halophytic plant communities, China (Pan et al. 1998)
Multivariate distance methods	DOMAIN: CIFOR (1999)	DOMAIN: marsupials in Australia (Carpenter et al. 1993) Other: wolf distribution in Italy (Corsi et al. 1999)
Generalized linear models	McCullagh and Nelder (1989) and Agresti (1996)	Arboreal marsupials in Australia (Lindenmayer et al. 1990a,b, 1991a), Eucalypts in Australia (Austin et al. 1983, 1984, 1990, 1994a), alpine plants in Switzerland (Guisan 1998), and a bird species in America (Akçakaya and Atwood 1997)
Generalized additive models	Hastie and Tibshirani (1990)	Trees in New Zealand (Yee and Mitchell 1991), Eucalypts in Australia (Austin and Meyers 1996), and wetland plants in The Netherlands (Bio et al. 1998)
Habitat suitability indices	Schamberger and O'Neil (1984)	Birds in United States (Van Horne and Wiens 1991), Bandicoots in Australia (Reading et al. 1996)
Machine learning methods		
Decision trees	Brieman et al. (1984)	Vegetation mapping in Australia (Keith and Bedward, 1999), forest type in United States (Lynn et al. 1995)
Neural networks	Aleksander and Morton (1990)	Fish in France (Mastrorillo et al. 1997) and land systems in Canada (Gong et al. 1996)
Genetic algorithms	Mitchell (1996)	GARP algorithm in Australia (Austin and Meyers 1995; Stockwell and Noble 1992), applied to plant species (Elith et al. 1998)

section of the program produces ranked predictions for each site of interest, and the ranks are based on user-specified bounds of the species profile. These predictions define the climatic suitability of the site for the species. The fact that a bounding box is used to describe the potential distribution rather than a more restricted polygon means that large areas are rated as suitable even though the species is likely to be absent, because on an ecological basis one would not expect species to occupy extreme combinations of the ranges (Nix 1986; Carpenter et al. 1993). The program only deals with climate variables, although the approach could theoretically be extended to a broader range of environmental variables. This may be especially important if the method is to be applied to animal species.

DOMAIN is one example of the application of multivariate distance measures to mapping. Conceptually, it takes the opposite approach to ANUCLIM, by defining sites of similarity rather than by determining bounds. It provides a measure of similarity for each site of interest, in which similarity is the environmental similarity between the site of interest and the most similar known record site, and is calculated from the Gower metric (Legendre and Legendre 1998). Environmental differences between sites are scaled by the range of each environmental variable at the sites at which the species has been recorded. DOMAIN can be used to specify an environmental envelope by selecting a lower threshold of similarity or to map similarities on a continuous scale. Only presence data are required as species records.

GLMs are a broad class of statistical models that include ordinary regression and analysis of variance. All GLMs have a random component (the response variable), a systematic component (the predictor or explanatory variables), and a link function that describes the relationship between the expected value of the response and the predictors. The predictors and their coefficients are always combined in a linear form, even if individual explanatory variables are presented in nonlinear form (see, e.g., Austin et al. 1994a). The particular forms of GLMs that are useful for modeling species distributions are logistic regression for presence-absence data and Poisson regression for abundance data. These methods estimate the probability of presence of the species. The predicted probability of presence may be used as an index of habitat quality.

GAMs are a nonparametric extension of GLMs, in which the linear or polynomial functions in a GLM are replaced by smoothed data-dependent functions in a GAM. They are considered a useful tool in modeling biological systems because the response is not limited to a parametric function, which means that the fitted response surface may be a more realistic representation of the true response shape. GAMs retain many of the features of GLMs, including additivity of the predictor effects—this means, in practice, that the roles of the different predictor variables can still be assessed in the model. GAMs can include both parametric and nonparametric terms. The predicted probability surface may be mapped and interpreted as an estimate of habitat suitability, as it may be for GLMs.

Genetic algorithms are general purpose optimization techniques based on a set of logical learning rules. The "genetic" analogy suggests evolution in which

- "chromosome" represents a string of variables, which in the current context are environmental variables, with an associated set of rules that describe the relationship between the variable and the presence or absence of the species
- a pool of chromosomes is created to initiate the algorithm
- random mutations alter the chromosome to a number of different states
- recombination (crossover) occurs between chromosomes
- resulting chromosomes are evaluated (how well do they predict the sample distribution?); the poorest are made "extinct" and replaced by new chromosomes.

GARP (genetic algorithm for rule set production) is a program developed by Stockwell and co-workers (ERIN 1995) to predict species distributions from environmental variables in the context of a geographical information system. The modeling component of the package uses a genetic algorithm to generate, test, and modify rules for predicting distribution. It develops the models on test and training sets of data that are resampled sets of the original presence-absence or presence-only species data. Predictions can be interpreted as relative likelihoods of the presence of the species and could be understood as an index of habitat quality for the species.

Comparative Features

There are differences between these methods that may define the most appropriate technique for a particular situation.

Scale and Definition

ANUCLIM uses only climate variables, which are estimates based on long-term climatic data. Although the resolution of these data is partly dependent on the resolution of the altitude values (usually derived from a digital elevation model) used to sample them, the climate estimates cannot provide surrogate measures of microhabitat features that may be important in some species modeling. Hence ANUCLIM can be viewed as a mesoscale modeling approach. Methods such as ANUCLIM and DOMAIN that only use species presence data will also provide less definition in their estimation of suitable habitat than those that include absence records.

Species Data

The form of species data required varies: ANUCLIM and DOMAIN only require records of presence, GARP can deal with either presence data or presence-absence data, and GLMs and GAMs require records of presence (or abundance) and absence. The requirement for presence-absence data is a serious limitation to the use of GLMs and GAMs, and there has been some exploration of their performance with pseudo-absence data generated from nonpresence sites (see, e.g., Ferrier and Watson 1996).

Predictor Variables

There is a range of approaches to selecting predictor variables. Variable selection is important when large amounts of environmental data are available but not necessarily relevant. GLMs and GAMs are the only methods with widely recognized approaches to selecting subsets of variables—the most common of these, forms of stepwise selection, are often available as options within the software used for modeling. For the other methods, there are no inbuilt procedures for variable selection. Nevertheless, there are several approaches including expert judgment, principal components analysis, and univariate regression that could be used to select variables prior to modeling with these methods.

GLMs and GAMs are the only methods with recognized practices for investigating interactions between variables. Several of the climate parameters in ANUCLIM combine data in an interaction-like relationship (e.g., mean temperature of the wettest quarter) and clearly new variables that represent interactions can be created for any of the methods, but there is no inbuilt capability for such analysis.

There are a variety of approaches to working with correlated (collinear) predictor variables. ANUCLIM recommends the use of all climate parameters even though the majority of them are often highly correlated. DOMAIN and GARP have no recommended approach. Most texts on regression advise that dependencies among the covariates should be investigated (see, e.g., Hosmer and Lemeshow 1989; Agresti 1996), but many statistical packages do not have inbuilt routines for correlation analysis that are part of the modeling component. This leaves it to the practitioners to develop their own good practice (see, e.g., Booth et al. 1994).

Goodness of Fit

GLMs and GAMs are the only methods that provide statistical estimates of error and deviance. These report the extent to which the model does not fit the data.

Usability and Transparency

All methods require some computing and statistical understanding. All can be mastered with some effort, and none requires expert programming knowledge. The time required to achieve proficiency will clearly vary with the user's experience. The more important consideration is that the method and its output are well understood by the user. Understanding is affected by the complexity of the underlying concepts, the transparency of the procedures, the form of the final model, the extent of documentation of the particular programs used, the number of alternative packages for implementing the method (can the results be replicated?), and the breadth of published applications of the method. Thus potential difficulties with the more complex statistical concepts of the GLMs and GAMs are balanced by the extensive documentation and literature associated with them (although GAMs are much newer and are less commonly implemented), and the user-friendly web interface of some versions of GARP is contrasted by the lack of an easily interpretable model, a complex and largely inaccessible background procedure, and few published applications. ANUCLIM and DOMAIN are based on simple concepts, and ANUCLIM has more documentation and published applications than DOMAIN. Nevertheless, it appears that none of the nonstatistical methods are accompanied by a comprehensive text, and they may require more effort or communication with the authors if they are to be thoroughly understood.

Assumptions

Austin and co-workers (1994b), McCullagh and Nelder (1989), and Hastie and Tibshirani (1990) discuss assumptions of each of the methods. All five methods assume that the species data represent an unbiased sampling of the full range of environmental conditions in the modeling and prediction regions and that the environmental variables are measured without error. Models and predictions based on data that do not conform to these assumptions will have unspecified errors that could only be investigated by some additional analyses (e.g., sensitivity analyses).

Models for Seven Plant Species

As an example of the use of these methods, predictions for seven Australian plant species are presented here from a larger study reported by Elith and associates (1998) and Elith and Burgman (in review) (see Table 4.2). The study area was the Central Highlands Regional Forest Agreement (RFA) area in Victoria, Australia (Fig. 4.1). The study aimed to investigate the suitability of existing species and environmental data for modeling plant distribution at a scale applicable to conservation reserve planning. Species data were presence-absence records from quadrat surveys completed by government agencies in the previous 20 years. There were 3,522 such quadrats in the RFA, with some spatial and environmental biases in their distribution (Elith et al. 1998). Quadrats tended to be located close to roads and away from ecotonal vegetation. They were clustered close to major population centers and in higher rainfall areas and occurred relatively frequently in particular vegetation types (e.g., in rainforest and alpine areas). The location of quadrats was estimated to be accurate to within 250 m of their true position for approximately 95% of the records (Elith et al. 1998). The full suite of 35 environmental variables described climate, topography, parent rock type, and vegetation class at a 9-second scale.

Modeling Approaches

Models were developed with five of the methods (ANUCLIM, DOMAIN, GAMs, GLMs, and GARP). Modeling was implemented for each method in a way that could be adapted to large numbers of species over several different regions.

TABLE 4.2. Details of seven of the modeled plant species.

Species name	Status[a]	Life form	Frequency in modeling quadrats	Range (within RFA)[b]	Frequency in full validation set	Records in validation subset[c]
Grevillea barklyana	Rr	Shrub	0.004	l	0.020	8/100
Tetratheca stenocarpa	Rr	Low shrub	0.015	m	0.043	10/101
Wittsteinia vacciniacea	r	Low shrub	0.037	m	0.146	54/24
Helichrysum scorpioides		Forb	0.058	(w)	0.066	17/70
Leptospermum grandifolium		Tall shrub	0.051	w	0.148	56/105
Nothofagus cunninghamii		Tree	0.145	w	0.253	79/73
Phebalium bilobum		Shrub	0.034	(w)	0.161	58/35

[a]Conservation status: species that are rare (r), with capitals indicating national status and lower case indicating state classification.
[b]Widespread (w), medium range (m), and localized (l), with parentheses indicating intermediate classes.
[c]Expressed as no. present/no. absent.

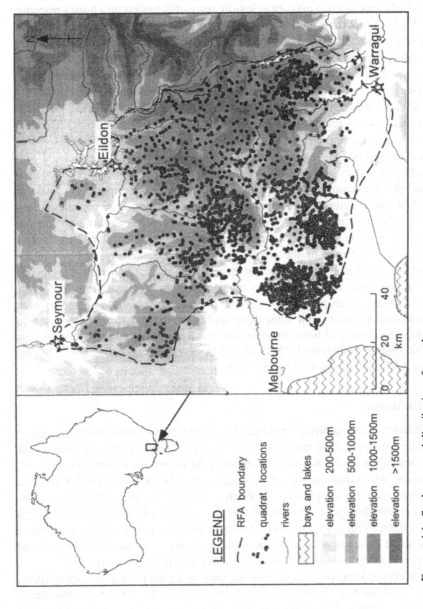

FIGURE 4.1. Study area and distribution of samples.

ANUCLIM used climate predictor variables and species presence data. Models developed in DOMAIN and GARP were constructed with the full set of environmental variables. Because there is noticable variation between runs of the genetic algorithm, spatial predictions were the average of ten iterations of the program.

GLMs were fitted to the data as logistic regression models on a binary response with a logit link function (Hosmer and Lemeshow 1989). All available predictor variables were initial candidates, with vegetation type handled as separate binary variables. For each species, the predictor candidate set was reduced to a subset through a univariate logistic analysis, in which a variable was retained for the subset if the change in deviance between the model containing the variable and the null model was significant at a p value of .2. In the subset, variables that were correlated were excluded through analysis of variance inflation factors (Booth et al. 1994; Sokal and Rohlf 1981). Models were developed with an automated forward selection procedure in which, at each step, the variable associated with the largest reduction in deviance was retained, and the process was repeated until none of the remaining variables significantly improved the model ($p < .05$). If latitude and longitude were included in the subset, they were considered last. Linear and quadratic terms were investigated for all continuous variables. Interactions were not considered. Most of these protocols were developed for species modeling being used by the government in land-use planning, and part of the intention of the study was to investigate modeling success under these protocols.

Variable selection for GAMs was approached in the same manner as for GLMs. GAMs were developed with 4 degrees of freedom initially assigned to each continuous variable, with subsequent testing of changes in deviance used to reduce the degrees of freedom when possible. Relationships were smoothed with cubic splines. Models for GLMs and GAMs were developed using S-PLUS (version 3.3 for Windows 95, MathSoft 1995).

Validation

Validation of the predictions was based on field surveys carried out by an experienced botanist in 1996 and 1997. Consistency between the original survey data and the validation set was achieved by using the same basic survey unit, a 900-m^2 quadrat, which was usually 30 × 30 m, or 15 × 60 m in riparian habitat. The botanist recorded the presence or absence of all 29 species, and searching continued until there was a reasonable certainty that any of the species would have been sighted if they were present. Quadrats were located on 1:100,000 mapsheets, and when these mapsheets did not appear consistent with the observed topography, 1:25,000 sheets were used to locate quadrats more precisely.

Validation sites were selected on the basis of model predictions. Two criteria were addressed: at least three sites from the top 0.1% of the distribution of predicted values for each species and each method were required, and for each species/method combination the full set of sites should span the full range of predictions. The choice of the first 100 sites was conditioned by expert judgment, to try to improve the probability of finding the rare species, and the remaining sites were selected randomly from locations not on private land, within the constraints of the criteria: 391 sites were selected. At each, the following species information was recorded:

- presence of each of the species within a 30 × 30-m quadrat located centrally within a 250 × 250-m target cell
- presence of each of the species anywhere within the cell
- presence of habitat suitable for the species (as assessed by the botanist) within the 30 × 30-m quadrat ("habitat probable")
- presence of habitat that is possible but not optimal for the species within the quadrat, or habitat suitable for the species outside the quadrat but still within the cell ("habitat possible").

Statistical Analysis

In this case, in which different forms of predictions (probabilities, relative likelihoods, and ranks) are produced and in which the ability of the models to predict distribution in new unsampled areas is critical, a measure of discrimination will provide the best comparison of methods. Confusion matrices are commonly used to assess this type of accuracy, but they require the selection of a threshold value that defines whether a prediction is for presence or absence. This is not necessarily a straightforward task (see Fielding and Bell 1997). Receiver operating characteristic (ROC) curves are threshold independent, because they assess the true-positive fraction against the false-positive fraction over numerous decision thresholds (Metz 1978). The area under the ROC curve indicates the percentage of correct decisions in paired comparisons (Swets 1988), which in a species modeling context will be an estimate of the probability of correctly ranking a presence-absence pair. For example, an area of 0.78 would indicate that, given a randomly selected pair of presence and absence observations, the model prediction at the presence site would be higher than at the absence site 78% of the time. This ROC area is equivalent to Wilcoxon's Mann-Whitney version of the nonparameteric two-sample statistic (Hanley and McNeil 1982; Ferrier and Watson 1996). The ROC area is more complex to compute and requires specialist software (for software listings, see Zweig and Campbell 1992; S-News 1999), but the Mann-Whitney statistic is a satisfactory alternative and is easy to compute. Standard errors can be calculated for either statistic (Zweig and Campbell 1992; Hanley and McNeil 1982; DeLong et al. 1988); the method of DeLong and associates (1988) was used in this study. The ROC area can vary from 0 to 1, although the convention in the medical field in which it originated is to constrain it to 0.5 or greater by reversing the decision rule if it is less than 0.5. This is not appropriate in the current context. An ROC area of 1.0 indicates perfect discrimination, whereas 0.5 indicates that discrimination is equivalent to that of a random set of predictions. Here, we view models with an ROC area greater than 0.75 as ones with sufficient discrimination to be potentially useful in reserve planning.

The full set of 391 validation sites was spread throughout the whole RFA, approximately 1.1 million ha in area. It has been shown (Elith and Burgman, in review) that discrimination between sites located across the whole region may be good even though discrimination between sites at a finer scale is poor. In this context, "finer scale" could be understood either in a geographic sense (compare sites within a subcatchment rather than within a whole region) or in an ecological sense (compare sites that are at least within possible ecological bounds rather than sites that are clearly outside the environmental tolerances of the species). Earlier publications of these data investigated the second concept (Elith et al. 1998), whereas here we focus on the first interpretation and create different validation sets for each species. The locations of the species in the full modeling set (3,522 presence-absence sites) were taken to indicate the geographic extent of the species in the RFA, and a mask was created that enclosed most presence locations and extended to the estimated bounds of the inhabited subcatchments. The masks were created through a visual interpretation of the region's digital elevation model and were intended to select subcatchments that might be expected to contain the species. The final validation sets for the seven species reported here vary in size from 78 to 161 sites, with species frequencies ranging from 0.07 to 0.62 (Table 4.2).

Results

Data presented here (Fig. 4.2) refer to observations of the presence of seven species within the 250-m cells (i.e., to the first two species types, as described under the heading Validation). The data for these species are representative of the data for all 29 species evaluated in the larger study (Elith et al. 1998). The majority of models do not discriminate well at a fine (subcatchment) scale. The predictions of any one method were not consistently better than the others, and there was no significant difference ($p < .05$) between the mean ROC areas for any of the methods across the seven species. In the larger data set, GAMs were more frequently represented among the models with ROC area greater than 0.75 than any other method (Elith et al. 1998). Species that occurred frequently in the original 3,522 quadrats were no better modeled than species that were relatively scarce.

Modeling success varied between species. Three species for which sufficiently discriminatory models could not be found were *Tetratheca stenocarpa*, *Wittsteinia vacciniacea*, and *Phebalium bilobum* (Fig. 4.2). These models were based on 54, 131, and 119 records, respectively. The species with the least successful models, *Wittsteinia vacciniacea*, is confined to high rainfall montane and subalpine areas. Although it is ubiquitous in some patches in the region, there are other areas where suitable habitat is restricted, and the available variables did not adequately describe the important variation in the environment. The "successful" models were for *Grevillea barklyana* (all methods), *Helichrysum scorpioides* and *Lep-*

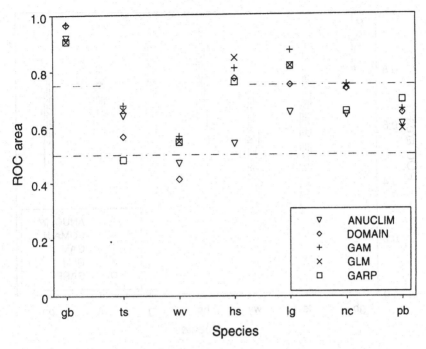

FIGURE 4.2. Discrimination (assessed by ROC area) of models of species distributions produced from five different modeling methods. Species are labeled by the first initials of their scientific name (see Table 4.2).

tospermum grandifolium (all methods except ANUCLIM), and *Nothofagus cunninghamii* (only GAMs). These models were based on 15, 204, 179, and 509 presence records, respectively, and the validation sets included 8, 17, 56, and 79 cell presence records, respectively. These models were able to discriminate between suitable and unsuitable habitat to a degree that would make them a useful aid in planning and decision making at a scale of subcatchments. The species that were modeled successfully shared no common factors with respect to their rarity, their range within the region, or the extent of their habitat. Results for the observations of the species at the quadrat and habitat level produce qualitatively similar results.

These results are true for the best estimates of the ROC areas. The 95% confidence intervals for these statistics (Fig. 4.3) depend on the value of the statistic (standard errors are smallest for very high ROC areas), the number of samples, and the balance of absence and presence records in the sample (standard errors are larger for small samples and for unbalanced samples) (Hanley and McNeil 1982). In most cases presented here, this means that the interpretation of modeling success for ROC areas between 0.65 and 0.85 has some uncertainty associated with it.

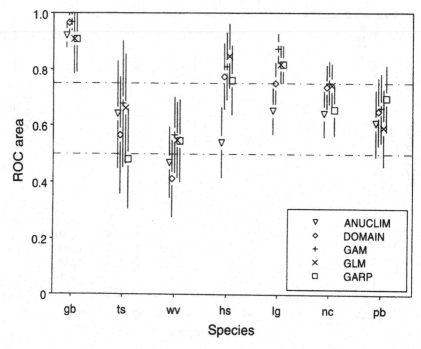

FIGURE 4.3. The 95% confidence intervals for ROC areas presented in Figure 4.2. Species are labeled by the first initials of their scientific name (see Table 4.2).

Discussion

Ferrier and Watson (1996) compared the performance of ANUCLIM with that of GAMs, GLMs, and decision trees for predicting species distributions, by testing predictions against an independent data set (in some cases obtained by jackknifing) with the Mann-Whitney statistic. Models derived from GAMs and GLMs performed significantly better than those from ANUCLIM and decision trees, although this was not consistent across all combinations of modeling techniques, biological groups, and species types (see Ferrier and Watson 1996). In the current study, there were no significant differences between methods across species, although there was a tendency for ANUCLIM models to perform more poorly than others. This is not surprising, given that ANUCLIM models were based on fewer variables (only climate data) than the other methods. Other studies (Austin and Meyers 1995; Bio et al. 1998) also identified the statistical models, and in some cases particularly the GAMs, as being the most promising. Published comparisons of GLMs and GAMs need to be interpreted in the light of the complexity of the response functions (e.g., linear, quadratic, beta) fitted in the GLMs, because GLMs may perform better given more complex parametric forms. The data sets in most comparative studies were collected specifically for modeling.

In the present study, the data were not particularly well suited to the modeling exercise. The 3,522 quadrats were compiled over many decades in response to a variety of research agendas and management objectives. The error in spatial location alone means that the finest scale at which they can confidently be used (250 m) is too coarse for a good representation of site topographic conditions (Elith et al. 1998). In my experience, the kinds of errors and gaps in the data used in this exercise are more representative of the kind of information likely to be available to management agencies than are the data collected specifically for modeling studies.

Under these data conditions, there was no clear distinction in modeling success between the methods. Variation in success between species was often greater than variation in success between methods within species (Fig. 4.2). The errors in the data were such that they masked most consistent differences between the performances of the methods. The lack of a clear difference between the methods is probably as much a reflection on the general lack of discrimination of all models as it is a revealing comparison of the methods. When data sets have more inherent "noise," there appears to be less advantage in using statistical models. Although there may be advantages in using GAMs to model species habitat, a result that is apparent in most comparative studies and even in these data, management agencies need to be aware that effort may be better invested in improving data quality than in more sophisticated modeling techniques. However, there is still sometimes a preference for the use of GLMs and GAMs because researchers like to be able to interpret the models on an ecological level, and the relationship expressed as a linear or additive model is easier to assess than, say, the complex rules evolved in a genetic algorithm.

It is apparent that the models for some species have good enough discrimination to be useful at a subcatchment scale, but it is difficult to predict a priori which species will be successfully modeled. This means that critical evaluation of any model is essential to its proper application. Data sets for testing models are commonly either the same as the modeling set or are partitioned subsets of the full data set (see Fielding and Bell 1997 for a summary of data partitioning methods). However, validation is more thorough if applied to a completely new data set (Chatfield 1995). In this study, the conclusions about modeling success will be dependent on the structure of the validation set. Issues of scale and scarcity need to be carefully considered, so that the validation results reflect the types of comparisons required by the models (Elith and Burgman, in review).

Variable selection is important, but not only in terms of the number of variables. Also important is the set of variables. In this study, the set was constrained by what was available. All methods would have done better with a set of variables that reflected the habitat requirements of the particular species being modeled. The competing argument is that in many circumstances, there will be no information on which to judge a priori which variables will be important. The problem of variable selection is frequently discussed in relation to the statistical methods (GLMs and GAMs), and specific approaches are advised for reducing the variable set (see, e.g., Harrell and Lee 1984; Booth et al. 1994; Ferrier and Watson 1996;

Harrell et al. 1996). Investigation of techniques for reducing the variable set for methods such as DOMAIN and GARP would be useful. It is likely that, for many species, expert knowledge could be used to guide the selection of a potential set of variables for a species. Most of the variables used in this study indirectly influence a species' distribution. Most are commonly available in digital form. Austin and Meyers (1995) demonstrated that the use of variables with a direct or causal effect on distribution improves modeling success. In any study, it is worth considering whether effort could be best spent developing variables appropriate to the modeling problem at hand, rather than (perhaps unsuccessfully) using variables that are conveniently available but only indirectly relevant.

The primary concern is predicting potential habitat for the species, because the objective is to provide advice on how best to protect the species in the long term. I assume implicitly that the presence of a species reflects suitable habitat. This is not necessarily so. For example, species may currently grow in places into which no further recruitment is possible because of changes in climate, the development of structural attributes in associated vegetation, or the presence of disease, predators, or competitors (including introduced species) that eliminate young plants. In this study, analyses based on presence records and on sites considered to be suitable habitat, regardless of absence, produce qualitatively the same results. Analyses of habitats that were apparently unsuitable even if the species was present were not explored.

The results suggest that the kinds of data that are available, their resolution and accuracy, and the spatial scale at which decisions are to be used will combine to determine the most appropriate method in any circumstance. In addition, it is clear that validation of models needs to be done in such a way that it tests predictions in the context in which they are to be applied.

Acknowledgments. Many people have contributed to the data, analyses, and ideas presented here. Among them are Jennie Pearce, Guy Carpenter, and David Stockwell (who helped with modeling), Graeme Watson, Simon Ferrier, and Elizabeth Atkinson (who allowed me to use their S-Plus routines), Paul Yates, Michele Arundelle, Fiona Young, Adrian Moorrees, Neville Walsh, Fons Vandenberg, Doug Frood, and Dale Tonkinson (who provided and interpreted data), and Andrew Taplin, David Barratt, Simon Ferrier, Brendan Wintle, Resit Akçakaya, and Michael McCarthy (who contributed ideas and comments). Mark Burgman provided invaluable ideas and support. I am very grateful to all of them for their contributions. The research was supported by project FB-NP22 of Environment Australia.

Literature Cited

Agresti A (1996) An introduction to categorical data analysis. John Wiley and Sons, New York

Akçakaya HR, Atwood JL (1997) A habitat-based metapopulation model of the California Gnatcatcher. Conservation Biology 11:422–434

Aleksander I, Morton H (1990) An introduction to neural computing. Chapman and Hall, London

Attiwill PM (1994) The disturbance of forest ecosystems: the ecological basis for conservative management. Forest Ecology and Management 63:247–300

Austin MP (1985) Continuum concept, ordination methods and niche theory. Annual Review of Ecology and Systematics 16:39–61

Austin MP, Meyers JA (1995) Modelling of landscape patterns and processes using biological data. Subproject 4: real data case study. CSIRO, Division of Wildlife and Ecology, Canberra, Australia

Austin MP, Meyers JA (1996) Current approaches to modelling the environmental niche of Eucalypts: implications for management of forest biodiversity. Forest Ecology and Management 85:95–106

Austin MP, Cunningham RB, Good RB (1983) Altitudinal distribution in relation to other environmental factors of several Eucalypt species in southern New South Wales. Australian Journal of Ecology 8:169–80

Austin MP, Cunningham RB, Fleming PM (1984) New approaches to direct gradient analysis using environmental scalars and statistical curve-fitting procedures. Vegetatio 55:11–27

Austin MP, Nicholls AO, Margules CR (1990) Measurement of the realised qualitative niche: environmental niches of five Eucalypt species. Ecological Monographs 60:215–228

Austin MP, Nicholls AO, Doherty MD, Meyers JA (1994a) Determining species response functions to an environmental gradient by means of a beta-function. Journal of Vegetation Science 5:215–228

Austin MP, Meyers JA, Doherty MD (1994b) Predictive models for landscape patterns and processes. Subproject 2: modelling of landscape patterns and processes using biological data. Division of Wildlife and Ecology, CSIRO, Canberra, Australia

Bio AMF, Alkemande R, Barendregt A (1998) Determining alternative models for vegetation response analysis—a non-parametric approach. Journal of Vegetation Science 9:5–16

Booth GD, Niccolucci MJ, Schuster EG (1994) Identifying proxy sets in multiple linear regression: an aid to better coefficient interpretation. Research Paper INT-470. Intermountain Research Station, USDA Forest Service, Ogden, UT

Box EO, Crumpacker DW, Hardin ED (1993) A climatic model for location of plant species in Florida, USA. Journal of Biogeography 20:629–644

Brieman L, Friedman JH, Olshen RA, Stone CJ (1984) Classification and regression trees. Wadsworth International Group, Belmont, CA

Burgman MA (1989) The habitat volumes of scarce and ubiquitous plants: a test of the model of environmental control. The American Naturalist 133:228–239

Busby JR (1986) A biogeographic analysis of Nothofagus cunninghamii (Hook) Oerst in south-eastern Australia. Australian Journal of Ecology 11:1–7

Carpenter G, Gillison AN, Winter J (1993) DOMAIN: a flexible modelling procedure for mapping potential distributions of plants and animals. Biodiversity and Conservation 2:667–680

Chatfield C (1995) Model uncertainty, data mining and statistical inference. Journal of the Royal Statistical Society Series A 158:419–466

CIFOR (1999) http://www.cgiar.org/cifor/research/intro_d.html. Centre for International Forestry Research, Bogor, Indonesia

Coates F (1996) Ecological and biogeographical correlates of rarity in two narrow endemics in Tasmania, *Spyridium microphyllum* (F Muell ex Reisseck) Druce and *Spyridium obcordatum* (Hook) WM Curtis. PhD thesis, University of Tasmania, Hobart, Tasmania, Australia

Corsi F, Duprè E, Boitani L (1999) A large-scale model of wolf distribution in Italy for conservation planning. Conservation Biology 13:150–159

Crance JH (1987) Guidelines for using the Delphi technique to develop habitat suitability index curves. U.S. Fish and Wildlife Service, Biological Report 82 10134. Department of the Interior, Washington, DC

CRES (1999) ANUCLIM documentation. http://cres.anu.edu.au/software/anuclim.html

DeLong ER, DeLong DM, Clarke-Pearson DL (1988) Comparing the area under two or more receiver operating characteristic curves: a non-parametric approach. Biometrics 44:837–845

Elith J, Burgman MA (in review) Predictions and their validation: rare plants in the Central Highlands, Victoria, Australia. In: Scott JM, Heglund P, Wall W, Samson F, Haufler J. (eds) Proceedings of "Predicting Species Occurrences: Issues of Scale and Accuracy" conference, Snowbird, Utah, USA

Elith J, Burgman MA, Minchin P (1998) Improved protection strategies for rare plants: consultancy report for Environment Australia Project FB-NP22. Environment Australia, Canberra, Australia

ERIN (1995) http://www.erin.gov.au/general/biodiv_model/ERIN/GARP/home.html. Environment Australia

Ferrier S, Watson G (1996) An evaluation of the effectiveness of environmental surrogates and modelling techniques in predicting the distribution of biological diversity. Consultancy report prepared by the NSW National Parks and Wildlife Service for Department of Environment, Sport and Territories, Australia, Canberrra, Australia

Fielding AH, Bell JF (1997) A review of methods for the assessment of prediction errors in conservation presence/absence models. Environmental Conservation 24:38–49

Gaston KJ (1994) Rarity. Chapman and Hall, London

Gong P, Pu R, Chen J (1996) Mapping ecological land systems and classification uncertainties from digital elevation and forest-cover data using neural networks. Photogrammetric Engineering and Remote Sensing 62:1249–1260

Gray PA, Cameron D, Kirkham I (1996) Wildlife habitat evaluation in forested ecosystems: some examples from Canada and the United States. In: DeGraaf RM, Miller RI (eds) Conservation of faunal diversity in forested landscapes. Chapman and Hall, New York, pp 407–533

Guisan A (1998) Predicting the potential distribution of plant species in an alpine environment. Journal of Vegetation Science 9:65–74

Hanley JA, McNeil BJ (1982) The meaning and use of the area under a receiver operating characterisitc (ROC) curve. Radiology 143:29–36

Harrell FE, Lee KL (1984) Regression modelling strategies for improved prognostic prediction. Statistics in Medicine 3:143–152

Harrell FE, Lee KL, Mark DB (1996) Multivariate prognostic models: issues in developing models, evaluating assumptions and adequacy, and measuring and reducing errors. Statistics in Medicine 15:361–387

Hastie TJ, Tibshirani RJ (1990) Generalized additive models. Chapman and Hall, London

Hosmer DW, Lemeshow S (1989) Applied logistic regression. John Wiley and Sons, New York

Keith DA, Bedward M (1999) Native vegetation of the South East Forests region, Eden, New South Wales. Cunninghamia 6:1–218

Jongman RH, ter Braak CJF, van Tongeren OFR (1995) Data analysis in community and landscape ecology, 2nd ed. Cambridge University Press, Wageningen, The Netherlands

Legendre L, Legendre P (1998) Numerical ecology, 2nd ed. Elsevier, New York

Lindenmayer DB, Cunningham RB, Tanton MT, Smith AP, Nix HA (1990a) The conservation of arboreal marsupials in the montane ash forests of the central highlands of Victoria, south-east Australia: I. Factors influencing the occupancy of trees and hollows. Biological Conservation 54:111–131

Lindenmayer DB, Cunningham RB, Tanton MT, Smith AP, Nix HA (1990b) The habitat requirements of the Mountain Brushtail Possum and the Greater Glider in the montane ash forests of the central highlands of Victoria. Wildlife Research 17:467–478

Lindenmayer DB, Cunningham RB, Tanton MT, Smith AP, Nix HA (1991a) The conservation of arboreal marsupials in the montane ash forests of the central highlands of Victoria, south-east Australia: III Models of the habitat requirements of Leadbeater's Possum and the diversity and abundance of arboreal marsupials. Biological Conservation 56:295–315

Lindenmayer DB, Nix HA, McMahon JP, Hutchinson MF, Tanton MT (1991b) The conservation of Leadbeater's Possum, *Gymnobelideus leadbeateri* (McCoy): a case study of the use of bioclimatic modelling. Journal of Biogeography 18:371–383

Lynn H, Mohler CL, DeGloria SD, McCulloch CE (1995) Error assessment in decision-tree models applied to vegetation analysis. Landscape Ecology 10:323–335

Main AR (1984) Rare species, problems of conservation. Search 15:93–97

Mastrorillo S, Lek S, Dauba F (1997) Predicting the abundance of minnow *Phoxinus phoxinus* (Cyprinidae) in the River Ariege (France) using artificial neural networks. Aquatic Living Resources 10:169–176

MathSoft (1995) S-PLUS guide to statistical and mathematical analysis, version 3.3. StatSci Division, MathSoft, Inc, Seattle, WA

McCullagh P, Nelder JA (1989) Generalized linear models, 2nd ed. Chapman and Hall, London

McIntyre S, Lavorel S (1994) Predicting richness of native, rare, and exotic plants in response to habitat and disturbance variables across a variegated landscape. Conservation Biology 8:521–531

Metz CE (1978) Basic principles of ROC analysis. Seminars in Nuclear Medicine 8:283–298

Mitchell M (1996) An introduction to genetic algorithms. MIT Press, Cambridge, MA

Mitchell ND (1992) The derivation of climate surfaces for New Zealand, and their application to the bioclimatic analysis of the distribution of Kauri (*Agathis australis*). Journal of the Royal Society of New Zealand 21:13–24

Nix H (1986) A biogeographic analysis of Australian elapid snakes In: Longmore R (ed) Atlas of elapid snakes of Australia. Australian Government Publishing Service, Canberra, Australia, pp 4–15

Pan D, Bouchard A, Legendre P, Domon G (1998) Influence of edaphic factors on the spatial structure of island halophytic communities, a case study in China. Journal of Vegetation Science 9:797–804

Prober SM, Austin MP (1990) Habitat peculiarity as a cause of rarity in *Eucalyptus paliformis*. Australian Journal of Ecology 16:189–205

Rand GM, Newman JR (1998) The applicability of habitat evaluation methodologies in ecological risk assessment. Human and Ecological Risk Assessment 4:905–929

Reading RP, Clark TA, Seebeck JH, Pearce J (1996) Habitat suitability index model for the Eastern Barred Bandicoot, *Perameles gunnii*. Wildlife Research 23:221–235

Richardson DM, McMahon JP (1992) A bioclimatic analysis of *Eucalyptus nitens* to identify potential planting regions in southern Africa. South African Journal of Science 88:380–387

Schamberger ML, O'Neil LJ (1984) Concepts and constraints of habitat-model testing. In: Verner J, Morrison ML, Ralph CJ (eds) Wildlife 2000: modeling habitat relationships of terrestrial vertebrates. Based on an international symposium held at Stanford Sierra Camp, Fallen Leaf Lake, CA. University of Wisconsin Press, Madison, WI, pp 5–10

S-News (1999) Website for functions available for use with MathSoft's S-PLUS statistics program. http://www.biostat.wustl.edu/s-news

Sokal RR, Rohlf FJ (1981) Biometry: the principles and practice of statistics in biological research, 2nd ed. WH Freeman and Co, New York

Stockwell DRB, Noble IR (1992) Induction of sets of rules from animal distribution data: a robust and informative method of data analysis. Mathematics and Computers in Simulation 33:385–390

Swets JA (1988) Measuring the accuracy of diagnostic systems. Science 240:1285–1293

ter Braak CJF (1986) Canonical correspondence analysis, a new eigenvector technique for multivariate direct gradient analysis. Ecology 67:1167–1179

USFWS (1980) Habitat evaluation procedures. ESM 102 Release 2–80. U.S. Fish and Wildlife Service, Department of the Interior, Washington, DC

Van Horne B (1983) Density as a misleading indicator of habitat quality. Journal of Wildlife Management 47:893–901

Van Horne B, Wiens JA (1991) Forest bird habitat suitability models and the development of general habitat models. Fish and Wildlife Research 8. U.S. Department of the Interior, Fish and Wildlife Service, Washington, DC

Wiser SK, Peet RK, White PS (1996) High-elevation rock outcrop vegetation of the southern Appalachian Mountains. Journal of Vegetation Science 7:703–722

Yee TW, Mitchell ND (1991) Generalized additive models in plant ecology. Journal of Vegetation Science 2:587–602

Zweig MH, Campbell G (1992) Receiver-operating characteristic (ROC) plots: a fundamental evaluation tool in clinical medicine. Clinical Chemistry 39:561–577

5

Risk Assessment of a Proposed Introduction of Pacific Salmon in the Delaware River Basin

Paul T. Jacobson

Introduction

Culture and introduction of nonindigenous species have long played an important role in fisheries management (Stroud 1986). Many fisheries biologists consider fish culture and stocking synonymous with fisheries management, and fisheries management has been a dominant consideration in management of the aquatic environment (Radonski and Martin 1986). Many species of fishes are highly fecund and experience high natural mortality in the wild during their early life stages. These factors make artificial propagation a feasible and often appealing way to ameliorate overharvesting and habitat degradation.

Hatchery rearing and stocking has also been used as a means of artificially extending the range of many fishes. Numerous species native to parts of North America have been widely introduced by resource management agencies and private citizens to provide additional sport fishing opportunities. Many of these species have become naturalized and are considered part of the native fish fauna by many persons. Examples of North American species transplanted outside of their native range include Smallmouth Bass (*Micropterus dolomieui*), Large-mouth Bass (*Micropterus salmoides*), Walleye (*Stitzostedion vitreum vitreum*), Rainbow Trout (*Oncorynchus mykiss*), Striped Bass (*Morone saxatilis*), and American Shad (*Alosa sapidissima*). In many rivers of the east coast, nonindigenous species have replaced native anadromous species that were extirpated following construction of dams and other blockages. Some interbasin transfers were undertaken to provide forage for indigenous and nonindigenous game fish (e.g., Rainbow Smelt [*Osmerus mordax*]); others have occurred inadvertently via anglers' bait buckets. Additional species were long ago introduced from Europe and beyond. Examples of exotic introductions include Brown Trout (*Salmo trutta*) and the Common Carp (*Cyprinus carpio*)

The aquarium trade has led to introductions of nonindigenous fish when such fish were released into suitable habitat, especially in Florida. More than 27 species of fish may have been introduced by that route (Courtenay and Williams 1992). The aquarium trade has led to introduction of at least seven species since 1980 (OTA 1993). Aquaculture also has led to introduction of nonindigenous fish. The likelihood of escape from aquaculture facilities is sufficiently high that the federal interagency group, the Aquatic Nuisance Species Task Force, considers eventual escape virtually guaranteed (OTA 1993).

Introductions have also occurred as a result of removal of natural barriers. Examples of this class of introduction include the Sea Lamprey (*Petromyzon marinus*) and the Alewife (*Alosa pseudoharengus*) which were able to bypass Niagara Falls and enter the upper Laurentian Great Lakes following construction of the Welland Canal. Finally, intercontinental transfers have occurred via ballast water. For example, the European Ruffe (*Gymnocephalus cernuus*) invaded the Laurentian Great Lakes via ballast water originating in Europe.

The consequences and public perceptions of these introductions are as varied as the means by which the introductions have occurred. The Sea Lamprey contributed to the demise of Lake Trout (*Salvelinus namaycush*), a large commercially valuable piscivore, and the abundance of non-native Alewife subsequently exploded in habitat that held few large predators. Alewife reached nuisance levels and underwent periodic die-offs that created offensive and unhealthy conditions requiring expensive cleanups. Beginning in 1966, Pacific salmonids were introduced into the upper Great Lakes in large numbers. Between 1966 and 1983, an average of 2.6 million Coho Salmon (*Oncorhynchus kisutch*) were stocked annually into Lake Michigan alone. Pacific Salmon have substantially reduced the abundance of Alewife and are the driving force behind a large recreational fishery in the Great Lakes. In 1980, the Great Lakes recreational fisheries comprised 54 million angler days and more than $1.7 billion in angler expenditures (Radonski and Martin 1986). Chemical treatment of streams flowing into the upper Great Lakes with lamprecide provides short-term control of lamprey abundance but is costly ($10 million annually for control and related research; OTA 1993). Native Lake Trout remain at low abundance, and the recreational fishery is maintained by massive stockings of Pacific salmonids.

The Common Carp was introduced to North America to provide a familiar food and sport fish for European immigrants (Courtenay and Kohler 1986; Lathrop et al. 1992). The first introduction occurred in the Hudson River, New York, in 1831; a second release was into a pond in California in 1872 (Moyle 1976; Courtenay and Kohler 1986). From 1877 until 1896, the U.S. Commission of Fish and Fisheries distributed Common Carp for culture and introduction (Courtenay and Kohler 1986; Lathrop et al. 1992). In 1897, the U.S. Fish Commission stopped distributing Carp because the species was well established and because of growing complaints by the public (Moyle 1969; Lathrop et al. 1992). Although the species has historically supported profitable commercial fisheries (Lathrop et al. 1992), the introduction of this species to North America is generally considered a fisheries management disaster. Adverse effects of this species extend beyond the lakes in which carp live because they destroy important submerged aquatic vegetation beds in staging lakes used by diving ducks on their annual migrations.

The Brown Trout was introduced from Europe for sport fishing and has become widely established in North America. It is prized as a sport fish, and widespread stocking continues. Although some adverse effects on native species have been reported (Fausch and White 1981; Waters 1983; Taylor et al. 1984; Kennedy and Strange 1986; Fausch 1988), the Brown Trout is generally considered a fisheries management success story (Courtenay and Kohler 1986).

At least 70 fish species have been introduced from outside the United States; most of these were introduced within the past 50 years (Fig. 5.1; OTA 1993); more than 168 species have been transplanted within the United States (Courtenay and Taylor 1984). The proportion of intentionally introduced or transplanted fishes causing harm (46%) is slightly higher than the percentage of unintentionally introduced species causing harm (38%), and intentional introductions outnumber unintentional introductions nearly three to one (OTA 1993). These findings suggest that researchers and resource managers have been ineffective in identifying and preventing the intentional introduction of harmful nonindigenous fish. Magnuson (1976) described intentional introductions of fish as a game of chance. Although the odds that an introduction will produce no harmful results may appear better than even, the alternative outcomes are, in fact, incommensurable, because introductions can lead to irreversible adverse effects on indigenous flora and fauna.

It was not until 1969 that the American Fisheries Society and the American Society of Ichthyologists and Herpetologists addressed exotic fish as an issue of concern (McCann 1984). Since that time, attitudes and approaches for dealing with exotic fish have slowly evolved. At its annual meeting in 1972, the American Fisheries Society adopted a position statement on introductions of exotic species (AFS 1972). The position statement called for an assessment of need, alternatives, and impacts and, if appropriate, experimental research prior to any introduction. The position statement received little attention, and it was rarely followed (Kohler and Courtenay 1986).

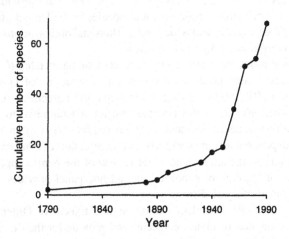

FIGURE 5.1. Cumulative number of fish species of foreign origin in the United States. Figure only includes those species with known introduction dates. (From OTA 1993.)

In 1973, the International Council for the Exploration of the Sea adopted a "Code of Practice to Reduce the Risks of Adverse Effects Arising from Introductions of Non-indigenous Marine Species" (Kohler and Courtenay 1986). In 1977, President Carter issued an executive order on exotic organisms (11987, 42 Fed. Reg. 26949 [25 May 1977]), directing federal agencies, to the extent authorized by law, to restrict introduction of exotic species into U.S. ecosystems. In practice, the U.S. Fish and Wildlife Service (USFWS) reviews each proposed introduction involving federal agencies or funding and issues a biological opinion. One of the criteria for determining the biological opinion is concurrence with the state responsible for management of the species and its habitat in the affected ecosystem (Versar 1992a).

In 1980, the Exotic Fish Section was formed within the American Fisheries Society to address exotic fish issues. Subsequently, the section changed its name to the Introduced Fish Section to encompass transplanted species, which far outnumber introductions of exotics and present many of the same risks (Kohler and Courtenay 1986). In 1986, the American Fisheries Society devoted an entire issue of its bulletin to the issue of introduced species (AFS 1986).

In 1987, the North American Commission of the North Atlantic Salmon Conservation Organization (NASCO) endorsed a set of recommendations that included a prohibition on intentional introductions of Pacific salmonids in eastern North America. The Non-indigenous Aquatic Nuisance Prevention and Control Act of 1990 (16 U.S.C.A. 4701–4751) created the interagency Aquatic Nuisance Species Task Force. The task force, co-chaired by representatives of the National Oceanic and Atmospheric Administration and the USFWS, is required to develop a program to prevent, monitor, and control unintentional introductions of nonindigenous aquatic nuisance species and to provide for related public education and research (OTA 1993).

The Office of Technology Assessment conducted a study of harmful nonindigenous species in the United States (OTA 1993). The report provided an overview of the status of harmful nonindigenous species in the United States and an examination of the technological issues and institutional organizations involved in dealing with harmful nonindigenous species.

The Chesapeake Bay Program, a joint federal–state partnership to protect and restore Chesapeake Bay, produced a policy on introduction of nonindigenous aquatic species (CBP 1993). Among other steps, the agreement requires participating jurisdictions (i.e., states and the District of Columbia) to submit any permit application that could ultimately be approved for review by an ad hoc panel made up of independent technical experts and representatives from each jurisdiction that is a party to the agreement. After review of the permit application and within 60 days of the date of submittal, the ad hoc panel provides nonbinding recommendations to the jurisdiction.

On February 3, 1999, President Clinton signed Executive Order 13112 "to prevent the introduction of invasive species and provide for their control and to minimize the economic, ecological, and human health impacts that invasive species cause." The order calls on each federal agency "to the extent practicable and permitted by law" and subject to the availability of funds to (1) identify actions that may affect invasive alien species; (2) prevent the introduction of alien species and control them in a cost-effective and environmentally sound manner; and (3) not authorize, fund, or implement actions that are likely to facilitate the introduction or spread of invasive species unless the agency involved has publicly determined that the benefits of such actions clearly outweigh the potential harm. An Invasive Species Council was also established to implement the executive order. The council comprises the Departments of State, Treasury, Defense, Interior, Agriculture, Commerce, and Transportation and the Environmental Protection Agency, as well as other invited federal agencies.

These and other state, regional, federal, and international agreements reflect a growing awareness of the potential hazard that non-native fish pose to aquatic ecosystems. Policies and attitudes are in a state of transition. During the early 1980s, in the midst of an evolving sociopolitical climate, the New Jersey Division of Fish, Game, and Wildlife (NJDFGW) initiated a program to introduce Pacific Salmon into the Delaware River watershed. At an earlier time, such a proposal might have received little or no opposition; however, because of its timing, New Jersey's proposal was destined to receive intense scrutiny by a large body of stakeholders. Interestingly, neighboring states with their own hatchery-supported Pacific Salmon fisheries opposed New Jersey's proposal; some angling groups opposed the introduction out of concern for their preferred quarry—another introduced Pacific salmonid that had become naturalized within the upper reaches of the Delaware River watershed.

Introduction of Pacific Salmon to the Delaware River Basin

In an effort to create additional sport fishing opportunities and to attract economic benefits to the region, the NJDFGW proposed an introduction of Pacific Salmon to the Delaware River. New Jersey applied for federal funding for the program, and the USFWS determined that the project could have significant impacts on environmental, cultural, or human resources. Thus, according to the National Environmental Policy Act of 1969 (NEPA), completion of an environmental impact statement (EIS) for the proposed program was necessary and appropriate. In response to concerns about the potential ecological and socioeconomic impacts of a large-scale introduction, NJDFGW proposed a small-scale, short-term experimental program to assess the feasibility and advisability of a larger-scale introduction. The proposed action consisted of a stocking of 300,000 Chinook Salmon (*Oncorhynchus tshawytscha*) in each of five consecutive years. Fish were to be reared on site at the Charles O. Hayford Fish Hatchery at Hackettstown and released into the adjacent Musconetcong River, a tributary of the Delaware River. Alternative proposals included stocking of 50,000 Steelhead Trout (*O. mykiss*) and 50,000 Coho Salmon. Reintroduction of the native Atlantic Salmon (*Salmo salar*) received little attention as an alternative, because the Delaware River historically was on the very edge of the Atlantic Salmon range and habitat conditions have since deteriorated. Attempts to reintroduce Atlantic Salmon earlier in this century failed; however, Pacific Salmon were considered a feasible option because they are more tolerant of water quality conditions to which they would be exposed in the lower Delaware River.

The proposed program included monitoring of returning adults. Monitoring was to begin 3 years after the first Chinook were stocked and to continue for 6 years after the last year of stocking. Elements of the monitoring/managerial program included marking or tagging of stocked fish, monitoring of movements of returning adults, creation of a public awareness program, seasonal closure of the lower Muscenetcong River, continuation of the statewide ban on snagging, and collection of resource use and socioeconomic data. Because monitoring would occur for at least 6 years beyond the last year of stocking, no action on a permanent program would occur for at least 11 years (Versar 1992a).

The experimental stocking and monitoring program was intended to provide the information needed to assess the feasibility and advisability of a large-scale, long-term program. If NJDFGW decided on the basis of the experimental program to proceed with a large-scale program, the data acquired from the monitoring program would be used to support a supplemental EIS.

Many of the potential environmental impacts identified by concerned citizens during the NEPA issue scoping process derive from straying by returning salmon. Examples of these impacts are colonization of streams, competition with native and previously introduced salmonids in the upper Delaware River watershed, and genetic depletion in a naturalized stock of Rainbow Trout in the upper Delaware River basin. The level of risk associated with these concerns is related to the probability and magnitude of straying. Thus, the risk assessment focused on straying. The preferred species, Chinook Salmon, and two alternative species, Coho Salmon and Steelhead Trout, were examined in the risk assessment.

Straying Model

Methods

Given the high degree of uncertainty and variability in factors that contribute to straying, a probabilistic approach is a useful means of evaluating potential levels of straying associated with stocking of anadromous salmonids. A Monte Carlo simulation approach was adopted to retain a high degree of flexibility in the model structure so that mechanistic detail could easily be added and deleted as needed. An additional benefit of this approach is its conceptual simplicity.

The model is driven by the following inputs: number of fish stocked each year, number of years in which stocking occurs, a return schedule characteristic of the species, statistical distributions for percentage return, percentage harvested, percentage straying (Table 5.1), and distribution of strays among streams (Table 5.2). Model parameters were derived from data in the scientific literature, as explained below. The model does not include natural production of smolts by returning stocked fish.

TABLE 5.1. Model parameter values.

Parameter	Chinook Salmon	Coho Salmon	Steelhead Trout
a	300,000	50,000	50,000
μ_h, σ_h	0.41, 0.19	0.41, 0.19	0.41, 0.19
μ_r, σ_r	0.0008, 0.0013	0.0096, 0.0161	0.0125, 0.0209
μ_s, σ_s	0.181, 0.101	0.0588, 0.0282	0.052, 0.032
p	0.427	0.424	0.620
$f_{\male,0}, f_{\female,0}$	0.000, 0.000	0.000, 0.000	0.000, 0.000
$f_{\male,1}, f_{\female,1}$	0.000, 0.000	0.000, 0.000	0.000, 0.000
$f_{\male,2}, f_{\female,2}$	0.001, 0.000	0.320, 0.000	0.041, 0.000
$f_{\male,3}, f_{\female,3}$	0.286, 0.010	0.680, 1.000	0.195, 0.062
$f_{\male,4}, f_{\female,4}$	0.463, 0.449	—[a]	0.479, 0.337
$f_{\male,5}, f_{\female,5}$	0.226, 0.500	—	0.207, 0.402
$f_{\male,6}, f_{\female,6}$	0.024, 0.041	—	0.077, 0.196
$f_{\male,7}, f_{\female,7}$	—	—	0.000, 0.004

[a]— Not applicable.

In each replication of the model, the annual number of strays is computed as follows:

$$N_i = a(1 - h_i)\{p\sum_j [r_j f_{(i-j)} s_j] + [1 - p]\sum_j [r_j m_{(i-j)} s_j]\} \qquad (5.1)$$

where

N_i = number of strays in year i
a = number of fish stocked each year
p = fraction of returning fish that is female
h_i = fraction of fish returning in year i that is harvested [a random variable, $N(\mu_h, \sigma_h)$]
r_j = fraction of cohort j that returns [a random variable, $N(\mu_r, \sigma_r)$]
s_j = fraction of cohort j that strays [a random variable, $N(\mu_s, \sigma_s)$]
$m_{(i-j)}$ = fraction of returning males that return at age $(i-j)$

TABLE 5.2. Probability density functions for distribution of strays among streams.

Stream	Chinook Salmon	Coho Salmon	Steelhead Trout
A	0.760	0.549	0.692
B	0.184	0.176	0.154
C	0.027	0.157	0.077
D	0.007	0.039	0.026
E	0.007	0.039	0.026
F	0.002	0.020	0.026
G	0.002	0.020	—[a]
H	0.002	—	—
I	0.002	—	—

[a]—Not applicable.

$f_{(i-j)}$ = fraction of returning females that return at age $(i-j)$

$j = \{0, 1, 2, \ldots, j_{max}\}$

$k = \{0, 1, 2, \ldots, k_{max}\}$

$i = \{0, 1, 2, \ldots, k_{max} + j_{max}\}$

j_{max} = number of cohorts stocked − 1

k_{max} = maximum age at which fish return

$N(\mu, \sigma)$ = a normal distribution having a mean of μ and a standard deviation of σ. Values of h_i, r_j, and s_j falling below zero were set equal to zero.

The return rate implicitly incorporates all sources of mortality that are not included in the harvest rate. Model year 0 begins when the first cohort of eggs hatches; thus, the first stocking of fish occurs in year 1. For each replication (iteration) of the model, the number of strays in each stream was tabulated. Frequencies for each combination of number of streams and number of strays were calculated on the basis of 50,000 replications of the model. Year-specific probabilities associated with a given outcome were then calculated as the proportion of model replications in which the outcome occurred.

There are several assumptions pertaining to the behavior and ecology of the fish that are explicit in the model. Within each replication of the model, straying rate is held constant for a given cohort, regardless of the year in which those fish return. This assumption is consistent with Shapovalov and Taft's (1954) observation that "the rate of straying from a given stream is fairly constant for a given year class (over all years in which returns occur), but may vary considerably from year class to year class." Return rate is also held constant for a given cohort in each replication of the model. This assumption is based on evidence that overall survival is most heavily influenced by factors operating relatively early in the cohort's marine life rather than by ecological conditions at the time of return (Mathews and Buckley 1976; Neilson and Geen 1986; Holtby et al. 1990).

Derivation of Model Parameters

Harvest Rate

The harvest rate mean, μ_h, and standard deviation, σ_h, for all three species were estimated from returns and harvest of the 1968–1984 brood years in the New Hampshire Coho fishery (R.S. Fawcett, New Hampshire Fish and Game Department, personal communication). Any fishery established in the Delaware River and estuary is more likely to be similar to the New Hampshire fishery than to those of the Great Lakes and the west coast of North America, where large numbers of many salmonid species are sought.

Return Rate

As with harvest rate, the mean return rate, μ_r, and standard deviation, σ_r, for each species was derived from the New Hampshire Coho fishery. Mean return rate for Coho Salmon was estimated from a linear regression of number of returns against the number of smolts stocked (Fig. 5.2; returns = 0.005002 (released) + 229.7,

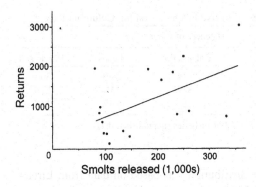

FIGURE 5.2. Number of smolts stocked and number of returns, New Hampshire Coho fishery, 1968–1984. Line is the least-squares best fit.

$R^2 = 0.28$, $p = .03$). Note that the assumed stocking level of 50,000 Coho smolts falls outside the range of stocking levels encompassed by the regression and that the expected number of returns is based on linear extrapolation. The standard deviation for the Coho return rate is the standard deviation of the prediction calculated from the regression statistics:

$$\sigma_r = \sqrt{\left(\frac{SE^2_{Estimate}}{MSE_{Regression}} + 1\right) MSE_{Regression}}$$

There are no comparable data for Chinook Salmon on the east coast of the United States. Thus, the return rate for Chinook relative to Coho on the west coast was used to obtain a return rate for Chinook on the east coast. West coast Chinook return at one-twelfth the rate of west coast Coho Salmon (Gross 1985; Lister et al. 1981a, b; Stauffer 1977; Bilton et al. 1982; Fedorenko and Shepherd 1986; Blackett 1979; Duke 1984). The mean and standard deviation for the Chinook return rate distribution for the Musconetcong stocking program was obtained by dividing the corresponding values for Coho Salmon by 12. Steelhead return rates were assumed to be comparable with those for Coho Salmon. However, the incidence of repeat spawning in winter run Pacific Coast Steelhead ranges from 5 to 31.3%, generally increasing from north to south (Hassinger et al. 1974). To incorporate repeat spawning in the Steelhead simulations, the mean return rate was set 30% higher than that for Coho Salmon. The coefficient of variation was assumed to be equal to that used for Coho.

Stray Rate

A stray rate distribution for each species was derived from the best available study for each species. The parameters for Chinook Salmon were derived from a study of straying encompassing 3 brood years and five hatchery release sites (Table 5.3). Stray rates ranged from 2.4 to 36%, with a mean of 18.1% and standard deviation of 10.1%.

The stray rate distribution for Coho Salmon was derived from four imprinting experiments conducted by Hasler and Scholz (1983). Those four experiments produced a mean stray rate of 5.88% with an estimated standard deviation of

TABLE 5.3. Chinook Salmon stray rates from five hatcheries on the Columbia River.[a]

	Hatchery of origin				
Brood Year	Abernathy	Cowlitz	Kalama Falls	Lewis	Washougal
1977	11%	28%	17%	2.4%	30%
1978	15%	19%	35%	15%	21%
1979	12%	36%	6.8%	9.4%	14%

[a]From T. Quinn, University of Washington School of Fisheries (unpublished data).

2.82%. The Steelhead straying rate distribution was based on data from Lirette and Hooton (1988). The mean of 5.2% and standard deviation of 3.2% were estimated from 5 brood-year and rearing-site combinations for which recoveries exceeded 100 adults (110–144 fish each). The selection of these studies used to develop the straying rates and variability of those rates was based on study design and amount of data available. In general, the best studies were those in which most available streams were surveyed for strays and in which multiple year classes were studied. Although the rates for the "best" studies for Coho and Steelhead are well within the range of rates reported in the literature, the Chinook mean rate of 18.1% from Quinn's Columbia River data is higher than all other straying rate values presented in the literature (T.P. Quinn, University of Washington, personal communication). For example, Major and co-workers (1978) report Chinook straying rates as being less than 8.4%. Quinn's data were selected for use here primarily because of their comprehensive nature, which allowed calculation of the variance of the straying rate (Table 5.3, Chinook). However, results of this analysis may be biased upward for Chinook Salmon.

Sex Ratio and Age at Return

The sex ratio and parameters for percentage return by age and sex of returning Chinook Salmon were computed from data in Fulton and Pearson (1981). The sex ratio and sex-specific percentage return-at-age parameters used for Coho were calculated from returns to Waddell Creek, California, between 1933 and 1939 (Shapovalov and Taft 1954). Steelhead parameters were calculated from spawning run data from two creeks on the Minnesota shore of Lake Superior in 5 consecutive years (Hassinger et al. 1974).

Distribution of Strays Among Streams

Strays are independently and identically distributed among streams in the model. This assumption is supported by Quinn's observations (personal communication, University of Washington School of Fisheries) that straying is exhibited by individuals and is not influenced by group behavior. Information adequate for the purpose of developing species-specific probability density functions (PDFs) used for assigning each stray to a stream was very limited. The Chinook PDF is based on data from a single natal stream and 4 brood years in the Columbia River basin (Quinn and Fresh 1984). That study was conducted by the same principal investigator within a subregion of the same research area providing the data on straying rates. Strays to a given stream were tabulated across recovery years. The straying data used for Chinook Salmon are geographically comprehensive in that every hatchery and all the spawning grounds available to Chinook Salmon in the Columbia River basin were surveyed. Furthermore, out-of-basin strays could be (and were) reported in the coded wire tag database (Quinn and Fresh 1984; Quinn, personal communication). Thus, the Chinook PDF can be considered an accurate representation of the distribution of strays from a single natal stream in 4 brood years.

Probability density functions for Coho Salmon and Steelhead Trout were derived from the same studies used to derive the respective stray rate distributions. These studies were chosen for development of the straying component of the model because, among all the literature reviewed, they provided the most comprehensive assessments of stray distribution and abundance. In all three species, strays entered a very small subset of the streams and rivers available to them. This observation is consistent with results from other less comprehensive studies of straying.

The PDF for Coho Salmon was derived from imprinting experiments conducted in Lake Michigan (Hasler and Scholz 1983). For each of four experimental treatments, streams were ranked by the number of strays they received. The PDF was then formed by summing the number of strays in streams with the same ranking.

For Steelhead Trout, data from 4 return years and two natal streams on Vancouver Island (Lirette and Hooton 1988) were used to approximate the distribution of strays among streams. For each natal stream, strays were tabulated by the stream in which they appeared, without regard to year. Two separate rankings were then made based on the number of strays received from a single natal stream. Counts from streams having the same ranking were summed to give the PDF. The normalized PDFs for the three species are presented in Table 5.2.

Results and Discussion

Adult Chinook Salmon would return to the Delaware River from 3 to 10 years after hatching of the first hatchery cohort (Fig. 5.3). Chinook stray densities were the highest among the three species examined. Stray densities reached a maximum during years 5–7 and subsequently declined. This is a consequence of the combined effect of five stocked cohorts reaching and then surpassing ages of peak return (see Table 5.1). Based on the mean return and stray rates, 240 fish were expected to return in the peak year, and 43 would be expected to stray. The probability of 50 or fewer strays appearing in a single stream is approximately 85%. If only females are considered, there is a 90% probability of 25 or fewer females appearing in a single stream (Fig. 5.4). Given the model assumptions, there is a 99% probability that the proposed stocking program would produce fewer than 100 strays in a single stream in a single year.

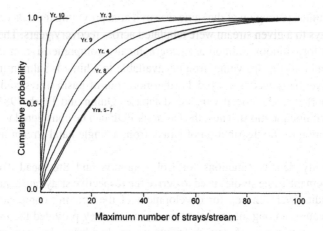

FIGURE 5.3. Year-specific distribution functions for the maximum number of Chinook Salmon strays per stream (males and females combined).

Coho Salmon return 2–7 years after hatching of the first cohort (Fig. 5.5). Coho Salmon produced the lowest stray densities of the three species examined. Coho have a shorter life span and return over just a 2-year period. Thus, the stocking program produces returns over a shorter period of time, and the stray density cumulative density functions (CDF) show less annual variation. In year 2, all the

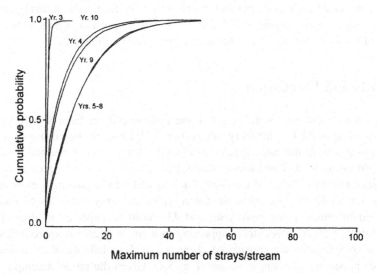

FIGURE 5.4. Year-specific distribution functions for the maximum number of Chinook Salmon strays per stream (females only).

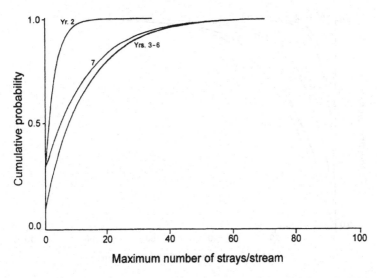

FIGURE 5.5. Year-specific distribution functions for the maximum number of Coho Salmon strays per stream (males and females combined).

returns, and thus all the strays, are males. The expected return in the peak year is 480 fish, of which 28 would be strays. The model indicates a 99% probability of 60 or fewer strays (35 or fewer females; Fig. 5.6) in a single stream.

The Steelhead Trout simulations produced intermediate stray densities (Fig. 5.7). The stray density CDFs display a large amount of annual variation as a

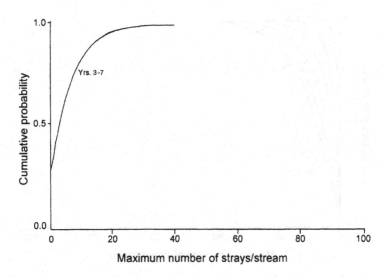

FIGURE 5.6. Year-specific distribution functions for the maximum number of Coho Salmon strays per stream (females only).

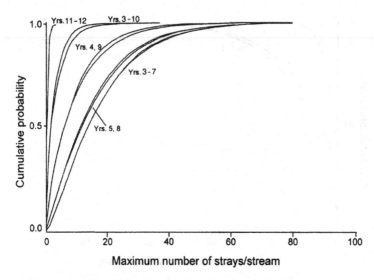

FIGURE 5.7. Year-specific distribution functions for the maximum number of Steelhead Trout strays per stream (males and females combined).

consequence of the Steelhead's complex life history. The expected number of returns in the peak year was 623, with 32 of them expected to stray. The model indicates a 99% probability that the maximum number of Steelhead strays occurring in a single stream would be 70 or fewer (Fig. 5.7), with 40 or fewer being females (Fig. 5.8).

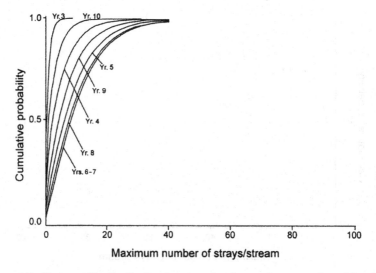

FIGURE 5.8. Year-specific distribution functions for the maximum number of Steelhead Trout strays per stream (females only).

Application of return rates and stray rates to cohorts rather than return years reduces variation in the number of strays occurring in any year, because more than one cohort contribute to each spawning run. Thus the model output is relatively insensitive to the variance in cohort-specific return and stray rates. This is true to a greater extent for Chinook and Steelhead, because in those species cohorts return over several years, versus 2 years in Coho.

The discrete finite PDF for each species defines the distribution of strays among streams and limits the number of streams to which fish can stray. That limit is the maximum number of streams observed to have received strays from a single natal stream. Missed strays and streams are unlikely, however, to seriously bias the estimates of maximum number of strays per stream, because the number of missed strays and the number of strays per missed stream are both probably very low. The effect of missed streams on model output is offset by the marginal reduction in stray rate attributable to the strays missed in those streams.

The set of values used to approximate the variance of the distribution for each variable includes an unknown amount of measurement error, which biases upward the variance of the model output, thereby broadening the stray density CDFs.

The proposed experimental introduction of Pacific salmonids is unlikely to result in large numbers of stocked fish entering streams other than the Musconetcong River. The short-term nature of the experimental program and the expected small number of strays per stream make it unlikely that stocked fish could seriously disrupt the streams into which they might stray. If any significant long-term effects arose from the experimental program, they would probably be a consequence of successful reproduction and establishment of a self-sustaining population.

The probability of establishment of a self-sustaining population by the limited experimental stocking program is very low but not zero. Spawning would depend entirely on strays—any fish returning to the carefully monitored Musconetcong River would be trapped and removed. The numbers of strays presented above are for entire years and streams. Successful spawning depends on the co-occurrence in time and space of heterosexual pairs, which may not happen with small numbers of strays returning to streams over a protracted period. If successful spawning were to occur, the eggs, larvae, and juveniles would be exposed to the suite of mortality factors that hatchery propagation attempts to minimize or eliminate. Wild smolts would also be more likely to migrate at a time of the year when environmental conditions in the middle and lower Delaware River estuary are hostile. These factors would all tend to diminish the chances that the number of fish completing the life cycle would be sufficient to establish a viable population. However, unlike the situation with stocked fish in which fish that return to their natal stream would be trapped and removed, all the wild fish that did return would be potential spawners, including the vast majority of them that would likely return to their natal stream.

A quantitative analysis to evaluate the likelihood of establishing a self-sustaining population was beyond the scope of the draft EIS, and it is beyond the scope of this chapter. If establishment of a self-sustaining population is unacceptable, an analysis encompassing the entire salmon life cycle should be required prior to any introduction; only then can there be an objective assessment of risk and a sound basis for designing steps to reduce that risk.

Postscript

In 1992, New Jersey Department of Environmental Protection and Enforcement announced a decision not to pursue the proposal to stock Pacific salmonids in the Delaware River watershed. Reasons cited in the press release included evaluation of the findings of the Draft EIS; consideration of public comment on the proposal; a salmon program would detract from a more immediate priority, promotion of a warm-water fisheries program; and a salmon program would require significant investment of resources over an extended period of time with no guarantee of success. Additional factors contributing to the decision to withdraw the proposal included (Versar 1992b) National Park Service (NPS) objections to introduction of any Pacific salmonid because of conflicts with NPS management plans for two parks on the upper Delaware River; a lack of support from other members of the Delaware River Fish and Wildlife Management Cooperative; a NASCO exemption was unlikely to be granted for any introduction of Pacific Salmon to the east coast; and future USFWS funding for an introduction of Pacific Salmon is unlikely because of Executive Order 11987 policy and the Nonindigenous Aquatic Nuisance Prevention and Control Act of 1990.

One of the findings of the draft EIS that was a major factor in the decision by NJDFGW to withdraw the stocking proposal was the low expected number of returns revealed by the straying model. Nonetheless, the proposal to introduce Pacific Salmon to the Delaware River basin was undoubtedly a casualty of the changing sociopolitical climate in which it was examined.

Literature Cited

AFS (1972) Position of American Fisheries Society on introductions of exotic aquatic species. Transactions of the American Fisheries Society 102:274–276

AFS (1986) The good, the bad, the ugly: introduced species. Fisheries 11(2)

Bilton HT, Alderdice DF, Schnute JT (1982) Influence of time and size at release of juvenile Coho Salmon (*Oncorhynchus kisutch*) on the returns at maturity. Journal of the Fisheries Research Board Canada 39:426–447

Blackett RF (1979) Establishment of Sockeye (*Oncorhynchus nerka*) and Chinook (*O. tshawytscha*) salmon runs at Frazer Lake, Kodiak Island, Alaska. Journal of the Fisheries Research Board Canada 36:1265–1277

CBP (1993) Chesapeake bay policy for the introduction of non-indigenous aquatic species. CBP/TRS 112/94 Chesapeake Bay Program, Annapolis, MD

Courtenay WR, Taylor JN (1984) The exotic ichthyofauna of the contiguous United States with preliminary observations on intranational transplants. European Inland Fisheries Advisory Commission Technical Paper 42:466–487, Rome, Italy

Courtenay WR, Kohler CC (1986) Exotic fishes in North American fisheries management. In: Stroud RH (ed) Fish culture in fisheries management. American Fisheries Society, Bethesda, MD, pp 401–413

Courtenay WR, Williams JD (1992) Dispersal of exotic species from aquaculture sources, with emphasis on freshwater fishes. In: Rosenfield A, Mann R (eds) Dispersal of living organisms into aquatic ecosystems. Maryland Sea Grant Program, College Park, MD, pp 49–81

Duke RC (1984) Anadromous fish marking and recovery 3/1/83–2/28/84. Idaho Department of Fish and Game, Fisheries Division, job performance report 52 (Article 08)

Fausch KD (1988) Tests of competition between native and introduced salmonids in streams: what have we learned? Canadian Journal of Fisheries and Aquatic Sciences 45:2238–2246

Fausch KD, White RJ (1981) Competition between Brook Trout (*Salvelinus fontinalis*) and Brown Trout (*Salmo trutta*) for positions in a Michigan stream. Canadian Journal of Fisheries and Aquatic Sciences 38:1220–1227

Fedorenko AY, Shepherd BG (1986) Review of salmon transplant procedures and suggested transplant guides. Canadian Technical Report Fisheries and Aquatic Science Report no. 1479 Fisheries and Oceans, Vancouver, BC, Canada

Fulton LA, Pearson RE (1981) Transportation and homing experiments on salmon, *Oncorhynchus spp.* and Steelhead Trout, *S. gairdneri*, in the Columbia River system: fish of the 1939–1944 broods. NOAA Technical Memorandum NMFS F/NWC-12. National Oceanic and Atmospheric Administration, National Marine Fisheries Service, Springfield, VA

Gross MR (1985) Disruptive selection for alternative life histories in salmon. Nature 313:47–48

Hasler AD, Scholz AT (1983) Olfactory imprinting and homing in salmon. Springer-Verlag, New York

Hassinger RL, Hale JG, Woods DE (1974) Steelhead of the Minnesota North Shore. Fisheries Technical Bulletin 11. Minnesota Department of Natural Resources Division of Fish and Wildlife

Holtby LB, Anderson BC, Kadowski RK (1990) Importance of smolt size and early ocean growth to interannual variability in marine survival of Coho Salmon (*Oncorhynchus kisutch*). Canadian Journal of Fisheries and Aquatic Sciences 47:2181–2194

Kennedy GJA, Strange CD (1986) The effects of intra- and inter-specific competition on the survival and growth of stocked juvenile Atlantic Salmon, *Salmo salar*, and resident trout, *Salmo trutta*, in an upland stream. Journal of Fish Biology 28:479–489

Kohler CC, Courtenay WR Jr (1986) Regulating introduced aquatic species: a review of past initiatives. Fisheries 11:34–38

Lathrop RC, Nehls SB, Brynildson CL, Plass KR (1992) The fishery of the Yahara lakes. Technical Bulletin 181. Wisconsin Department of Natural Resources, Madison, WI

Lirette MG, Hooton RS (1988) Coded-wire tag recoveries from Vancouver Island sport caught Steelhead 1982–1986. Fisheries Management Report 92. Ministry of Environment, Nanaimo, British Columbia, Canada

Lister DB, Hickey DG, Wallace I (1981a) Review of the effects of enhancement strategies on the homing, straying, and survival of Pacific salmonids, vol 1. Department of Fish and Oceans, Vancouver, BC, Canada

Lister DB, Thorson LM, Wallace I (1981b) Chinook and Coho Salmon escapements and coded-wire tag returns to the Cowichan-Koksilah River system, 1976–1979. Canadian Manuscript Report of Fisheries and Aquatic Sciences 1608. Division of Fisheries and Oceans, Pacific Biological Station, Nanaimo, BC Canada

Magnuson JJ (1976) Managing with exotics—a game of chance. Transactions of the American Fisheries Society 105:1–9

Major RL, Ito J, Ito S, Godfrey H (1978) Distribution and abundance of Chinook Salmon (*Oncorhynchus tshawytscha*) in offshore waters of the North Pacific. International North Pacific Fishery Commission Bulletin 38

Mathews SB, Buckley R (1976) Marine mortality of Puget Sound Coho Salmon (*Oncorhynchus kisutch*). Journal of the Fisheries Research Board Canada 33:1677–1684

McCann JA (1984) Involvement of the American Fisheries Society with exotic species, 1969–1982. In: Courtenay WR Jr, Stauffer JR Jr (eds) Distribution, biology, and management of exotic fishes. Johns Hopkins University Press, Baltimore, MD, pp 1–7

Moyle PB (1969) Ecology of the fishes of a Minnesota lake with special reference to the cyprinidae. PhD thesis, University of Minnesota, St. Paul, MN

Moyle PB (1976) Fish introductions in California: history and impact on native fishes. Biological Conservation 9:101–118

Neilson JD, Geen GH (1986) First-year growth rates for Sixes River Chinook Salmon as inferred from otoliths: effects on mortality and age at maturity. Transactions of the American Fisheries Society 115:28–33

OTA (1993) Harmful non-indigenous species in the United States. US Congress, Office of Technology Assessment, OTA-F-565. U.S. Government Printing Office,Washington, DC

Quinn TP, Fresh K (1984) Homing and straying in Chinook Salmon (*Oncorhynchus tshawytscha*) from Cowlitz River Hatchery, Washington. Canadian Journal of Fisheries and Aquatic Sciences 41:1078–1082

Radonski GC, Martin RG (1986) Fish culture is a tool, not a panacea. In: Stroud RH (ed) Proceedings of a symposium on the role of fish culture in fisheries management, Lake Ozark, Missouri. American Fisheries Society, Bethesda, MD, pp 7–13

Shapovalov L, Taft AC (1954) The life histories of the Steelhead Trout (*Salmo gairdneri*) and Silver Salmon (*Oncorhynchus kisutch*) with special reference to Waddell Creek, California, and recommendations regarding their management. California Department of Fish and Game Fisheries Bulletin 98:1–375

Stauffer TM (1977) Number of juvenile salmonids produced in five Lake Superior tributaries and the effect of juvenile Coho Salmon on their numbers and growth 1967–1974. Fisheries Research Report 1846. Michigan Department of Natural Resources, Lansing, MI

Stroud RH (ed) (1986) Fish culture in fisheries management. Proceedings of a symposium on the role of fish culture in fisheries management, Lake Ozark, Missouri. American Fisheries Society, Bethesda, MD

Taylor JN, Courtenay WR Jr, McCann JA (1984) Known impacts of exotic fishes in the continental United States. In: Courtenay WR Jr, Stauffer JR Jr (eds) Distribution, biology, and management of exotic fishes. Johns Hopkins University Press, Baltimore, MD, pp 322–373

Versar (1992a) Introduction of Pacific salmonids into the Delaware River watershed. Draft Environmental Impact Statement prepared for the US Fish and Wildlife Service and New Jersey Division of Fish, Game, and Wildlife. Versar, Inc., Columbia, MD

Versar (1992b) Notice of withdrawal of Pacific salmonid stocking proposal for the Delaware River basin. Prepared for the New Jersey Division of Fish, Game, and Wildlife. Versar, Inc., Columbia, MD

Waters TF (1983) Replacement of Brook Trout by Brown Trout over 15 years in a Minnesota stream: production and abundance. Transactions of the American Fisheries Society 112:137–146

6

Likelihood of Introducing Nonindigenous Organisms with Agricultural Commodities: Probabilistic Estimation

Michael J. Firko and Edward V. Podleckis

Introduction

When nonindigenous organisms become established in new areas, biodiversity is altered, and there may be direct impacts on indigenous species as well as disruption of communities and ecosystems. The frequency of species introductions into the United States has increased dramatically during the past 100 years, primarily because of human activities (OTA 1993). The advent of regular long-range human movements and increased trade in international markets has added to the global dissemination of organisms. Plant Protection and Quarantine (PPQ) within the Animal and Plant Health Inspection Service (APHIS) of the U.S. Department of Agriculture (USDA) intercepts tens of thousands of nonindigenous plant pests every year in passenger baggage, in ships' stores, and in agricultural commodities—both legal and smuggled.

Many introductions that occur are difficult to manage. For example, it is believed that the Zebra Mussel (*Dreissena polymorpha* Pallas) was introduced when commercial ships in North American waterways discharged ballast water taken on in foreign ports. Sailer (1978) discussed exotic insect fauna in the United States. Well-known introduced plant pests include Gypsy Moth, Japanese Beetle, Chestnut Blight, Dutch Elm Disease, and Karnal Bunt of wheat.

Commercial trade in living nonindigenous organisms has led to many introductions. The U.S. Congress Office of Technology Assessment (OTA) found that importations by the nursery, pet, and aquaculture industries constitute a major source of introductions (OTA 1993). And although nonindigenous biological control agents such as parasitic wasps and plant pathogens have been imported and released for many years, activity in this area has increased markedly during the past decade with the development of international markets for mass-produced biological control agents. Simberloff and Stiling (1996) discussed risks associated with purposeful introductions of biological control organisms. Release of nonindigenous biological control organisms—as well as other biologically based technologies for pest control—was also considered by OTA (1995). Bennett (1993) discussed replacement of native parasitoids by exotic species, and several specific issues related to release of nonindigenous biological control agents for control of both native and exotic plant pests were discussed by Lockwood (1993) and Carruthers and Onsager (1993), who examined these issues from different perspectives. Other arthropods have been introduced as crop pollinators; significant international trade in nonhoneybee crop pollinators began during the past decade with technical innovations for mass rearing of bumblebees (see discussion below).

Walk-through displays of live butterflies and other arthropods have become popular worldwide. Currently, there are about 35 facilities in the United States displaying live nonindigenous butterflies, and many more are currently in the construction or certification phase. Most nonindigenous butterflies imported for display are imported as pupae, many of which are collected from the wild or "farmed" in open areas where host plants are cultivated to provide oviposition sites for wild females. A variety of other taxa such as walking sticks, leaf cutter ants, beetles, and cockroaches are also imported and displayed. Although strict guidelines are in place to prevent escape of displayed organisms, escapes have occurred. Importation of any of these organisms may lead to introductions of associated species such as endo- and exoparasites, phoretic organisms, and diseases.

International trade in agricultural commodities such as fresh produce is also a potential pathway for movement of nonindigenous organisms. Vast quantities of fresh produce are traded on international markets. Although the United States enjoys a large agricultural trade surplus, millions of tons of fresh produce are imported annually, especially during the winter. Virtually all the bananas consumed in the continental United States are imported, and for each of the 5 years between 1992 and 1996, the United States has imported at least 30 million lugs (boxes) of table grapes from Chile alone.

The responsibility for regulating most of the activities listed above falls on federal and state government agencies. USDA-APHIS is the lead regulatory agency for many of these issues. Regulating the movement of nonindigenous organisms, primarily plant pests, via agricultural trade, passenger traffic, and other human activities is the fundamental mission of PPQ. APHIS also considers requests for release of a variety of organisms with potential plant pest characteristics including genetically modified organisms and nonindigenous biological control organisms and pollinators. USDA has deliberated the risk of introducing nonindigenous organisms with agricultural trade for more than 100 years. However, over the past two decades, analysis of phytosanitary risks has attracted increased scrutiny by governments, trading partners, and academic institutions around the world. The advance of free trade agreements such as the General Agreement on Tariffs and Trade and the North American Free Trade Agreement has fueled the increased emphasis on pest risk analysis by removing tariffs as tools for restricting trade while dictating that phytosanitary decisions be technically justified. Technical justification is most often achieved through pest risk analysis conducted according to accepted international standards (FAO 1996).

APHIS' heuristic model for risk analysis (Gipson 1991) identifies three distinct components: risk assessment, risk management, and risk communication. Risk management includes consideration of risk mitigation measures and the decision-making process. Risk communication can assume a variety of forms such as consultation with stakeholders during preparation of risk assessments, public hearings on proposed actions, and publication of risk assessments. This concept of risk analysis and the associated definitions conform with most international agreements concerning trade in agricultural commodities and the primary literature in the field of risk analysis. In this chapter, we focus on the assessment component of risk analysis.

Plant Pest Risk Assessments for Nonindigenous Organisms

Risk assessments focus on three basic questions: what can go wrong, how bad would it be, and what is the likelihood that it will happen? Each of these questions is discussed in the following sections. Risk assessments are science-based evaluations; however, they are not original scientific research. Risk assessments form a link between scientific data and decision makers and express risk in terms appropriate for decision making. Risk assessments are conducted to support decisions that must be made regarding agricultural trade, the quarantine status of various organisms intercepted at U.S. borders, and domestic regulation of agricultural commodities. Assessments are considered necessary but not sufficient on their own to provide the basis for regulatory decisions. The process for pathway (commodity-based) plant pest risk assessments (USDA 1995a) has nine steps. It starts by documenting the purpose and scope of the assessment, identifying any previous risk assessments or decisions regarding the entry status of the commodity, and describing the current status of commodity importations. The biological portion of the assessment starts by considering the "weediness potential" of the commodity. The focus of these biological assessments is on the three key issues in any risk assessment, summarized by the three following questions.

What Can Go Wrong?

Frequently, we state this question as "What is the bad event?" This phase of a risk assessment typically is referred to as hazard identification. The answer to the question determines the endpoint of the risk assessment. The bad event for most of our assessments is the introduction or spread of a nonindigenous plant pest. *Introduction* means the entry and establishment of an organism (Hopper 1995; FAO 1996) in a new area. Entry and establishment are distinct phenomena and should be considered separately. Although entry of nonindigenous organisms is frequently considered undesirable, entry does not necessarily imply establishment. It is probably safe to assert that many more nonindigenous organisms enter an area than ever become established. Establishment can be defined as growing and reproducing successfully in a given area (Lincoln et al. 1982).

The hazard identification phase of the assessment typically involves identifying the species of nonindigenous plant pests that may be moved with the commodity. The process of determining which pests are quarantine pests follows internationally accepted criteria (Hopper 1995; FAO 1996) and is based largely on the presence or absence of the pest in the area at risk. Because not all quarantine pests in the export area can reasonably be expected to accompany the commodity, some quarantine pests may not be considered in detail in the risk assessments. For example, consider a soil-dwelling nematode that is a pest of orange trees in the production area: if only fresh fruit—which must be free of all leaves, stems, and other plant parts—are to be moved, there is little risk that the soil nematode will be moved.

Note that the hazard considered is human-assisted spread of pest organisms to areas outside of their current ecological range. Although organisms spread naturally, our risk assessments typically deal only with human-assisted spread because the assessments are prepared to support decisions about regulated activities. USDA sometimes conducts assessments that focus on the persistent problem of introductions via smuggling of agricultural commodities. Occasionally, natural spread is considered along with human-assisted spread. Assessments are also limited by the scope of the regulatory authority. Some types of organisms, such as mammals, birds, reptiles, many spiders, and scorpions, fall under the jurisdiction of other federal agencies regardless of whether they can be considered plant pests. The regulatory focus is on human activities that may disseminate invertebrate animals and plant pathogens with the potential to act as direct or indirect plant pests.

How Bad Would It Be?

This question can be restated as "What are the consequences?" We consider the biological potential of the quarantine pests to have negative economic (e.g., lowering yields or decreasing the value of the crop) and environmental impacts in the area at risk. We also consider the pest's potential to cause direct environmental impacts or impacts on species listed as endangered or threatened. We do not consider quantitative measures of the consequences of plant pest introductions. Although risk managers typically are informed of the economic ramifications of pest introductions, focus on monetary values is outside the scope of biological risk assessments. Although we and other biologists work with economists conducting monetary assessments, the methods and tools are the domain of economics. Biological risk assessments use qualitative methods to consider the biological potential of pests to cause economic damage.

What Is the Likelihood That It Will Happen?

For our assessments, this question becomes "What is the likelihood that nonindigenous plant pests will be introduced into the area at risk as a result of the regulated activity?" Most USDA decisions regarding movement of agricultural commodities are based on qualitative risk assessments in which estimates of risk are expressed in terms such as *high* or *low*. But increasingly, quantitative approaches for estimating the likelihood of "bad events" such as pest introductions are being used. Our quantitative assessments are probabilistic and estimate the likelihood of introduction using scenario analysis and Monte Carlo simulation. Probabilistic assessments contain more detailed and rigorous appraisals of the likelihood that nonindigenous species will be spread. We conduct probabilistic assessments—as opposed to less rigorous, qualitative assessments—for a variety of reasons including trade with a bilateral trading partner (a country with which the United States has special trading relationship), controversy, large domestic industry, or potential for significant impacts.

We have estimated the biological risks associated with trade in a variety of agricultural commodities. Because different exporting areas typically have unique collections of agricultural pests, separate assessments may be needed for each unique combination of commodity and export area. To illustrate the use of the concepts described below, examples are provided from four probabilistic assessments conducted to estimate the likelihood of bad events. The first addresses risks associated with the importation and sale of self-contained bumblebee hives for crop pollination, two involve importation of fresh fruits, and the last involves a set of assessments conducted to support decisions regarding domestic quarantines and regulations to limit spread of Karnal Bunt disease of wheat. Although probabilistic methods had been used previously by other staffs in USDA, their first use in PPQ was in the 1994 assessment conducted in support of issuance of permits for importation of bumblebee hives (unpublished).

Bumblebees

Wild bumblebees play a significant role in pollination of both wild plants and crops. In the past decade, there has been increased interest in managed colonies of bumblebees, *Bombus* spp. (Hymenoptera: Bombidae), for crop pollination. This development followed a breakthrough in the ability to mass produce bumblebee colonies. In the early 1990s, APHIS began to receive requests for permission to import European species of bumblebees such as *B. terrestris*. Releases of nonindigenous species have been authorized in various countries; *B. impatiens,* a species native to eastern North America, is now established in Europe following releases for crop pollination. Agricultural officials in the United States and Canada have worked together to ensure that only *Bombus* species native to North America have been authorized for release here. Bumblebees for crop pollination are delivered as self-contained cardboard "hives" that are simply opened within a greenhouse or grove to allow exit and entrance of worker bees; we estimate that at least 1 million hives were shipped worldwide during 1995.

Qualitative risk assessment conducted in support of permit decisions regarding importation of nonindigenous bumblebees indicated that both economic and environmental risks were too high to allow introduction of bumblebees not native to North America. Not only do workers compete for limited resources, queens compete for nest sites and sometimes fight to the death over suitable nest sites (Heinrich 1979). Aggressive nonindigenous species could place pollination of crops and native plants, as well as indigenous species of bumblebees, at significant risk. The risk of introducing nonindigenous bumblebee pests was also considered to be unacceptably high.

Permission was never granted for release of *Bombus* species not native to North America. After adoption of this policy, foreign producers of bumblebee hives collected live bumblebee queens from North America and used them to establish colonies in Europe. They then requested authorization to import hives of foreign-produced *B. impatiens* back into the United States. For a short time, APHIS granted permission for importation of these hives into the United States. But after bumblebee and pollination scientists and stakeholders raised concerns about this activity, APHIS expanded its risk assessment and conducted a probabilistic assessment to consider the likelihood that nonindigenous bumblebee pests such as parasitic wasps and diseases might be imported with the foreign-produced hives. We concluded that this activity placed native species at risk and revoked permission for importation of bumblebee hives produced outside of North America. Current federal policy only allows the release of native species of *Bombus* that have been produced in North America.

Importation of Fresh Unshu Orange Fruit from Japan

Quarantine 56 (Title 7 Code of Federal Regulations § 319.56) prohibits entry of fruits and vegetables grown in foreign countries. Fresh produce is allowed entry only if a risk assessment finds that there is no significant risk of introducing a nonindigenous plant pest of quarantine significance. A series of risk assessments conducted over recent decades have provided the technical basis for the current status of Japanese Unshu Orange fruit that may enter the United States. Their distribution is limited to states that do not produce commercial citrus crops.

The primary concern has always been a bacterial disease of citrus, *Xanthomonas axonopodis* pv. *citri* (Syn. *X. campestris* pv. *citri*), Citrus Canker A, and *Xanthomonas axonopodis* pv. *aurantifolii* (Syn. *X. campestris* pv. *citri*), Citrus Canker B. After several years of importation with no apparent pest introductions, Japanese officials requested access to all 49 continental states. We conducted an assessment that focused on risks to citrus-producing states (USDA 1995b) and considered all potential plant pests that might be associated with shipments. We identified several quarantine significant arthropods, pests, and diseases of citrus in Japan. The assessment was forwarded to Japan, and we are currently awaiting their risk mitigation proposals.

Importation of Fresh Avocado Fruit from Mexico

Mexico has been requesting access to the U.S. avocado market for at least 90 years. An assessment conducted by USDA led to a decision in 1914 that fresh intact avocado fruit grown in Mexico would be denied entry into the United

States. Over the past few decades, avocado producers in Mexico, along with Mexican agricultural officials, have developed comprehensive pest control programs, and Mexico has been exporting avocado fruit to many countries around the world for several years. In the early 1990s, Mexican agricultural officials again requested authorization to export fresh avocados to the United States, and shipments have been allowed into Alaska under certain conditions since 1993. Then in 1994, the Mexican government formally requested that APHIS further amend its import regulations to allow importations into the northeastern United States. We conducted a complete risk assessment including a probabilistic assessment of the likelihood of introduction for nine species of arthropod pests. Fresh avocado fruit from Mexico can now enter the United States, subject to certain restrictions. For example, fruits are enterable only from certain regulated export groves, they may be imported only during the months of November through February, and they may only be imported into 19 northeastern states.

Domestic Regulation of Wheat to Limit the Spread of Karnal Bunt Disease of Wheat

In March 1996, Karnal Bunt, a disease of wheat caused by the smut fungus *Tilletia indica*, was first detected in the wheat-growing regions of Arizona. The wheat harvest in Arizona was scheduled to begin in May. Because much of the grain and seed produced in Arizona is shipped to other wheat-producing areas of the United States, we had to consider the consequences of various regulatory actions. For example, much of the grain produced in Arizona is shipped to flour mills in the Midwest that produce semolina for pasta. Some of the seed is shipped to northern states for planting, and some is shipped to grain elevators in other states, where it is stored until it is exported to foreign countries. Regulation of grain destined for mills is complicated by the fact that roughly 25% of mill products are used as animal feed, and spores of *T. indica* are capable of surviving the digestive systems of both birds and mammals. Manure from these animals is used to fertilize fields, including wheat fields. In a series of probabilistic risk assessments (USDA 1996a–d), we considered risks associated with 24 different program options by describing each option with a specific scenario.

Probabilistic Risk Assessment: Estimating the Likelihood of "Bad Events"

Our probabilistic assessments focus on the likelihood that plant pests will be spread if particular activities are authorized. However, the word *probabilistic* in this context refers to the fact that the input parameters and model output are expressed in probabilistic terms. That is, estimates of the likelihoods of bad events are expressed in probabilistic terms, with each estimate having an associated probability. In simple quantitative assessments, point estimates are used to calculate estimates of risk (a single number resulting from a single calculation), and estimates are expressed in quantitative terms such as "a probability of 0.01." Usually, the point estimate represents a best estimate. However, quantitative assessments that rely on point estimates can neither account explicitly for uncertainty in the estimated input values nor can they express uncertainty in the final risk estimate.

The primary reason for conducting a probabilistic assessment is to provide a definitive mechanism to account for the uncertainty in the model inputs. Examples of "bad events" (risk assessment endpoints) include the frequency or probability of contamination, pest entry, pest establishment, or pest outbreaks. For most of these probabilistic risk assessments, the bad event—the endpoint of the risk assessment—is the frequency of pest establishment. Our probabilistic risk assessments have four basic steps: (1) scenario analysis, (2) development of the mathematical model, (3) construction of probability density functions as model inputs, and (4) Monte Carlo simulations.

Scenario Analysis and Development of a Mathematical Model

Scenario analysis is essentially model building; the risk scenario represents a risk model. Our scenario analyses involve identifying the events (nodes) that must occur before some "bad outcome" can result. The scenario provides a visual representation of the risk model. The nature of the scenario dictates the appropriate mathematical model to use for risk calculations. For example, if all nodes are independent and if each of the events must occur before the endpoint can be reached, the appropriate mathematical model is a simple, linear, multiplicative model.

Figure 6.1 shows the scenario used in the Mexican avocado risk assessment (USDA 1995c), a simple, linear, multiplicative model. The nodes shown for program option A (imports with no additional risk mitigations beyond those already in place) are the same as for program option B (a specific systems approach for risk mitigation). Both scenarios are displayed to show node designations that are useful for keeping track of the progress of the subsequent calculations. The models used in the Karnal Bunt assessments are similar to but more complex than those shown for the Avocado assessment. When events (primary nodes) can occur as a result of more than one event (subnodes), the model becomes more elaborate as illustrated in Figure 6.2 by a scenario from one of the Karnal Bunt risk assessments (USDA 1996a). Note that in this scenario, although the primary nodes share a simple, linear, multiplicative relationship, some of the primary nodes are composed of multiplicative and additive subnodes and subsubnodes.

In the Japanese Unshu orange assessment (USDA 1995b), four scenarios representing four different program options were examined. In the risk assessments for

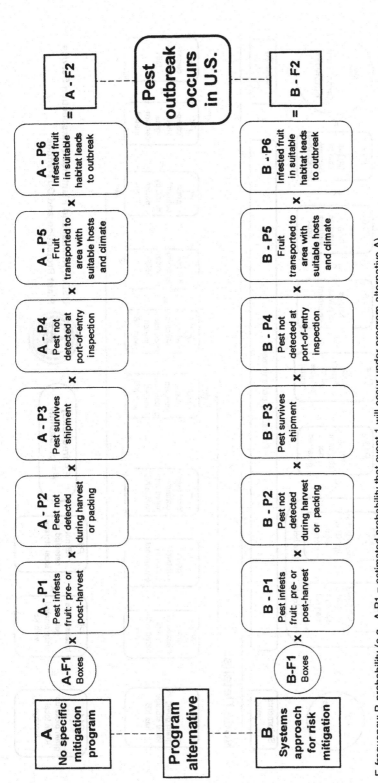

F, frequency; P, probability (e.g., A-P1 = estimated probability that event 1 will occur under program alternative A).

FIGURE 6.1. Risk scenario from APHIS's Mexican avocado risk assessment (USDA 1995c).

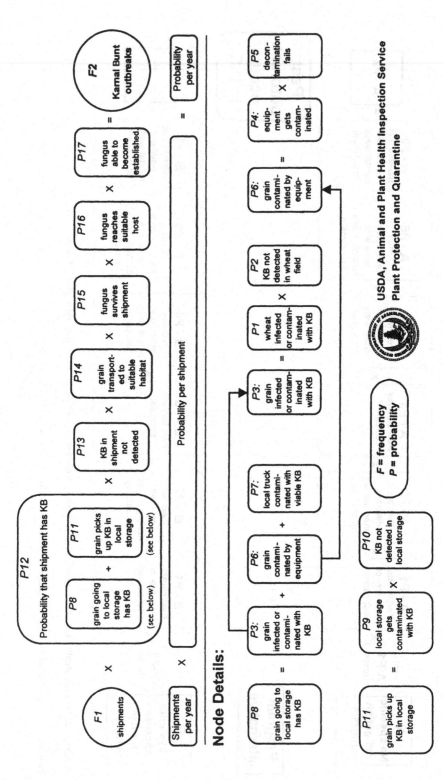

FIGURE 6.2. Risk scenario from APHIS's Karnal Bunt (KB) plant pest risk assessment (USDA 1996a).

domestic regulation of wheat shipments (USDA 1996a–d, 1997) to limit the spread of Karnal Bunt, each of the 24 scenarios represented a different program option. We construct the probabilistic risk assessment models as fault models. Each independent event (primary node) in the model represents a "failure" with respect to risk management (the probabilities of the subnodes may be additive, conditional, and so on). The final estimate of risk depends on the mathematical relationship among the nodes. In developing the risk assessment model, it is important to conduct a units analysis to ensure that the model inputs, together with the mathematical relationship among the nodes, result in the appropriate output. The scenario analysis provides a conceptual framework for estimating risk.

Estimation of Input Values: Choosing Distributions and Distribution Parameters to Represent Estimated Frequencies and Probabilities

Estimates of risk are only as precise and accurate as the estimates used as input values. We use the best scientific information available for assessments. Ideally, existing data would provide the basis for direct estimation of model inputs. However, scientific experiments are seldom conducted specifically to provide these estimates for risk assessments, and results are seldom provided that can be used directly in our models. In addition, because most of our risk assessments are conducted to support decisions that must be made within relatively narrow time frames, research programs can seldom be conducted to provide data specifically for our assessments. Fortunately, this situation has already started to change as risk-based decisions for trade in agricultural commodities become the international standard. Agricultural research has already started in support of this type of risk assessment.

For now, however, a variety of biological data usually is available that is pertinent to the needed probability. These data are reviewed, and professional judgment is used to represent the available data regardless of whether estimates are characterized as point estimates or distributions of possible values. We base estimates on pest interception records, the known biology of the organism being assessed (or the known biology of related taxa), expert judgment based on laboratory experience with the pest or related organisms, expert judgment based on field experience with the pest or related organisms, expert judgment based on experience conducting commodity inspections at ports of entry or in the exporting country, and experience working with export programs and export-quality commodities.

Probabilistic risk assessments are based on Monte Carlo simulations. Input values are characterized as probability density functions (PDFs) of possible values. Normal and lognormal distributions are familiar examples of PDFs. Using PDFs provides an explicit and transparent method to account for the uncertainty inherent in estimated probabilities for events. The two basic components of a PDF are the shape of the distribution and the values of distribution parameters such as the minimum, mean, and standard deviation. Typically, we construct PDFs by assembling a group of specialists in the appropriate field (e.g., entomology, mycology, plant pathology, virology) to review and discuss the available data. We use the "expert information" approach described by Kaplan (1992). With this approach, all pertinent information is shared among the group of specialists before any estimates are made. This approach increases objectivity and helps avoid personal ownership of estimates that are made before all pertinent data are shared. In every biological system, there is uncertainty about pest biology, how well risk management systems will perform, the level of compliance with the program, and future ecological and environmental conditions. Uncertainty in estimated values results from natural biological variation, climate variation, lack of precision in the model, data gaps, poor data, multiple components in a node, and a variety of other sources.

The basic approach is to use whatever PDF best represents the available information. It is always tempting to overinterpret available data and specify distributions that are not supported by the data. For example, specification of distributions such as normal and lognormal requires assumptions concerning the central tendency of the distribution and characteristics of the data and the processes that generated them. Most risk assessment programs available on the market today have 20–30 different PDFs from which to choose. However, the following five are among those chosen most often.

Uniform Distribution

Uniform distributions are the simplest PDF; only a minimum and maximum value are needed to specify the distribution. Every value between the minimum and maximum has an equal probability of being selected by the sampling algorithm. Uniform distributions may be appropriate when there is little justification for assuming that some values are more likely than others or when data do not suggest a central tendency. Values beyond the minimum and maximum value are not used for calculations. Figure 6.3 shows an example of a uniform distribution used in the assessment for importation of Mexican avocado fruit (USDA 1995c). Uniform distributions used for probability values that range over an order of magnitude should be used with caution. If the analysts were thinking on a log scale, results could be overly conservative. For example, consider a uniform distribution covering two orders of magnitude (minimum = 0.0002, maximum = 0.02). Values between the minimum and the geometric mean (0.002) will be chosen only about 9% of the time, and values between the geometric mean and the maximum will be chosen about 91% of the time. Values between 0.01 and 0.02 will be chosen for about 50% of the calculations.

Triangular Distribution

Triangular distributions are specified by three values: minimum, most likely, and maximum. The relative frequency of the various values in the distribution is

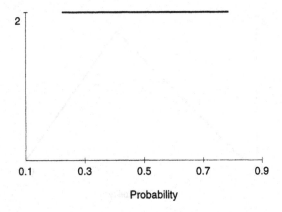

FIGURE 6.3. Uniform distribution from APHIS's Mexican avocado risk assessment (USDA 1995c, Table 8) used to estimate the probability that fruit would be transported to a habitat suitable for establishment of the avocado seed moth (*Stenoma catenifer*) under program option A. Distribution minimum = 0.25, maximum = 0.75.

indicated by the shape of the curve, which is determined by the relative position of the three parameters. Values at or near the most likely value are selected for calculations more often than values at or near the minimum and maximum. Triangular distributions can be used when it is fair to assume that some values are more likely than others but have the disadvantage of not selecting values below the minimum and maximum. Choice of the minimum and maximum values can be considered more important than choice of the most likely value because the minimum and maximum imply a high degree of assurance about the limits of the underlying distribution. We use triangular distributions occasionally but avoid them because biological systems are seldom well represented by distributions with strict limits. The arbitrary shape of the triangular distribution and its strict limits are indications that the precise nature of the distribution is not known. Figure 6.4 shows an example of a triangular distribution used when assessing the risk of importing Japanese Unshu orange fruit into citrus-producing states (USDA 1995b).

Normal Distribution

Normal distributions have the advantage of being representative of many types of biological phenomena. They also have the advantages of not being limited by minimum or maximum values. However, when the parameter being estimated is a probability—which has a domain from 0 to 1—care must be taken to avoid including "probability" values below 0 and above 1. The risk is greatest with probability values greater than 0.1. Figure 6.5 shows an example of a normal probability distribution used during assessment of risks associated with importation of the Japanese Unshu orange (USDA 1995b, Table 11). This distribution was

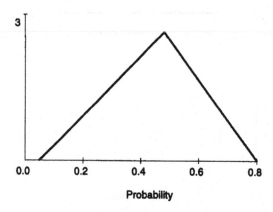

FIGURE 6.4. Triangular distribution from APHIS's Japanese Unshu orange risk assessment (USDA 1995b, Table 5) used to estimate the probability that quarantine species of mites, *Tetranychus kanzawi* and *Eotetranychus asiaticus,* would escape detection at the port of entry. Distribution minimum = 0.05, most likely value = 0.5, maximum = 0.8, mean = 0.45, mode = 0.5, variance = 0.025, kurtosis = 2.4, skewness = –0.19.

used to estimate the probability that the Citrus Fruit Fly (*Bactrocera tsuneonis* [Diptera: Tephritidae]) would be able to find host material if transported to a citrus-producing state. Although the mean of this normal distribution is 0.5, values beyond 1 are extremely unlikely because the specified standard deviation is only 0.1. Nonetheless, in these cases it is prudent to use the software options to verify that values below 0 and above 1 were not included in the calculations.

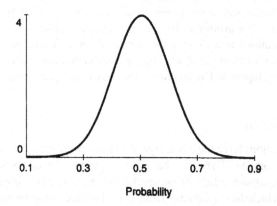

FIGURE 6.5. Normal distribution from APHIS's Japanese Unshu orange risk assessment (USDA 1995b, Table 11) used to estimate the probability that the Citrus Fruit Fly (*Bactrocera tsuneonis* [Diptera: Tephritidae]) would be able to find host material if transported to a citrus-producing state. Distribution mean = 0.5, mode = 0.5, variance = 0.01, skewness = 0, kurtosis = 3.

Lognormal Distribution

Lognormal distributions have many of the same advantages and disadvantages as normal distributions: they are representative of many biological phenomena, and they are not limited by minimum and maximum values, but there is a chance that values greater than 1 will be chosen for probability estimates. Lognormal distributions have the advantage of not allowing negative values (i.e., 0 is the minimum value). The problem of sampling values greater than 1 can be managed with many risk assessment programs by specifying a "truncated lognormal distribution" with a maximum value of 1. Figure 6.6 shows an example of a lognormal distribution used to represent that likelihood that a bumblebee queen with a parasite infection would be able to establish a new hive (unpublished risk assessment).

Beta Distribution

Beta distributions are especially useful when estimating probabilities because the domain of the distribution (0 to 1) is the same as for probabilities. Beta distributions are specified with two parameters, $\alpha1$ and $\alpha2$ (sometimes referred to as α and β). The shape of the beta distribution can vary significantly and depends on the values of $\alpha1$ and $\alpha2$. Many beta distributions resemble a lognormal distribution. For example, a beta distribution with $\alpha1 = 1.1$ and $\alpha2 = 50$ (mean = 0.02, mode = 0.002, variance = 0.0004) has the same general shape and specifications similar to a lognormal distribution with a mean and standard deviation of 0.01 (mean = 0.01, mode = 0.003, variance = 0.0001). But beta distributions can be constructed so that very low or very high values are chosen less frequently than

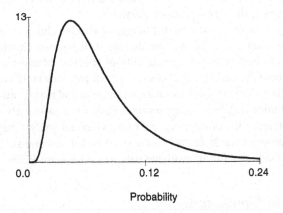

FIGURE 6.6. Lognormal distribution from APHIS's bumblebee assessment (unpublished) used to estimate the probability that a bumblebee queen with a parasite infection would be able to successfully establish a new hive. Distribution mean = 0.075, mode = 0.0432, variance = 0.0025, skewness = 2.3, kurtosis = 13.6, 5th percentile values = 0.023, 95th percentile value = 0.17.

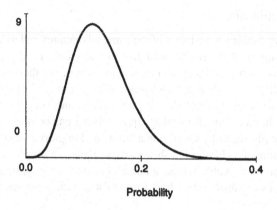

FIGURE 6.7. Beta distribution from APHIS's Karnal Bunt risk assessment (USDA 1996d) used to estimate the probability that spores would not be detected in the preharvest grain sample. In this beta distribution, $\alpha1 = 6$ and $\alpha2 = 40$. Distribution mean = 0.13, mode = 0.11, variance = 0.002, skewness = 0.63, kurtosis = 1.5.

with a similar lognormal distribution, and with certain types of beta distributions, values beyond certain limits will probably not be sampled. For example, we used a beta distribution with $\alpha1 = 6$ and $\alpha2 = 40$ to estimate the likelihood that spores of *Tilletia indica* (Karnal Bunt fungus) would not be detected in an infected wheat field (USDA 1996d, Table 5) (Fig. 6.7). The mean (0.13) and mode (0.11) of this distribution accurately reflect the available data concerning the efficacy of the field test for spores. But available information made it clear that the likelihood of not detecting spores was both greater than 0 and less than 1. With this distribution, values below 0.01 and above 0.4 will almost certainly not be included in the calculations even with 10,000 or more iterations.

Although it can be demonstrated that in many cases the choice of PDF has little impact on the final result of the simulation, distributions should be chosen carefully so as to best reflect the availability of specific information or specific assumptions about the underlying characteristics of parameters. If available information only allows the analyst to estimate a minimum and maximum value, then a uniform distribution may be most appropriate. Risk assessment software packages provide an extensive list of distributions from which an analyst may choose, as well as user-programmable options for specialized distributions. Curve-fitting software also facilitates fitting available data to well-known PDFs.

Monte Carlo Simulations

We use Monte Carlo and Latin hypercube sampling methods as the basis for probabilistic risk calculations and refer to these calculations as Monte Carlo simulations. In a typical Monte Carlo simulation, we run 10,000 iterations (i.e.,

TABLE 6.1. Example of output from Monte Carlo simulation.[a]

Risk estimate parameter	Outbreak frequency (per year)	Number of years between outbreaks
Minimum	1.48×10^{-9}	>Million
Mode	1.11×10^{-7}	>Million
Median	2.14×10^{-6}	467,290
Mean	8.56×10^{-6}	116,822
95th Percentile	3.74×10^{-5}	26,738
Maximum	2.72×10^{-4}	3,676

[a]From USDA (1996d, Table 3a).

10,000 sets of input values are selected for each of 10,000 output calculations). For each individual calculation, the software uses Monte Carlo, Latin hypercube, or some other sampling algorithm to select a value randomly from the input distributions according to the specified distribution. The program then collects the output from all iterations and presents the statistics of the output PDF. Some programs also provide graphs of the output PDF. Although the basic calculation is typically presented as the probability per year of a bad event, the output is also expressed in terms of one chance in X of the bad event occurring, where X is equal to the inverse of the probability per year. Table 6.1 shows a typical table of results from a Monte Carlo simulation designed to estimate the frequency/probability of pest outbreaks.

Conclusion

The mission of APHIS and its predecessor agencies within the USDA is consistent with a key goal of conservation biology: preventing the introduction of harmful nonindigenous organisms. Realizing this goal requires informed decisions by regulatory risk managers regarding importations of agricultural commodities. The technical basis for these informed decisions is most often a plant pest risk assessment. Past plant pest risk assessments and, indeed, the majority of current assessments have used qualitative measures of the risk posed by commodity pathways. Increasingly, probabilistic assessments have emerged as the risk assessment method of choice to support difficult regulatory decisions. The United States has taken the lead in using this type of assessment for agricultural trade. But assessments of this type are becoming more common, and New Zealand, Australia, and Canada have conducted risk assessments of this type in support of decisions on international trade in agricultural commodities. Biological systems in general and plant pest–commodity interactions in particular are difficult to assess both because of their inherent complexity and the frequent dearth of data available to describe them. Probabilistic assessments provide the best available tools to account for the variability and uncertainty that biological complexity and data gaps introduce into the plant pest risk assessment process.

Literature Cited

Bennett FD (1993) Do introduced parasitoids displace native ones? Florida Entomologist 76:54–63

Carruthers RI, Onsager JA (1993) Perspective on the use of exotic natural enemies for biological control of pest grasshoppers (Orthoptera: Acrididae). Environmental Entomology 22:885–903

FAO (1996) International standards for phytosanitary measures, Part 1—Import regulations: guidelines for pest risk analysis. Secretariat of the International Plant Protection Convention, Food and Agriculture Organization (FAO) of the United Nations, Rome, Italy

Gipson P (1991) Enhancing credibility and delivery of APHIS programs through risk communication: special report to the APHIS administrator and the APHIS management team. APHIS/USDA Risk Communication Work Group, Washington, DC

Heinrich B (1979) Bumblebee economics. Harvard University, Cambridge, MA

Hopper BE (ed) (1995) NAPPO compendium of phytosanitary terms. North American Plant Protection Organization (NAPPO), Nepean, Ontario, Canada

Kaplan S (1992) "Expert information" versus "expert opinions." Another approach to the problem of eliciting/combining/using expert knowledge in PRA. Reliability Engineering and System Safety 35:61–72

Lincoln RJ, Boxshall GA, Clark PF (1982) A dictionary of ecology, evolution and systematics. Cambridge University Press, Cambridge, UK

Lockwood JA (1993) Environmental issues involved in biological control of rangeland grasshoppers (Orthoptera: Acrididae) with exotic agents. Environmental Entomology 23:503–518

OTA (1993) Harmful non-indigenous species in the United States. Office of Technology Assessment, US Congress, OTA-F-565. U.S. Government Printing Office, Washington, DC

OTA (1995) Biologically based technologies for pest control. Office of Technology Assessment, US Congress, OTA-ENV-636. U.S. Government Printing Office, Washington, DC

Sailer RS (1978) Our immigrant insect fauna. ESA Bulletin 24:3–11

Simberloff D, Stiling P (1996) How risky is biological control? Ecology 77:1965–1974

USDA (1995a) Pathway-initiated pest risk assessment: guidelines for qualitative assessments, version 4.0. U.S. Department of Agriculture, USDA-APHIS-PPQ, Riverdale, MD

USDA (1995b) Importation of Japanese Unshu orange fruits (*Citrus reticulata* Blanco var. *unshu* Swingle) into citrus producing states, pest risk assessment. U.S. Department of Agriculture, USDA-APHIS-PPQ, Riverdale, MD

USDA (1995c) Importation of avocado fruit *(Persea americana)* from Mexico, supplemental pest risk assessment. U.S. Department of Agriculture, USDA-APHIS-PPQ, Riverdale, MD

USDA (1996a) Karnal Bunt: likelihood of spread via conveyances, harvest equipment and wheat shipments, preliminary probabilistic risk assessment. U.S. Department of Agriculture, USDA-APHIS-PPQ, Riverdale, MD

USDA (1996b) Karnal Bunt: special risk assessment addendum, risk of outbreak in states receiving grain for milling from quarantine areas where seed and fields test negative and millfeed is not treated. U.S. Department of Agriculture, USDA-APHIS-PPQ, Riverdale, MD

USDA (1996c) Karnal Bunt: special risk assessment addendum II, risk of new outbreaks via seed harvested and planted in Arizona. U.S. Department of Agriculture, USDA-APHIS-PPQ, Riverdale, MD

USDA (1996d) Karnal Bunt: special risk assessment addendum III, risk of outbreak in states receiving grain for milling from quarantine areas when millfeed is not treated. U.S. Department of Agriculture, USDA-APHIS-PPQ, Riverdale, MD

USDA (1997) Karnal Bunt: likelihood of spread associated with proposed 1997 program options. Plant pest risk assessment. I. Grain shipped from 1996 program areas to domestic mills, millfeed not treated. II. Seed shipped from 1996 program areas for domestic planting. U.S. Department of Agriculture, USDA-APHIS-PPQ, Riverdale, MD

7

"Best" Abundance Estimates and Best Management: Why They Are Not the Same

Barbara L. Taylor and Paul R. Wade

Introduction

In *New Principles for the Conservation of Wild Living Resources,* Holt and Talbot (1978) give their second principle as, "Management decisions should include a safety factor to allow for the facts that knowledge is limited and institutions are imperfect." Inclusion of uncertainty in management has been difficult partly because of the failure of scientists to explain adequately the importance of incorporating estimates of uncertainty and the consequences of not accounting for this uncertainty in management decisions to policy makers and managers. On the surface, using the best estimates of abundance for management purposes seems sound. This chapter explains why using the mean estimate of abundance (N_{MEAN}, often referred to as the "best" estimate) counterintuitively can result in poor management practices. We illustrate the benefits of using estimates of uncertainty with a management scheme incorporated in the 1994 amendments to the Marine Mammal Protection Act (MMPA) and will therefore use marine mammal examples, although the lessons generalize to many species.

Why should management be concerned with uncertainty in abundance estimates? Consider the following cases: animals from two populations are incidentally killed in fishery interactions. Population A is well known, but considerable uncertainty exists about population B. How should management proceed in the short term when decisions must be based on best current information? Relatively good estimates may be made for the number of animals that could be killed by the fishery without depleting A. Difficulties arise, however, with population B. Let us assume that the best abundance estimates for both populations are about the same. Confidence in the estimates, however, differs greatly. Is the best management strategy to limit incidental kill based on no more than some small fraction of the best estimate (N_{MEAN}) or to somehow incorporate the level of uncertainty concerning the populations into our management decision? Using a lower percentile of an abundance distribution (N_{MIN}) incorporates uncertainty. It embodies Holt and Talbot's (1978) statement: "The magnitude of the safety factor should be proportional to the magnitude of risk." To compare management strategies using N_{MEAN} and N_{MIN}, we must first understand what these terms mean.

Background

Abundance Estimation

To understand the effect of human-caused mortality on a population, we need to know the size of the population. No population of marine mammals can be counted in its entirety (a true census). Instead, the population is sampled, and mathematical techniques are used to estimate the absolute abundance. The precision of abundance estimates depends not only on the effort made to make the estimate but also on properties inherent to the populations themselves. To understand the concepts of precision and bias, consider the analogy of archery (Fig. 7.1; White et al. 1982). For the small remaining population of Hawaiian Monk Seals, nearly every individual is identified. Thus, the estimate should be both precise and unbiased (Fig. 7.1a). Seals and sea lions are photographed and counted during maximum abundance on land (breeding or molting). Seasonal counts from animals on land are accurate (coefficients of variation in abundance (CV) often <10%). Because some unknown proportion of the population is at sea, the estimate would be precise but biased (Fig. 7.1b). Estimates of the proportion at sea could be made to correct for the bias. Most whale and dolphin populations and some seal populations must be estimated with distance sampling techniques. Obtaining precise estimates is frequently difficult. An uncommon but highly visible species, such as the Killer Whale, would have an imprecise estimate (Fig. 7.1c). Both imprecision and bias (Fig. 7.1d) would be expected for an animal such as a Sperm Whale, which is both uncommon and easily missed even when close because it typically dives for 40 minutes. A few examples will illustrate the difficulties in estimating abundance.

The most common technique for abundance estimation is line-transect (Buckland et al. 1993). Observers on ships or planes traveling along survey lines record number of animals seen, species identification, and perpendicular distance (Fig. 7.2). Not all animals are seen, and observers have a better chance of seeing animals that are close than those that are more distant. Data are used to estimate the total number of animals. For a small population, few sightings will be made. If the survey were replicated, the resulting abundance estimate would be different (possibly substantially) due to many random factors. If you could repeat the survey many times, the distribution of resulting estimates would be relatively wide for rare species and would be narrow for common species. For Vaquita, an endangered porpoise, Taylor and Gerrodette (1993) showed that the precision of the abundance estimate drops sharply with decreasing population size. Thus, one reason for poor precision is small population size.

A second reason for poor precision is that the species may be difficult to see. Consider again populations A and B. Assume A is Blue Whales and B is Beaked Whales. Blue Whales are conspicuous. Not only are they large but blows can be seen for great distances. Thus, the probability of sighting does not decrease with distance until distance becomes large. The smaller Beaked Whales surface quickly, often erratically, and have no conspicuous blow. Sighting probability

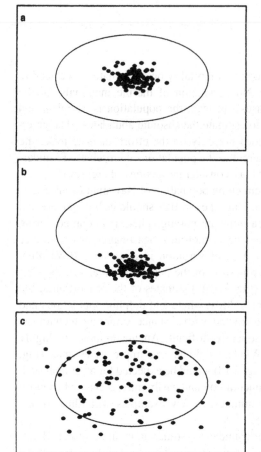

FIGURE 7.1. Archery targets demonstrating the meaning of precision and bias. Unlike the archer, an abundance estimate is like having one shot with no target to compare the shot with. Biologists must estimate the precision and bias of abundance estimates statistically.

decreases rapidly with distance. If numbers of these two species were equal, we would be able to estimate Blue Whales more accurately than Beaked Whales. Figure 7.3a shows two distributions that illustrate good precision (CV = 0.2) and poor precision (CV = 0.8). For the example, assume the true population size for A and B is 1,000 (the mean of the lognormal distributions of abundance estimates).

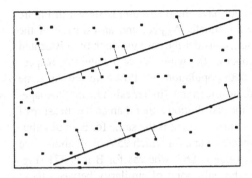

FIGURE 7.2. Schematic of a line-transect survey. Squares indicate animals (50). Sighted animals are connected with perpendicular lines to the survey trackline (12). Total abundance in the area is estimated from the sighting data and would differ if the survey was conducted again.

For any single survey, there is a single best estimate of abundance (N_{MEAN}) that comes from the appropriate distribution with the probabilities shown. The "best" estimate tells us nothing about the confidence we have in our estimate. Best estimates for Beaked Whales will vary more than best estimates for Blue Whales. For Blue Whales, the estimated population would usually be between 600 and 1,600. For Beaked Whales estimates are less certain. There is a good chance of estimating abundance anywhere from 200 to 3,000.

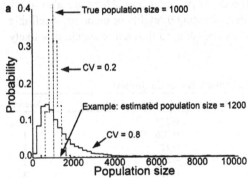

FIGURE 7.3. Probability distributions for abundance estimates where N (true population size) = 1,000 and distributions are assumed to be lognormal. (a) Distributions for the true population, (b) distributions centered on an estimated abundance (N_{MEAN}) of 1,200.

After conducting a survey, we do not have the distributions shown in Figure 7.3b. Instead, we have an abundance estimate (N_{MEAN}) and an estimate of the precision of our survey. For illustration, consider the case in which both Blue and Beaked Whales are estimated to number 1,200, with CVs of 0.2 and 0.8, respectively. Because N_{MEAN} is the same, both populations would be treated the same under the N_{MEAN} strategy. To incorporate uncertainty in our estimate into management, we need to focus on the tails of the distribution rather than on the measure of central tendency (the mean or best estimate), which is the same for both distributions. Take, for example, the abundance estimate for which 95% of all abundance estimates will be greater. For A this value is 867, whereas for B it is 371 (Fig. 7.3b). The mean or best estimate is the only point of similarity between these distributions. If we want to give importance to the difference in our degree of certainty, we should consider something other than the mean. The actual percentage of the distribution chosen depends on our management objectives. A sample of precision estimates for marine mammals is given in Table 7.1.

Line-transect abundance estimates can also be biased. Usually, it is assumed that all animals in the path of the ship (or plane) are seen. For most animals, especially those that can dive for long periods, this assumption is false. If this problem goes uncorrected, the estimate would be too low (negatively biased). Animals that are attracted to or repelled from the ship will also bias abundance estimates. If abundance estimates are thought to be low and fisheries are being threatened with closure because incidental mortality is thought to be excessive, then there would be pressure to correct for potential bias. There are also likely sources of positive bias, such as underestimating mortality or incorrectly defining population structure. Although bias can be reduced through research, it is likely

TABLE 7.1. Sample CVs for estimates of abundance in California.[a]

Species	Coefficient of variation (CV)	Source[b]
Short-beaked Common Dolphin	0.275	1
Long-beaked Common Dolphin	0.706	1
Northern Right Whale Dolphin	0.41	2
Bottlenose Dolphin	0.472	1
Harbor Porpoise	0.31	3
Baird's Beaked Whale	1.004	1
Mesoplodont Beaked Whale	0.924	1
Cuvier's Beaked Whale	0.864	1
Pygmy Sperm Whale	0.813	1
Risso's Dolphin	0.396	1
Killer Whale	1.207	1
Humpback Whale	0.409	1
Blue Whale	0.363	1
Fin Whale	0.591	1
Minke Whale	1.100	1
Sperm Whale	0.472	1

[a]The lowest current CVs are given.
[b]1, Barlow (1993); 2, Forney and Barlow (1993); 3, Barlow and Hanan (1995).

that decisions will have to be made while biases of unknown magnitude still exist. We treat the problem of bias by incorporating a safety factor parameter (F_R) that we set by simulating various levels and types of bias.

Qualitative Management Objectives

Management regimes can be evaluated only in the context of specific objectives. The Marine Mammal Commission (Robert Hofman, testimony to Senate Committee on Commerce, Science and Transportation, July 14, 1993) defined objectives for marine mammal management: (1) maintain the fullest possible range of management options for future generations, (2) restore depleted species and populations of marine mammals to optimum sustainable level with no significant time delays, (3) reduce incidental take to as near zero as practicable, and (4) as possible, minimize hardships to commercial fisheries while achieving the previous objectives. These objectives are based on the recommendations of Holt and Talbot (1978). Before we can define these qualitative objectives into quantitative objectives that can be used to measure the performance of the management scheme, we must examine the proposed management scheme and choose measurements that correspond to the above objectives.

Management Scheme

The basic idea is to ensure that populations recruit enough members to both make up for human-caused mortality and maintain a certain "safe" population level. To account for basic population dynamics, we begin with a simple model of how populations grow. We know that many marine mammal populations have been reduced to a small fraction of their historical abundance, and most have been recovering at an exponential rate. Some, such as Gray Whales, have population growth rates that have recently started to slow (Wade in press) perhaps because they are reaching the capacity of the environment to sustain them. A simple model that describes this density-dependent growth is a θ-logistic model (Equation [7.1]):

$$N_{t+1} = N_t + rN_t\left[1 - \left(\frac{N_t}{K}\right)^\theta\right]$$ (7.1)

where N = population size, t = time, r = maximum growth rate (near $N = 0$), K = carrying capacity (for illustrative purposes set at 10,000), and θ = shaping parameter that controls the level of maximum net growth.

With density-dependent growth, the fastest growth rate is near $N = 0$, and at K the births are equal to the deaths. For marine mammals, θ is thought to occur such that maximum growth is $0.5K$ or greater (Wade in press). Somewhere in the middle range of abundance then, net recruitment is the greatest. The term often used for this level is *maximum net productivity level* (MNPL). It is perhaps, then, in this range that populations will be safe (at >50% of historical levels) while minimizing hardships to fisheries because net recruitment is highest.

Management may be governed by the following equation:

$$\text{PBR} = N_{\text{MIN}} \tfrac{1}{2} R_{\text{MAX}} \, F_R \tag{7.2}$$

where PBR = potential biological removal, N_{MIN} = minimum population estimate, R_{MAX} = maximum population growth rate (r in Equation [7.1]), and F_R = recovery factor.

With these basic terms laid out, we can return to the next step of choosing quantitative management objectives. This process will serve to illustrate how the equation is intended to work.

Quantitative Management Objectives

After passage of the 1994 amendments to the Marine Mammal Protection Act, a group of scientists was convened and agreed to a set of quantitative objectives that met the qualitative objectives specified in the law. Wade (1998) lists these objectives as (1) the percentile of the abundance estimate used for N_{MIN} will be chosen to satisfy both of the following conditions: (a) a population will be above MNPL with a 95% probability after 100 years in the absence of biases and using $F_R = 1$ and (b) a population starting at MNPL will remain at or above MNPL after 20 years; (2) a default value for F_R will be chosen such that the above criteria are met during simulation trials that include plausible levels of bias; (3) a value of F_R will be chosen for populations listed as endangered (under the U.S. Endangered Species Act) such that the recovery time of a population depleted to just 5% of K will not be delayed by more than 10% (relative to the recovery rate of a population with no human-caused kills) with a 95% probability. For species in which the maximum growth rate (R_{MAX}) is unknown, we use conservative default values (0.04 for cetaceans and 0.12 for pinnipeds).

With definitions of the quantitative objectives, we can now "tune" the performance of the management model to achieve the objectives. Because we cannot actually test the performance of a management scheme on real animals (and we would not want to even if we could), we use computer simulations to achieve the task.

Methods

To test the management schemes, we used a technique developed by the International Whaling Commission in testing the Revised Management Plan (Donovan 1989; Cooke 1994). Simulations are used to project the population sizes through time. We separate the tuning of the model into two steps. The first step is to find the percentile of the abundance distribution (N_{MIN}) to meet objective 1. During this step, we will also compare the difference between using N_{MEAN} and N_{MIN}. The N_{MEAN} strategy explored here uses Equation (7.2) with N_{MEAN} substituted for N_{MIN}. Use of N_{MEAN} rather than N_{MIN} is not merely a change of a number in the

TABLE 7.2. Base cases for management analysis.[a]

Base case	Starting N	R_{MAX}	Survey CV
1	K	0.04	0.2
2	**K/3**	**0.04**	**0.2**
3	K	0.04	0.8
4	**K/3**	**0.04**	**0.8**
5	K	0.12	0.2
6	**K/3**	**0.12**	**0.2**
7	K	0.12	0.8
8	**K/3**	**0.12**	**0.8**

[a]Cases shown in figures are given in **bold** type.

equation but constitutes a very different management strategy because it does not incorporate uncertainty about the level of precision in the abundance estimate.

Analysis of management schemes must consider different types of populations (different growth rates, initial population status, and different levels of abundance precision). These types are called base cases (Table 7.2) and are used to find the appropriate percentile of the abundance distribution for N_{MIN}. Base case trials are also used to compare the N_{MIN} to the N_{MEAN} strategy and assume that there are no errors in any of the other parameters.

The second step is to simulate different types of errors (bias) and find the value of F_R to meet objective 2 (Table 7.3). Choice of levels for unknown bias is difficult and will depend on the special problems likely for the suite of species to be managed. Wade (1998) provides detailed discussion of the choice of bias levels.

For each time step, the N_{MIN} strategy follows these steps: (1) N_{t+1} determined (Equation [7.1]), (2) N_{MIN} drawn from lognormal distribution with mean = N_{t+1}, CV as specified, (3) N_{MIN} calculated as a lower percentile of a lognormal distribution with mean = N_{MEAN}, CV as specified, (4) PBR calculated from Equation (7.1) (every 4 years), (5) kill drawn from a normal distribution with mean = PBR, and CV as specified, and (6) N_{t+1} adjusted by subtracting kill. The value of the percentile for N_{MIN} is found iteratively by finding the percentile that exactly satisfies objective 1. The N_{MEAN} strategy omits step 3 and uses N_{MEAN} in Equation (7.1) for step 4. Note that step 5 contributes to a worst-case scenario as it assumes

TABLE 7.3. Robustness trials for management schemes.

Problem type	Symbol	Description
Data	D1	Estimated N twice actual N
	D2	Estimated abundance CV 1/4 actual CV
	D3	Estimated mortality 1/2 actual mortality
	D4	Estimated mortality CV 1/4 actual CV
Criteria	C1	Estimated R_{MAX} twice actual R_{MAX}
	C2	Classified as within OSP ($F_R = 1$) when actually below ($F_R = 0.5$)
Research	R1	Abundance estimated every 8 years

that all PBRs are taken. For each trial, 2,000 replications are done and the mean, 5th, and 95th percentiles of the distribution are calculated and saved.

Results

We present results that illustrate the special properties of this management scheme. More detailed results of each type of trial are in Wade (1998). Samples of the simulated population trajectories (Figs. 7.4 and 7.5) show that N_{MEAN} and N_{MIN} differ little when CVs are low but perform very differently for high CVs. The N_{MIN} strategy manages the population more conservatively, whereas the N_{MEAN} strategy always results in populations attaining lower population sizes in 100 years and is highly variable. Performance of N_{MEAN} was unacceptably poor for all the measured performance statistics.

The population level achieved in 100 years is shown for the most stringent base cases (highlighted in bold in Table 7.2) for different percentiles of N_{MIN} (Fig. 7.6). Although a different percentile could be chosen for each potential case, the scientists convened to implement the MMPA decided a simple formula that was more easily understood by the affected parties (primarily fisherman and environmental groups) would be preferable. Therefore, the lower 20th percentile for N_{MIN} was chosen to meet objective 1.

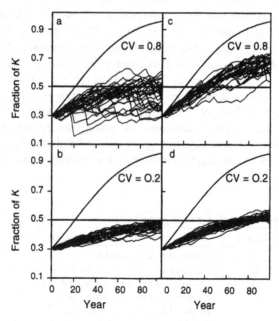

FIGURE 7.4. Sample of 30 simulations for base cases for cetaceans ($R_{MAX} = 0.04$), the N_{MEAN} strategy (a, b) and N_{MIN} strategy (c, d).

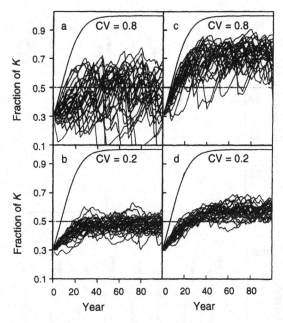

FIGURE 7.5. Sample of 30 simulations for base cases for pinnipeds ($R_{MAX} = 0.04$), the N_{MEAN} strategy (a, b) and N_{MIN} strategy (c, d).

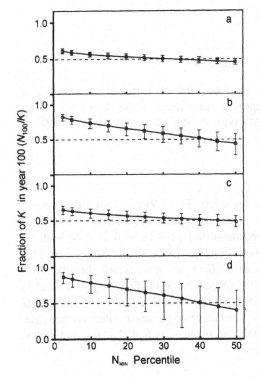

FIGURE 7.6. Fraction of K attained in 100 years for the cases in bold in Table 7.2 and for different percentiles of N_{MIN} (a–d refer to base cases 2, 4, 6, and 8, respectively, in Table 7.2).

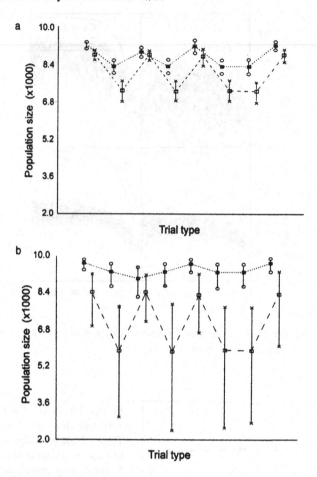

FIGURE 7.7. Distributions of population size after 100 years. The N_{MIN} strategy is symbolized by solid squares, the N_{MEAN} strategy by open squares. (a) Base case 2 (CV = 0.02); (b) base case 4 (CV = 0.8). Trial type symbols as in Table 7.2. Vertical bars include 95% of the distribution.

A sample of the results of robustness trials (Fig. 7.7) also shows the superior performance of N_{MIN}. The base trial (no errors) appears on the left. The magnitude of the effect of different errors (robustness trials) can be viewed along with the outcomes of the different management strategies (N_{MIN} versus N_{MEAN}). For both strategies, errors that have the greatest effect are those which directly affect parameters multiplied in Equation (7.2) (N_{MIN} [D1], PBR [D3], R_{MAX} [C1], and F_R [C2]). The first two can be minimized through scientific scrutiny of abundance and mortality estimation techniques. The latter two parameters caution that criteria for changing from default growth rate parameters and changing population status should be carefully considered.

Conclusion

Given the objective of preventing population declines while minimizing restrictions on fisheries, this modeling exercise has shown that including uncertainty in the management regime (the N_{MIN} strategy) is superior. Abundance estimation is difficult for many marine mammal species (Table 7.1). Management objectives are not met for species with high CVs using the N_{MEAN} strategy. If CVs could be reduced to low levels for all species, the N_{MIN} and N_{MEAN} strategies would be similar. Unfortunately, it is often difficult to reduce CVs. As population size decreases, CVs increase (Taylor and Gerrodette 1993). Therefore, threatened, endangered, or depleted populations may be managed poorly by using N_{MEAN}. Species that are difficult to sight (e.g., those with long dive times, that surface with little splashing and no visible blow) may also suffer from chronically high CVs and therefore poor management by using N_{MEAN}. Using N_{MEAN} can also contribute to economic uncertainty for fisheries. For example, the probability of depleting a population is much higher using the N_{MEAN} strategy, which could severely affect fisheries if closures were required for population recovery.

The creation of a model to meet management objectives must consider the data available (past, present, and future). Species managed by the International Whaling Commission have historical catch data that allow a calculation of population status, and a more sophisticated model was created to use these data (Cooke 1994). Such data are not available for most marine mammals in U.S. waters. Given less data, a model that requires less data to drive the management regime is appropriate. Although the models must differ, the principle of creating a model to meet performance standards based on quantitative management objectives remains sound.

Acknowledgments. This research was supported with a National Research Council Associateship (Taylor) and by the National Marine Fisheries Service's Office of Protected Resources (Wade). Improvements were made thanks to reviews by Jay Barlow, Robert Brownell, Doug DeMaster, Tim Gerrodette, Lloyd Lowery, Robert Hofman, Steve Reilly, and Michael Tillman.

Literature Cited

Barlow J (1993) The abundance of cetaceans in California waters estimated from ship surveys in summer/fall 1991. Administrative Report LJ-93-09. National Marine Fisheries Service. Southwest Fisheries Science Center, La Jolla, CA

Barlow J, Hanan D (1995) An assessment of the status of harbor porpoise in central California. Report of the International Whaling Commission, Special Issue 16:123–140

Buckland ST, Anderson DR, Burnham KP, Laake JL (1993) Distance sampling: estimating abundance of biological populations. Chapman and Hall, London

Cooke JG (1994) The management of whaling. Aquatic Mammals 20:129–135

Donovan GP (1989) The comprehensive assessment of whale stocks: the early years. Report of the International Whaling Commission, Special Issue 11

Forney K, Barlow J (1993) Preliminary winter abundance estimates for cetaceans along the California coast based on a 1991 aerial survey. Report of the International Whaling Commission 43:407–415

Holt SJ, Talbot LM (1978) New principles for the conservation of wild living resources. Wildlife Monographs 59:6–33

Taylor BL, Gerrodette T (1993) The uses of statistical power in conservation biology: the Vaquita and Northern Spotted Owl. Conservation Biology 7:489–500

Wade, PR. (in press) A Bayesian stock assessment of the eastern Pacific grey whale using abundance and harvest data from 1967 to 1996. Journal of Cetacean Research and Management, Special Volume

Wade, PR (1998) Calculating limits to the allowable human-caused mortality of Cetaceans and pinnipeds. Journal of Marine Mammal Science 14(1):1–37

White GC, Anderson DR, Burnham KP, Otis DL (1982) Capture-recapture and removal methods for sampling closed populations. LA-8787-NERP NOAA. Technical Information Service, 5285 Port Royal Road, Springfield, VA 22161

8
Whaling Models for Cetacean Conservation

Mark S. Boyce

The moot point is, whether Leviathan can long endure so wide a chase, and so remorseless a havoc; whether he must not at last be exterminated from the waters, and the last whale, like the last man, smoke his last pipe, and then himself evaporate in the final puff.
—Herman Melville, *Moby Dick* (1851)

Introduction

Norway recently resumed commercial whaling, and Japan may follow suit. The Scientific Committee of the International Whaling Commission (IWC), the body charged with the management of large whales, has put forward a quantitative management plan to set quotas for the commercial harvest of whales. This plan is the result of 6 years of modeling efforts by numerous teams of scientists. My objective is to review the historical background for the modeling, describe the final accepted model, and question its adequacy in the broader context of ecosystem management.

Historical Context

The near extirpation of the world's great baleen whales (rorquals) during the 20th century ranks among the most tragic consequences of human technology. In particular, global stocks (populations) of Blue (*Balaenoptera musculus*), Fin (*B. physalus*), Sei (*B. borealis*), Humpback (*Megaptera novaeangliae*), Bowhead (*Balaena mysticetus*), and Right (*Eubalaena glacialis* and *E. australis*) Whales were severely depleted (Table 8.1). Since 1946, monitoring and regulation of whaling have been the purview of the IWC, presently comprising 40 member-nations including several that have never had commercial whaling interests.

Initially, gross quotas for whale harvest were set in Blue Whale units whereby catches were scaled according to the value of the whales (e.g., two Fin Whales or six Sei Whales were equivalent to one Blue Whale unit) (Gambell 1976). Quotas were set high enough that stocks were rapidly depleted. In 1975, a new

TABLE 8.1. Commercial whale stocks.[a]

Species	Stock	Prewhaling	Present	Source	Latin binomial
Blue Whale	Global	250,000	3,000–10,000	Schmidt 1994	Balaenoptera musculus
Blue Whale	Antarctic	250,000	210–1,000	Myers 1993; IWC 1994	
Bowhead	Arctic	57,000	8,000	Scheffer 1994	Balaena mysticetus
Bryde's Whale	Global	100,000	90,000	Myers 1993	Balaenoptera edeni
Fin Whale	S Hemis[b]	548,000	15,000	Myers 1993; IWC 1994	Balaenoptera physalus
Fin Whale	N Hemis	58,000	20,000	Evans 1987	
Gray Whale	NE Pacific	12–15,000	21,000	Buckland et al. 1993; Schmidt 1994	Eschrichtius robustus
Humpback	Global	115,000	20,000	Myers 1993	Megaptera novaeangliae
Humpback	S Hemis[b]		22,000	IWC 1994	
Minke Whale	Global	140,000	725,000	Myers 1993	Balaenoptera acutorostrata
Minke Whale	Antarctic		600,000	IWC 1994	
Northern Right	N Atlantic	?	350	Schmidt 1994	Eubalaena glacialis
Southern Right	S Hemis[b]	100,000	3,000	Myers 1993	Eubalaena australis
Sei Whale	Global	256,00	54,000	Myers 1993	Balaenoptera borealis
Sei Whale	S Hemis[b]	98–135,000	8,300	Horwood 1987; IWC 1994	
Sperm Whale	Global	2,400,000	400,000	Myers 1993	Physeter catodon
Sperm Whale	S Hemis[b]		128,000	IWC 1994	

[a]The primary source for many of these data is the International Whaling Commission (IWC 1993, 1994). Note that the prewhaling estimates are likely to be increased resulting from the increased take of whales that was previously unreported by the USSR because prewhaling stock estimates are based on the cumulative recorded harvests.
[b]South of 30°S.

management procedure (NMP) was adopted by the IWC that based harvest quotas on a harvest model adapted from fisheries (Young 1993). With this model, quotas were set for each species at levels thought to be sustainable.

In some instances, the apparent mismanagement of whale stocks may partly relate to the modeling approach used to recommend quotas for taking whales. The underlying structure of the NMP was a model of maximum sustained yield (MSY) (i.e., harvesting that achieves a maximum harvest that can be sustained indefinitely). Flaws in the IWC's application of the MSY model included (1) poor or insufficient data with which to parameterize the model, and (2) inappropriateness of the model because it is a single-species equilibrium model (Botkin 1990). Further problems plagued implementation of the NMP that could be attributed to inflexible management procedures on the part of the IWC and their refusal to heed advice of the Commission's Scientific Committee (Frost 1979; Smith et al. 1994). Perhaps most important, insufficient attention had been given to the role that uncertainty should play in shaping harvest quotas.

Finally, the IWC agreed in 1982 to establish an indefinite moratorium on commercial whaling effective in 1986. To conclude the moratorium, the IWC's Scientific Committee developed a revised management procedure (RMP) that offers a more conservative and robust method for setting whaling quotas. The RMP requires data on stock size and harvests of sufficient accuracy and precision to ensure that populations are not overexploited before implementation can go forward.

Japan and Norway are forcing the issue to lift the moratorium on whaling. Indeed, Norway began whaling again, in 1993 taking 226 Minke Whales (*Balaenoptera acutorostrata*). Japan has asked to harvest 2,000 Minke Whales annually from Antarctic seas and has spent millions of dollars on sighting surveys to justify future whaling.

Management Objectives

In 1989, the IWC provided its Scientific Committee with priority criteria for the RMP. These included (1) perpetuation of stocks, (2) maximum sustained harvest, and (3) relative constancy of catch limits. The IWC gave clear priority to conservation (i.e., the first criterion). To optimize the harvest under these seemingly conflicting objectives, the IWC proposed to use a model, the catch-limit algorithm (CLA), to set harvest quotas for whaling that are encumbered with numerous cushions so that harvesting will occur only when it poses no threat to the persistence of populations of whales. Accomplishing this balance among objectives requires careful consideration of various sources of uncertainty in data and the model. If used according to the guidelines developed by the IWC, the CLA is designed to accomplish all objectives by setting consistent harvests of whales, while still protecting whale stocks from overexploitation. For the near future, this implies that none of the severely depleted stocks will be harvested.

Catch-Limit Algorithm

To formulate a new CLA for the RMP, the Scientific Committee of the IWC solicited harvesting algorithms from a number of scientists. Because whales are long-lived and growth rates are slow, empirical validation of a CLA may not be feasible. Instead, competing models were evaluated by using extensive simulation trials of other more detailed models that included various forms of complexity to see how well the CLA performed. Although the models had to meet each of the three objectives established by the IWC, the principal criterion in the selection of a CLA was that harvests would be conservative under the algorithm and afford security of persistence for the stocks. The selected algorithm was one formulated by Justin Cooke.

Model Structure

Cooke's CLA is essentially an ad hoc statistical device that uses historical time series of harvests and estimates of abundance to calculate an allowable current harvest. The model underlying the CLA has no age structure and assumes that the population was at equilibrium prior to exploitation,

$$N_{t+1} = N_t - C_t + rN_t[1 - (N_t/N_0)^2] \qquad (1)$$

where N_t = population size in year t, C_t = catch in year t, and N_0 is the population size prior to exploitation in year 0. In practice, $N_0 = N_T/D_T$ such that stock depletion, D_T, is the ratio of the population size at the beginning of the catch quota period, T ($0 \le t \le T$).

At low population density, the potential growth rate for the population, $r = 1.4184\mu$, where μ is a productivity parameter and 1.4184 is a constant assigned to whale productivity by the IWC for arcane historical reasons (because μ can be adjusted, the 1.4184 is of no real consequence). The density-dependent response that permits sustainable harvests is labeled "sustainable yield" in Figure 8.1. The maximum sustained yield rate (MSYR) is 0.9456μ. Faithful application of the CLA should eventually result in stock sizes no less than about 75% of preexploitation levels.

One of the most difficult problems associated with application of the CLA, and a key criterion used in selection of a CLA, is how to deal with uncertainty in parameter estimates. In general, fitting ecological models to data is difficult because when sufficient ecological structure is included in the model, data are seldom sufficient to estimate all parameters reliably. The IWC has taken the approach of keeping the structural model as simple as possible, so that few parameters need to be estimated.

The model is fitted by using a joint likelihood function similar to a Bayesian procedure, but the CLA does not "learn" quickly (Anon. 1993) (i.e., catch limits do not respond quickly to new data on stock estimates). The model is governed by prior distributions for parameters, and the underlying population model is used to

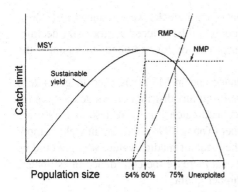

FIGURE 8.1. Yield model that forms the basis for the IWC's revised management procedure. Structural differences in the catch-control law between the new management procedure (NMP) of 1975 are compared with the current revised management procedure (RMP) (from Young 1993). Maximum sustained yield is assumed to occur when the population is maintained at 60% of the unexploited stock size, N_0. Both the RMP and the NMP proscribe curtailed harvests if populations drop below 54% of the unexploited stock size but, as illustrated, differ substantially in the catch limits allowed for larger populations.

generate posterior distributions that reflect the model as well as the prior distributions. The prior distributions require assumptions (e.g., the CLA assumes that the stock sizes are lognormally distributed). If, instead, the underlying distribution were bimodal due to multiple equilibria, the Bayesian assumptions could yield a spurious distribution of allowable catches.

For some species, the IWC has been willing to adopt more complex models (e.g., age structure has been built into a Bayesian model for aboriginal harvest of Bowhead Whales) (Raftery et al. 1995). Bayesian statistics afford a powerful way to estimate model parameters if one is comfortable with the underlying model and the assumed distributions. These Bayesian-like procedures bring state-of-the-art statistics to the problem of interfacing models and data. (Readers of this volume will appreciate potential applications to population viability analysis.)

Protecting Against Overexploitation

Several features of the RMP are designed to protect against overharvests.

1. Although whale populations can achieve population growth rates in excess of 3% (Horwood 1990), the CLA assumes a base yield rate of only 1%. This yield rate can be increased if supported by sufficient data.
2. As in the NMP, harvest rates are further backed off from MSY by an additional 10%. This can be seen in Figure 8.1, in which harvests are always below the peak in the yield curve.
3. In contrast to the NMP, the RMP sets quotas as a function of current stock size if sufficient data exist to justify the harvests. As illustrated in Figure 8.1, quotas of zero harvest will be set for stocks at or less than 54% of N_0, and harvest will be increased with increasing population size. The NMP set constant quotas that

could be maintained at the same level until stocks were reduced to 60% of stock size (see Fig. 8.1), whereas quotas will be reduced as stock size declines under the RMP. This affords greater protection for exploited stocks under the RMP.

4. Uncertainty in estimates of abundance and in MSYR is used to generate a distribution of possible catch limits. From this distribution, an actual quota is selected in which 41% of the calculated catch limits are less and 59% are greater than the recommended number (Young 1993). This again yields a more conservative quota than choosing the mean or median value, which would be closer to 50%. As more and better data increase the confidence in estimates, the CLA typically recommends a higher quota.

5. Before annual quotas can be set for a 5-year period, data are required on previous harvests and sighting surveys to estimate current stocks.

6. Periodic sighting surveys must be conducted to monitor the efficacy of the CLA for each stock. If sighting data are not collected within a prescribed period, the RMP stipulates that the harvest quota is automatically reduced by 20% per year and set at 0 if survey data are not completed within 5 years.

7. Extensive robustness trials were performed to evaluate performance under various assumptions. For each application of the CLA (i.e., before whaling begins), additional "implementation trials" will be required in which computer simulations will be performed that incorporate what is known of spatial organization of the stock.

Populations most likely to be harvested under the CLA will be Minke Whales in the north Atlantic, the north Pacific, and possibly the Antarctic. Minke Whales are by far the most abundant rorquals, with population estimates greater than 750,000 animals, including more than 600,000 Minke Whales in Antarctic seas (IWC 1994). Because stocks are large and proposed harvests are small, implementing the CLA probably will not threaten the future persistence of Minke Whales. Japan's request to harvest 2,000 Southern Hemisphere Minke Whales annually constitutes a kill of approximately 0.3% of current stock size. Sustainable harvests of other animal populations have been possible given that rates of exploitation are not excessive (Rosenberg et al. 1993).

There is evidence that rorqual stocks can recover, even after severe depletion and populations of several species of whales appear to be increasing (Best 1993; Schmidt 1994). In particular, the Grey Whales (*Eschrichtius robustus*) in the northeastern Pacific Ocean have staged a remarkable comeback from near extirpation at the turn of the century to approximately 21,000 animals today (Buckland et al. 1993). This population is thought to be near or even greater than pre-exploitation levels (Schmidt 1994). Estimates of Bowhead Whales in the Arctic have increased from some 1,500 whales in 1976 to at least 7,800 today, although part of this increase is probably due to improved estimates (J. Zeh, personal communication). Likewise, under protection, Southern Right Whales were thought to have achieved rates of increase in excess of 7% per annum, and Sei Whales off South Africa increased at annual rates of 9–13% (Horwood 1987);

such extraordinary rates of increase of 7% or more are likely to be inaccurate estimates. Nevertheless, these many observations of population resilience imply that other baleen whales can recover after having been overexploited, although many years may be required given the low reproductive rates of these large species.

Testing the CLA with Simulations

The IWC Scientific Committee conducted extensive computer simulation trials to evaluate the CLA and to ensure that it met the IWC's three objectives. Models were constructed that included considerable detail about whale life histories including age and sex structure, variable vital rates, and a variety of assumptions about depensation and density dependence. In addition, thousands of simulations were conducted to evaluate the consequences of biases and errors in survey data and prior catch data. To simulate changes in future environments, simulations were conducted in which the carrying capacity declined or underwent cyclic oscillations.

The Cooke procedure was selected because it was most robust to the various modeling scenarios. The simplicity of the Cooke procedure was a benefit because estimation of fewer parameters was required, and therefore the power associated with input data was higher. The many conservative cushions incorporated into the CLA outlined above would seem to be sufficient to prevent harvests of whales in depleted stocks.

Of the algorithms evaluated by the Scientific Committee, the Cooke procedure best met the objectives provided by the IWC. However, the IWC has not provided guidelines on precisely what constitutes acceptable performance of the CLA, and the simulations demonstrate that there are some circumstances under which it performs poorly (Taylor et al. 1994). Using the population size at the end of 100 years to evaluate the CLA, model performance appears poor when current stock size is overestimated (Taylor et al. 1994). Deriving reliable estimates of whale populations over extensive areas of ocean is a daunting task and one that can be highly sensitive to observer bias and sampling protocol. If estimates were made that indicated a number higher that the true population size, quotas set by the CLA will be excessive and drive the population to low levels. The slow rate at which the CLA learns is partly responsible for this weakness in model performance (Young 1993).

Simulations also showed that the CLA performs poorly when historical catches are underreported (Taylor et al 1994). The CLA uses historical catch to back-calculate preexploitation stock size, and if only a fraction of the actual harvest is reported, the model will target a harvested population that is far too low. For some stocks, this can be a serious problem. Right Whales have been protected since the 1930s, yet recent reports document Soviet exploitation of Right Whales in the Okhotsk Sea, near Tristan da Cunha in the south Atlantic, near Kurile Island in the Pacific, and from factory ships in the Antarctic (Best 1988; Yablokov 1994).

Large numbers of Humpback and Blue Whales were harvested by Soviets well after their protection by the IWC (Yablokov 1994), prompting a scheme requiring international observers on all whaling vessels. The reported kill of Humpback Whales from one of four Soviet factory ships in the Southern Hemisphere was 152, a number that recently has been corrected to 7,207. Computer simulations performed by the IWC Scientific Committee found the CLA to be robust to (i.e., criteria for evaluation not greatly affected by) 50% underreporting of true catches. But recent data released by the Russian Federation show that the USSR was underreporting catches by some 90%. The CLA cannot be expected to perform adequately with such poor data.

Japanese whaling operations also have been accused of underreporting of Sperm Whale catches (Kasuya 1991). Norway has been exposed for underreporting of Minke Whale kills in the mid-1980s and has recently resumed whaling in the northeast Atlantic without IWC permission. We must trust whaling countries to conform to quotas and to provide reliable data on harvests. Resolving such issues is beyond the scope of the CLA but is included in the RMP. The adopted approach, the International Observer Scheme, has proved to be effective at reducing the extent of such underreporting of whale harvests (Young 1993).

Problems of Multistock Management

The spatial structure of whale populations may constitute the most serious risk to implementation of the CLA. Intended for a single population, Cooke's CLA can lead to local stock depletion if spatial distribution is not carefully considered. Imagine two Minke Whale populations: a near-shore population and a pelagic offshore population. If quotas for harvest were allocated for the entire area, but all harvests were near-shore, we would expect to see serious depletion of near-shore stocks. The Scientific Committee of the IWC is using an approach of dividing ranges into small areas that will be the focus for CLA application (i.e., "disjoint areas small enough to contain whales from only one biological stock" [Smith and Polacheck 1994]). Genetic information or radiotelemetry data can be used to designate medium-sized areas if data show that a population ranges over a larger area. Otherwise, quotas will be set within each small area (although this rule is being contested).

Current guidelines for the RMP have quotas determined by one of two protocols: catch cascading or catch capping. For catch cascading, data among multiple small areas are pooled and the CLA applied; harvests for individual small areas are proportional to the estimated population of whales in the respective small areas. Catch capping, however, involves comparison of the summed quotas over all small units with CLA quotas from the pooled data. If pooling the quotas from small areas yields higher total quotas than application of the CLA to the sum of the parts, the original individual-area quotas are used. But if pooled quotas are less than those from the sum of the individual areas, harvests are reduced proportionally so that the sum equals the pooled CLA quota.

Catch cascading appears to work well for Minke Whales in the north Atlantic, but choice of method must be evaluated by simulation for each area. Catch capping is the safest approach when high variance exists among small areas. Pooling offers advantages for catch cascading when sample sizes are small for individual areas such that increased power associated with pooling samples can increase yields over an area.

Even if commercial whaling were to follow the RMP, there may be errors in stock identification that could lead to local extirpation. Uncertainty of distribution and movement patterns is high and may vary considerably. For example, Alaskan Humpback Whales normally winter off Hawaii, but individuals documented off Hawaii have been seen in subsequent years off Baja (Evans 1987). Also, errors in identification of species may occur by whalers; in particular, the status of the Pygmy Right Whale (*Caperea marginata*) is poorly known, but it may be subject to accidental harvest as a Minke Whale because it is difficult to distinguish the two species (Evans 1987). Likewise, because of their similar appearance, whalers often have not recognized Bryde's (*Balaenoptera edeni*) and Sei Whales to be distinct species (Gambell 1976).

Species-level problems of identification, however, are relatively minor compared with the difficulty of identifying small genetically isolated subpopulations within a species. Recall my example that if whales from small coastal populations are likely to be caught in conjunction with whaling of larger pelagic populations, and if these coastal stocks are not recognized to be distinct, they easily could be extirpated. Insufficient data exist on the spatial distribution and composition of most whale populations to eliminate such a threat. Furthermore, due to problems of population estimation, most estimates of stock size are burdened by enormous confidence intervals.

Unfortunately, no monitoring is required that would ensure appropriateness of small area definitions nor to evaluate the efficacy of harvest implementations (Smith et al. 1994).

Should We Use the CLA?

The immediate issue at hand is whether to implement the CLA as a component of the RMP to set quotas for the commercial harvest of Minke Whales. Strong arguments can be made for either position. Many reasons exist to be concerned about the reauthorization of whaling, even if primarily sanctioned for the abundant Minke Whale (plans also exist to continue with whaling for Fin Whales off Iceland and Brudas Whales in the north Pacific). The primary arguments against future whaling include the following: (1) Many people find whaling objectionable on ethical grounds (Freeman and Kellert 1994) irrespective of the strength of the stocks. (2) Nations proposing to reinstate commercial whaling have a poor track record with documented violations of past IWC regulations. Rather than risking future violations of quotas that could further threaten stocks perhaps the safest approach is to ban all commercial whaling. (3) Because we do not understand the complexity of ocean ecosystems and cannot anticipate the full consequences of a whale harvest, the safest approach most compatible with conservation may be one of strict protection. (4) Uncertainty in the CLA (for reasons outlined here) suggests that the scientific basis for whaling may be on shaky ground, and the precision with which we can monitor whale stocks is poor. Thus we are unlikely to detect negative consequences of whaling until considerable damage has been done to stocks.

Arguments in favor of reinstating commercial whaling guided by the CLA include the following: (1) Whales are a protein-rich food source acceptable to many people of the world. Foregoing whales as food may result in greater environmental costs to replace that lost food (e.g., land-based agriculture has severe consequences for biodiversity, pollution, erosion, and global change) (Freeman 1994). (2) Harvest of Minke Whales might reduce competition with Blue, Fin, and Sei Whales, thereby enhancing recovery of the great whales that have been so severely decimated. The possible existence of multiple equilibria due to competitive interactions might require harvest of Minke to permit recovery of reduced whale populations. If stocks of the great whales are increasing, reducing Minke stocks may hasten their recovery. (3) Conservative measures built into the CLA minimize the risk that Minke Whale stocks could be overexploited as were the other species of rorquals. (4) Without commercial whaling as an incentive to estimate whale stocks, we would have no support for monitoring, which is very expensive. Without monitoring data, no responsible management is possible.

Ecosystem Management

My primary concern with the proposed use of the CLA for commercial whaling is the failure to recognize the context of the whale stocks in ocean ecosystems. Most obvious are the severe disruptions already imposed on the Antarctic marine ecosystems, which will require hundreds of years to recover, if recovery is possible. The issue is one that needs to be evaluated by using the evolving principles of ecosystem management (Boyce and Haney 1997; Vogt et al. 1997).

Freeman (1994) argued that the environmental costs to replace lost human food associated with suspension of whaling would be much greater if this food came from land-based agriculture. However, simulations of stocks of the great whales suggest that restoration of the southern ocean ecosystems will require substantially more than 500 years (see below), suggesting that the environmental costs of whaling simply have not been measured adequately. There are few baseline data on marine ecosystems, so it is difficult to evaluate the long-term consequences of actions. Freeman's conclusion that whaling may be conducted with few environmental consequences may be based on the "out of sight, out of mind" view that has tolerated abuse of aquatic ecosystems worldwide.

The implicit assumption of the RMP is that the safest way to ensure the recovery of Blue, Fin, Sei, Humpback, and Right Whales in Antarctic waters is to offer protection for these species. A step in this direction was taken with the 1994 resolution by the IWC for the establishment of a whaling sanctuary in Southern Hemisphere oceans. But protection might not be sufficient, and in many circumstances, conservation management must include intervention (Belovsky et al. 1994).

Interactions between whales and their prey and other competing taxa are largely unknown. This fact must raise concerns about the utility of applying a single-species whaling model to species embedded in complex ecosystems (May et al. 1979). Sound ecosystem management requires that we confront this complexity in resource management (Grumbine 1994; Boyce and Haney 1997). After all, application of the RMP can only occur in the context of a complex ecosystem.

One argument to support the "release-from-competition" hypothesis is that despite protection since 1962 (Gambell 1976), Southern Blue and Humpback Whales have not increased at rates that might have been expected (Schmidt 1994). Yet, not until November 1993 was it learned that the USSR did not abide by their IWC agreements to curtail whaling of Right, Blue, and Humpback Whales and that the harvests of these species were seriously underreported (Yablokov 1994). Given the weakness of understanding of the Antarctic ecosystem, justification for whaling based on the competitive-release hypothesis easily could be challenged. But if nothing is done, we are unlikely to improve our management.

Multispecies Context

The CLA is driven by a single-species population model; yet all whale populations are embedded in multispecies ecosystems. Interspecific interactions including competition and predation surely have consequences for the dynamics and management of whale stocks. The complex dynamics that can emerge from multispecies and multistock systems would defy the existence of simple dynamic equilibria as assumed by the CLA (May 1987; Botkin 1990). Of particular significance for the krill-based southern ocean stocks are the trophic-level interactions between krill (*Euphausia* spp.) and whales and the competitive relations among whales feeding on krill.

Competitive interactions among Antarctic species feeding on krill have changed as a consequence of whaling earlier in this century (Horwood 1987). Stocks of Blue Whales were depleted first, followed by heavy harvests of Fin Whales. When Fin Whale stocks declined, an eruption of harvests of Sei Whales occurred in Antarctic waters, presumably curtailed by the ban on whaling in 1986. Only in recent years has a significant number of Minke Whales been taken from Antarctic waters (Fig. 8.2).

Before whalers shifted from their focus to taking Sei Whales in the 1960s, increased pregnancy rates and decreased age of maturity were documented for Sei Whales, suggesting that they were prospering from reduced competition for krill stocks resulting from depressed Blue and Fin Whale stocks (Gambell 1973). Similarly, first age of maturity for Minke Whales has decreased since the 1930s despite limited take of this species, again implying that the species is responding

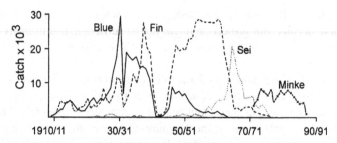

FIGURE 8.2. Historical pattern of exploitation for four species of baleen whales in the Antarctic during this century (from Horwood 1990). These data do not include substantial unreported take of whales by the USSR. For example, Soviets took at least 1,433 Blue Whales during the 1960s, but only 156 were reported (Yablokov 1994). Historical records such as these are required inputs for the CLA.

to competitive release from the depletion of stocks of larger rorquals (Laws 1977). The age at maturity is particularly important demographic information for Minke Whales because their 1-year breeding cycle suggests limited scope for population response in pregnancy rates (Horwood 1987, 1990). These lines of evidence are more likely to be valid interpretations for Minke Whales than for Sei Whales because the spatial and temporal niche overlaps with Blue and Fin Whales are greater for Minke Whales than for Sei Whales (Laws 1977).

Baleen Whales can have major effects on the structure of marine ecosystems (Katona and Whitehead 1988). Estimates of the quantity of krill consumed annually in the Antarctic seas approach 190 million metric tons, which is a substantial fraction of the estimated annual production of 250 million metric tons (Laws 1977). Prior to severe stock depletion, cetaceans consumed more biomass of prey than humans take in the entire world's fisheries (Kanwisher and Ridgway 1983). Because whale stocks have been so severely depleted, however, now only some 45 million metric tons of krill are being eaten by whales.

Subsequent to the decimation of stocks of the great whales, several nonwhale species also increased in abundance, apparently partly in response to the increased availability of krill (May 1984; Nybakken 1993). Crabeater Seals (*Lobodon carcinophagus*) from the Antarctic Peninsula appear to have increased based on reduced age at maturity (Laws 1977; Horwood 1987). Under protection, Fur Seals (*Arctocephalus gazella*) on South Georgia increased from fewer than 100 to more than 1.1 million (Gentry and Kooyman 1986). Also, Chinstrap (*Pygoscelis antarctica*), King (*Aptonodytes patagonicus*), Gentoo (*P. papua*), Macaroni (*Eudyptes chrysolophus*), and Adelie (*P. adeliae*) Penguins have increased in abundance in recent years. These examples of changes in bird and seal populations are not without problems of interpretation (e.g., some of these population increases are also associated with release from exploitation) (Horwood 1987). Yet the consistency in the pattern over the species listed lends support to this theory. There can be no question that the community composition in krill-based ecosystems has been drastically altered (cf. Fig. 8.2).

The Japanese government has suggested that harvests of Minke Whales may actually release the larger rorquals from competition and thereby hasten their recovery. Although data supporting the Japanese hypothesis are weak, if true, this may make a defensible case for commercial harvest of Minke Whales in the Antarctic. To examine the hypothesis, a population trajectory was calculated for Southern Hemisphere Blue Whales by using the population model underlying the CLA and compared with a trajectory from a Lotka-Volterra model of two-species competition between Blue Whales and Minke Whales (Fig. 8.3). In addition, population growth for Blue Whales was plotted with an annual harvest of 5,000 Minke Whales. Note that the CLA model's projection is much more optimistic than would occur under competition, yet still requiring 575 years for Blue Whales to recover to 50% of their prewhaling stock size. Also, note that although harvest of Minke Whales indeed hastens recovery for Blue Whales, because growth rates are low the benefit appears small: recovery to 50% of prewhaling stock size requires 765 years under a complete moratorium, whereas the same level of recovery still requires 711 years under an annual harvest of 5,000 Minke Whales. Under competition, however, a continued harvest of 5,000 Minke Whales exceeds the CLA and is not sustainable.

The simulation results suggesting more than 700 years to achieve one-half of prewhaling stock size is discouraging. Also discouraging is a comparison of the effectiveness of Minke harvest as a means of hastening Blue Whale recovery. Note, for example, that after 20 years and a total harvest of 100,000 Minke Whales, there will be only four additional Blue Whales above the no-harvest alternative. In other words, harpooners would only have to err in their identification of a young Blue Whale (or Pygmy Blue) on four occasions out of 100,000 strikes to eliminate any conservation advantage to the harvest of Minke Whales.

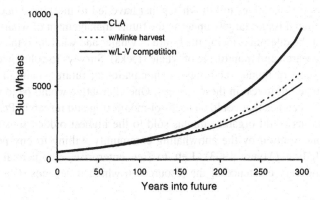

FIGURE 8.3. Population recovery for Blue Whales under the population model that underlies the CLA (top curve) versus a simplistic two-species Lotka-Volterra competition model between Blue and Minke Whales with a 5,000/yr harvest of Minke Whales (middle curve) and without harvesting (bottom curve).

Recovery of the great whales will not be rapid because of their low fecundity and long time to maturation. Thus the great whales will be slow to respond, and any krill resources released by the exploitation of Minke Whales are likely to be used by smaller species with greater reproductive potential (e.g., penguins, Crab-eater Seals, and fish). It is by no means certain that the krill-based community is an equilibrium system, an issue that also plagued attempts to anticipate conse-quences of proposed human harvests of Antarctic krill over a decade ago (May 1984). Therefore, the food resource as well as consumer populations may be expected to undergo highly dynamic fluctuations. Or multiple equilibria may exist, say, due to Allee effects or diffuse competition, such that some stocks may never recover from the excesses of the 1930–1970 slaughter. Thus, it is not certain that this complex multispecies system can return to some preexploitation equi-librium, or if it can, that Minke Whale harvests will facilitate restoration.

Economic Considerations

Effective ecosystem management must consider human values, and economics certainly has played a primary role in the management of whales in the past. For example, capital investment in whaling ships is so great that economic moti-vation for curtailing whaling will not occur even if stocks dwindle toward extinc-tion (Clark 1975). Fundamental to frustrations of managing whale stocks is the "tragedy of the commons" associated with resource management of the ocean, whereby economic incentives invariably lead to overexploitation (Hardin 1968; Myers 1993).

For several reasons developed above, we cannot be completely confident that the RMP can be used effectively for the conservation of whale stocks. The fact that Iceland and Norway have decided to ignore the IWC's direction calls to question the ability of the IWC to implement regulations. Rather whale conserva-tion may be better driven by the creative use of economic incentives. Indeed, it has been economic incentives to kill whales that have led to the severe depletion of stocks. Financial issues largely underlie the future management of whales as well (e.g., one of the dilemmas facing the future of whales and whaling is the source of funds for research and monitoring of whale stocks). Surveys to count whales cost enormous sums of money, and only in anticipation of future whaling has Japan been convinced to invest in these surveys. One alternative would be to impose a tax on future catches that could be used for monitoring and research. Perhaps the United Nations could organize quotas sold to the highest bidder, thus allowing financial intervention by the antiwhaling community willing to buy permits to prevent whaling (Myers 1993). Left to free enterprise, commercial whaling threatens the very existence of the resource on which it depends (Clark 1975).

Adaptive Management

The IWC is not proposing to begin use of the RMP on a strictly experimental basis to evaluate the CLA's efficacy. Rather, if implemented, it may be used as a management tool for the exploitation of whale stocks into the indefinite future. Because of the many uncertainties surrounding its use, one might argue that the CLA should not be implemented until it has been rigorously evaluated. Such field trials would require several years or decades to complete. The IWC Scientific Committee is divided on the issue of monitoring the performance of the CLA (IWC 1993). Some members argue that intensive surveys of whale populations are essential for the purpose of monitoring the effectiveness of the CLA (Smith and Stokes 1993). Others argue that, given the poor precision of current estimates of abundance, it is impractical to monitor the CLA in real time. Alternatively, perhaps we should be monitoring other components of the ecosystem.

Active adaptive management (Walters and Holling 1990) may afford the only sound approach to structuring the management of the ecosystems that are home to the world's whale stocks. Adaptive management involves first using a model (e.g., the CLA) to synthesize how we think the system works and to formulate hypotheses. This is followed by active management intervention such as an experimental harvest or management by protection. The system is monitored to see how well we are able to predict the system's behavior. The monitoring data are used to evaluate the model predictions, followed by reformulation of the model. Adaptive resource management is structurally like the modern scientific method but iterated through time and in different areas to accommodate the fact that complex ecosystem processes cannot be replicated with a rigorous experimental design. The idea of integrating science and management is not a new one, indeed; such an experimental approach to evaluating the management of whale stocks was outlined 20 years ago (Allen 1980).

Conclusions

The IWC's RMP makes use of a CLA with a simple density-dependent model that only requires data on previous catch and current stock size from which to set catch quotas. Although the underlying model is simple in structure, the algorithm is designed to reduce harvests to compensate for uncertainty in a way that makes it unlikely that overexploitation will occur when following the CLA. Careful consideration for uncertainty ensures that the RMP will be a safer model for governing whale harvests than its predecessor, the New Management Procedure. The RMP's Bayesian approach to interfacing data with the model affords a powerful way to design ecological models. Nevertheless, problems with implementation may remain due to excessive harvests under the CLA if there are unreported harvests, overestimates of stock size, and failure to identify spatially structured stocks.

To my mind, however, these problems pale in comparison to the fundamental flaw that the RMP is a single-species approach to management that fails to consider ecosystem structure and function in setting management objectives. Despite knowledge that the krill-based ecosystem of the southern ocean has been seriously disrupted by the decimation of the great whales, the IWC has recently established a sanctuary making the entire region off limits to whaling. This may not be sufficient, and recovery of the great whales may require intervention by selective removal of Minke Whales as proposed by Japan. If taken in the context of the ocean ecosystems in which it would be applied, the RMP could provide a useful baseline for adaptive management.

Acknowledgments. My service on a scientific panel assembled by the National Marine Fisheries Service to review the IWC catch-limit algorithm provided background for this chapter. The panel was chaired by A. Rosenberg with the support of G. Kirkwood, D. Pelka, and T. Smith. Committee members included J. Cushing, R. McKelvey, R. Myers, A.M. Starfield, G.L. Swartzman, and B. Taylor. Thanks to E. Slooten, T. Smith, R. Spill, A.M. Starfield, and three anonymous reviewers for comments on the manuscript.

Literature Cited

Allen KR (1980) Conservation and management of whales. University of Washington Press, Seattle, WA

Anonymous (1993) The revised management procedures (RMP) for baleen whales. IWC Report 45/4, Annex H. International Whaling Commission, Cambridge, UK

Belovsky GE, Bissonette JA, Dueser RD, Edwards TC Jr, Luecke CM, Ritchie ME, Slade JB, Wagner FH (1994) Management of small populations: concepts affecting the recovery of endangered species. Wildlife Society Bulletin 22:307–316

Best PB (1988) Right Whales *Eubalaena australis* at Tristan da Cunha—a clue to the "non-recovery" of depleted stocks? Biological Conservation 46:23–51

Best PB (1993) Increase rates in severely depleted stocks of Baleen Whales. ICES Journal of Marine Science 50:169–186

Botkin DB (1990) Discordant harmonies. Oxford University Press, Oxford, UK

Boyce MS, Haney A (1997) Ecosystem management: applications for sustainable forest and wildlife resources. Yale University Press, New Haven, CT

Buckland ST, Breiwick JM, Cattanach KL, Laake JL (1993) Estimated population size of the California Grey Whale. Marine Mammal Science 9:235–249

Clark CW (1975) Mathematical bioeconomics. J Wiley, New York

Evans PGH (1987) The natural history of whales and dolphins. Facts on File, New York

Freeman MMR (1994) Science and trans-science in the whaling debate. In: Freeman MMR, Kreuter UP (eds) Elephants and whales: resources for whom? Gordon and Breach Science, Basel, Switzerland, pp 143–157

Freeman MMR, Kellert SR (1994) International attitudes to whales, whaling and the use of whale products: a six-country survey. In: Freeman MMR, Kreuter UP (eds) Elephants and whales: resources for whom? Gordon and Breach Science, Basel, Switzerland, pp 293–315

Frost S (1979) The whaling question. Friends of the Earth, San Francisco, CA

Gambell R (1973) Some effects of exploitation on reproduction in whales. Journal of Reproduction and Fertility Supplement 19:531–551

Gambell R (1976) Population biology and the management of whales. In: Coaker TH (ed) Applied biology, vol 1. Academic Press, New York, pp 247–343

Gentry RL, Kooyman GL (1986) Fur Seals: maternal strategies on land and at sea. Princeton University Press, Princeton, NJ

Grumbine RE (1994) What is ecosystem management? Conservation Biology 8:27–38

Hardin G (1968) The tragedy of the commons. Science 162:1243–1248

Horwood J (1987) The Sei Whale: population biology, ecology and management. Croom Helm, London

Horwood J (1990) Biology and exploitation of the Minke Whale. CRC Press, Boca Raton, FL

IWC (1993) International Whaling Commission Report of the Scientific Committee. International Whaling Commission, Cambridge, UK

IWC (1994) International Whaling Commission Report of the Scientific Committee. International Whaling Commission, Cambridge, UK

Kanwisher JW, Ridgway SH (1983) The physiological ecology of whales and porpoises. Scientific American 248:110–121

Kasuya T (1991) Density-dependent growth in north Pacific Sperm Whales. Marine Mammal Science 7:230–257

Katona S, Whitehead H (1988) Are cetaceans ecologically important? Oceanography and Marine Biology Annual Review 26:553–568

Laws RM (1977) Seals and whales of the southern ocean. Philosophical Transactions of the Royal Society of London B 279:81–96

May RM (1984) Exploitation of marine communities: report of the Dahlem Workshop. Springer-Verlag, Berlin

May RM (1987) Chaos and the dynamics of biological populations. Proceedings of the Royal Society of London A 413:27–44

May RM, Beddington JR, Clark CW, Holt SJ, Laws RM (1979) Management of multi-species fisheries. Science 205:267–277

Myers N (1993) Sharing the earth with whales. In: Kaufman L, Mallory K (eds) The last extinction. MIT Press, Cambridge, MA, pp 179–194

Nybakken JW (1993) Marine biology: an ecological approach. 3rd ed. HarperCollins New York

Raftery AE, Givens GH, Zeh JE (1995) Inference from a deterministic population dynamics model for Bowhead Whales. Journal of the American Statistical Association 90:402–416

Rosenberg AA, Fogarty MJ, Sissenwine MP, Beddington JR, Shepherd JG (1993) Achieving sustainable use of renewable resources. Science 262:828–829

Scheffer VB (1994) Book review: the Bowhead Whale. Journal of Mammalogy 75:558–559

Schmidt K (1994) Scientists count a rising tide of whales in the seas. Science 263:25–26

Smith T, Stokes A (1993) SC/45/Mg11. International Whaling Commission, Cambridge, UK

Smith TD, Polacheck T (1994) Information requirements for multi-population application of the catch limit algorithm. SC/46/Mg8. International Whaling Commission, Cambridge, UK

Smith TD, Polacheck T, Swartz S (1994) The role of science in resource management: the International Whaling Commission's revised management procedure. SC/46/Mg9. International Whaling Commission, Cambridge, UK

Taylor BL, Smith TD, Palka D (1994) Towards understanding the performance of the catch limit algorithm. SC/46/Mg6. International Whaling Commission, Cambridge, UK

Vogt KA et al (1997) Ecosystems: balancing science with management. Springer, New York

Walters CJ, Holling CS (1990) Large scale management experiments and learning by doing. Ecology 71:2060–2068

Yablokov AV (1994) Validity of whaling data. Nature 367:108

Young NM (ed) (1993) Examining the components of a revised management scheme. Center for Marine Conservation, Washington, DC

9

Assessing Land-Use Impacts on Bull Trout Using Bayesian Belief Networks

Danny C. Lee

Introduction

Management agencies responsible for public lands within the United States increasingly find their activities circumscribed by considerations of threatened, endangered, or sensitive species (TES species). In the western states, where the U.S. government owns a majority of the land base, many TES species are critically dependent on habitat conditions within federal lands. Federal regulations provide specific guidelines when dealing with TES species on these lands. For example, U.S. Department of Agriculture Forest Service (Forest Service) regulations require the Forest Service to maintain viable populations and promote recovery of TES species throughout their native range. Commonly, the Forest Service prepares biological assessments for all management activities that may affect TES species. Such activities include timber sales, road construction, mining leases, grazing allotments, and recreational enhancements. It is the responsibility of the forest supervisor and staff to assess the threat posed by each activity; those that may adversely affect TES species are either modified to mitigate the impact or abandoned. Species protected under the Endangered Species Act additionally require that the federal oversight agency (either the U.S. Fish and Wildlife Service or the National Marine Fisheries Service, depending on the species) concurs with the impact assessment and proposed mitigation measures.

The contentious nature of species protection combined with the large number of assessments required by the agencies has prompted a huge demand for information and methods that can provide quantitative estimates of the risks to TES species posed by land-use activities. Linkages between management activities and population risk are complex, however, and the level of information available about them inevitably is incomplete. Frequently, biological assessments are based largely on the judgment and experience of forest biologists because specific information on local populations simply is unavailable and unlikely to be obtained within a reasonable timeframe. Managers need methods that recognize the uncertainty inherent in decisions made in information-poor environments, yet are efficient, rigorous, and defensible. Such tools must be capable of incorporating complexity, while simultaneously acknowledging uncertainty.

The key to developing suitable management tools is to find a way to link conventional well-understood viability models with the more nebulous reality of habitat impacts on population parameters. Management concerns for Bull Trout (*Salvelinus confluentus*) motivate the following demonstration of an approach to risk assessment that combines a stochastic logistic model of population abundance with conditional probability matrices that tie habitat conditions and proposed activities to model parameters. The model parameters, model outputs, and the conditional probability matrices are all combined within a Bayesian belief network (BBN), which provides the analytical structure for examining possible implications of proposed actions.

The example network presented here was constructed in 1993 as a prototype to explore the potential for BBN technology and to help identify research needs. It was not developed for real-world decision making, nor has it been used as such. Because BBNs are a relatively recent development in the field of artificial intelligence, a brief introduction to the structure and purpose of BBNs is provided.

Bull Trout: A Species in Jeopardy

The plight of the Bull Trout is illustrative of the problems facing many native trout and char of western North America and of the challenges facing land-use managers that consider the needs of sensitive species in their management activities. Twenty-four endemic species and subspecies of the genera *Oncorhynchus, Salvelinus,* and *Thymallus* join the Bull Trout on the American Fisheries Society's list of endangered, threatened, or fishes of special concern (Williams et al. 1989). Heightened concern for the viability of the Bull Trout prompted the U.S. Fish and Wildlife Service recently to review the status of the Bull Trout and list it as a threatened species.

A principal threat to Bull Trout and other native salmonids is loss or disruption of habitat. Such losses result from a combination of changing land-use patterns and instream alterations over the past 150 years (Meehan and Bjornn 1991; Nehlsen et al. 1991; Behnke 1992; Rieman and McIntyre 1993). Logging, mining, agricultural practices, and urbanization have degraded watersheds, whereas irrigation and hydroelectric dams or diversions have reduced the quantity and quality of available habitat. In addition, overharvest and hybridization with introduced species and stocks have compounded the problems. The decline in suitable habitat has fragmented and isolated native trout populations. Many of the strongest populations supported by natural reproduction are found in undisturbed or wilderness areas (Rieman and Apperson 1989; Rieman and McIntyre 1993; Lee et al. 1997).

Bull Trout pose a special challenge to land managers because they are widely distributed throughout their range (Fig. 9.1), but they are seldom locally abundant due to demanding habitat requirements (see Howell and Buchanan 1992; Rieman and McIntyre 1993). This combination increases both the likelihood of the presence of a Bull Trout population within a given watershed and the sensitivity of that population to watershed disruption. In the United States, much of the best remain-

FIGURE 9.1. Distribution of Bull Trout in North America (Meehan and Bjornn 1991).

ing habitat lies within public lands administered by the Forest Service and the U.S. Department of Interior's National Park Service and Bureau of Land Management. Outside of special wilderness areas and national parks, proposals to eliminate all detrimental activities are untenable, given that current federal law requires public lands be managed for multiple purposes, including commodity extraction. The public debate over balance between wildlife needs and economic and social factors is becoming increasingly strident, and Bull Trout are caught in the middle.

Bayesian Belief Networks

The formal reasoning behind BBNs is complex, yet elegant. Excellent introductions can be found in the work of Pearl (1988), Olson and co-workers (1990), and Haas (1991). Pearl (1988) and Haas (1991) provide not only the conceptual underpinnings but compare BBNs to other reasoning schemes and describe BBNs within the context of current graph theory. Whittaker (1990) provides a thorough introduction to conditional dependencies that are inherent in graphical models, including BBNs. Influence diagrams (Clemen 1996), which are common in formal decision analysis, are a logical extension of belief networks. Bayes (1763) first articulated the conditional probability relationship between a prior hypothesis, evidence, and a posterior hypothesis that is at the core of current belief network methodology.

I do not cover the details of BBNs here nor discuss all the ramifications of their use. Readers are referred to the references herein for such information. Rather, I would hope to stimulate by way of example, sufficient interest in BBNs that readers might pursue more advanced understanding on their own.

Pearl (1988) defines a BBN as

A directed acyclic graph in which each node represents a random variable that can take on two or more possible values, and the arcs signify the existence of direct influences between linked variables. The strengths of these influences are quantified by forward conditional probabilities.

To illustrate, consider a simple example represented by a directed acyclic graph containing three nodes (Fig. 9.2). Each node represents a system property that is characterized by using a discrete random variable. The nodes represent (A) watershed history, (B) habitat condition, and (C) population trend. The directed arcs signify that watershed history influences habitat condition directly, which, in turn, affects the population trend.

Each random variable, or node, can take on a range of values. To simplify matters, each variable is divided into three or four discrete values, each of which may represent a range of conditions. At any given point in time, the state of the system is reflected by the set of random variables, A_i, B_j, C_k, a subset of all possible combinations of A, B, C. Although there is only one true state of the system, one may be uncertain of its nature. This uncertainty is expressed as a belief vector, represented here by the histogram at each node. The belief (probability) attached to each level within each node represents the degree to which one believes each level is the true state of the system. In this example, I am certain that the watershed is undisturbed but less certain about habitat conditions and recent population trend; the most likely combination is favorable habitat and a stable population trend.

Simply stated, the purpose of a BBN is to track belief vectors and update them systematically as information is added to the system. The terminology associated with trees is often used to describe BBNs because of their branching structure. Nodes that have no directed arcs entering them are referred to as root nodes, whereas those with no directed arcs originating from them are called leaf nodes. In this simple example, watershed history is a root node, and population trend is a leaf node. Once the structure of a causal network has been defined, the bulk of the remaining effort is dedicated to developing reasonable estimates for the link matrices. Link matrices are conditional probability matrices that define the relationships between nodes (Table 9.1). The values within the matrices come from combinations of empirical evidence and expert opinion or may be generated by ancillary models as demonstrated below.

Updating the BBN involves entering findings (i.e., changing the belief vectors of nodes in which you have information germane to the issue at hand). Findings

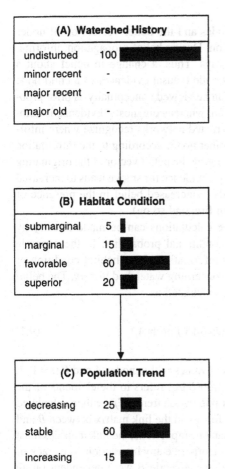

FIGURE 9.2. Three-node example Bayesian belief network.

TABLE 9.1. Link matrices of conditional probabilities for the example Bayesian belief network shown in Figure 9.2.

		Probability that habitat condition is			
Given:		Submarginal	Marginal	Favorable	Superior
Watershed disturbance history	Undisturbed	0.05	0.15	0.60	0.20
	Minor recent	0.08	0.22	0.55	0.15
	Major recent	0.20	0.50	0.28	0.02
	Major old	0.10	0.41	0.42	0.07

		Probability that population trend is		
Given:		Decreasing	Stable	Increasing
Habitat condition	Submarginal	0.85	0.13	0.02
	Marginal	0.70	0.25	0.05
	Favorable	0.15	0.65	0.20
	Superior	0.07	0.83	0.10

can be entered at any combination of nodes and the belief vectors of all nodes updated accordingly via belief propagation. During belief propagation, information is passed along two separate pathways. Thus, a change in belief about a certain node can originate from a parent node (causal evidence) or a child node (diagnostic evidence). Causal evidence can be viewed conceptually as prior probabilities in a classical Bayesian application, whereas diagnostic evidence is comparable with sample data. Properly constructed networks recognize where information enters the system and update all other nodes according to the information received but prevent the updates from changing the belief vector of the originating node. This prevents circular reasoning (e.g., evidence for smoke leads to increased belief in the presence of fire, which leads to increased belief in the presence of smoke, which leads to increased belief in fire, and so on).

In the simple three-node example, the calculations can be made directly by using Bayes theorem and the rules of conditional probability. In the example shown in Figure 9.2, I assume no prior information about habitat condition or population trend. Information is entered concerning watershed history. The belief vectors for habitat condition can then be calculated as

$$\text{prob}(B_j) = \sum_{i=1}^{a} \text{prob}(B_j|A_i) \, \text{prob}(A_i) \qquad (9.1)$$

where the subscripts i and j denote different values within nodes A and B ($i = 1, 2, \ldots, a; j = 1, 2, \ldots, b$), and the term $\text{prob}(B_j|A_i)$ refers to the elements of the conditional link matrix. Belief vectors for population trend (C) are then calculated in like manner by using the belief vector for B and the link matrix between B and C. When a node is conditional on more than one input node, the link matrices must be specified for the joint probability and incorporate any interactions (i.e., lack of independence) that might exist. For example, if node B were dependent on an additional fourth node (D), the equation would be

$$\text{prob}(B_j) = \sum_{i=1}^{a} \sum_{l=1}^{d} \text{prob}(B_j|A_i, D_l) \, \text{prob}(A_i) \, \text{prob}(D_l) \qquad (9.2)$$

The calculations can become more complex as information flows in both directions. For example, if we know something about both watershed history and population trend and want to calculate a belief vector for habitat condition, the formula becomes

$$\text{prob}(B_j) = \frac{\sum_{k=1}^{c} \text{prob}(C_k) \left[\text{prob}(C_k|B_j) \sum_{i=1}^{a} \text{prob}(B_j|A_i) \, \text{prob}(A_i) \right]}{\sum_{j=1}^{b} \text{prob}(C_k|B_j) \sum_{i=1}^{a} \text{prob}(B_j|A_i) \, \text{prob}(A_i)} \qquad (9.3)$$

As the networks become more complex, with multiple pathways between nodes, exact solutions become intractable. Various analytical routines for belief updating in complex networks that overcome this limitation have been developed by Henrion (1986), Pearl (1988), Lauritzen and Spiegelhalter (1988), and Andersen and associates (1989).

Relatively simple BBNs can be programmed by using spreadsheets or other software packages that easily manipulate matrices. More complex networks generally are beyond the programming skills or patience of many potential developers. Sophisticated software such as the Hugin system (http://www.hugin.dk), Netica (http://www.norsys.com), and Analyticia (http://www.lumina.com) provide convenient shells for developing and implementing BBNs with a minimum of programming. Timothy Haas (University of Wisconsin-Milwaukee, personal communication) developed a simpler program, BAYES, that is suitable for many applications. The Hugin system was used for the network presented below; a more limited spreadsheet version was also developed using Excel.

Bull Trout Network

Three basic steps may be used in assessing land-use impacts on the population viability of Bull Trout. First, a population viability model is needed that provides a quantitative assessment of the risk of extinction (or quasi-extinction) given some combination of population parameters. Second, one must be able to identify physical changes in habitat conditions that result from land-use activities. The third and perhaps most problematic step is to link the changes in physical habitat to the parameters of the viability model.

The overall structure of the Bull Trout BBN is as shown in Figure 9.3. The intention of introducing it at this point is to help guide the reader through the following discussion. In essence, I want to use information about the status of a watershed and the nature of proposed activities to judge the future viability of a local Bull Trout population. The following sections elaborate on the rationale behind each node and the linkages among them.

Underlying Viability Model

Some of the best available population data for Bull Trout are time series of counts of spawning beds (redds). Bull Trout spawn in the fall when stream flows are generally low. Thus it is relatively easy to observe their spawning redds and obtain convenient indices of adult population abundance. Rieman and McIntyre's (1993) analysis of time series data from northern Idaho and Montana by using the exponential trend model of Dennis and co-workers (1991) suggests high probabilities of extinction for many of the 19 populations examined. I fit a density-dependent variant of this model to these same data. The density-dependent model gives a more optimistic view of the chances of population persistence and is the underlying model used in the Bull Trout network. The basic structure of the viability model is

$$N_{t+1} = N_t \exp \left(\frac{\tilde{N} - N_t}{\gamma \tilde{N}} + \varepsilon \right) \tag{9.4}$$

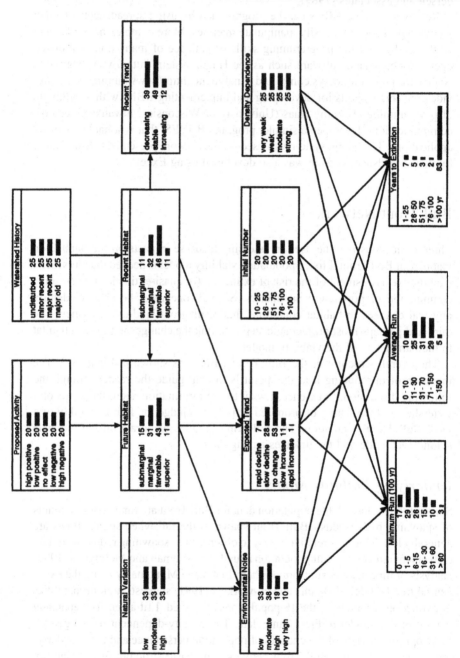

FIGURE 9.3. Bull Trout Bayesian belief network at initialization.

where N_i = the redd count in year i, $\tilde{N}$ = the equilibrium attraction point (i.e., the count at which the expectation of $N_{t+1} = N_t$), g = density-dependent scaling parameter, and ε = normally distributed error term with zero mean and standard deviation, $s(N_t)$. For simulation purposes, $s(N_t)$ is defined as

$$s(N_t) = \frac{\sigma}{1 + \left(\dfrac{N_t - \tilde{N}}{\gamma\tilde{N}}\right)} \tag{9.5}$$

if $N_t > \tilde{N}$, and $s(N_t) = \sigma$ otherwise. This adjustment in the error term imposes higher fidelity to the density-dependent control at high population levels and tends to suppress large unreasonable values from occurring during simulation.

Parameter estimates were obtained by using linear regression and Dennis and Taper's (1994) version of the model:

$$\ln\left(\frac{N_{t+1}}{N_t}\right) = a + bN_t + \varepsilon \tag{9.6}$$

where $a = 1/\gamma$, and $b = -1/\gamma\tilde{N}$, and ε is a normally distributed error term with zero mean and standard deviation, $s(N)$. As Dennis and Taper (1994) note, point estimates obtained in this manner are appropriate, although hypothesis testing based on linear regression estimates of the error terms would be invalid due to the autoregressive structure of the model.

The data from northern Idaho (Nelson et al. 1992) and western Montana (Weaver 1992) produced a range of parameter estimates for γ, σ, and initial values (N_0) that were used to help define the parameter space used for Monte Carlo simulation (Table 9.2). Model fits were variable, with r^2 values ranging from 0.13

TABLE 9.2. Viability model parameter ranges used in Monte Carlo simulation.

Description	Parameter symbol	Observed range[a]	Level	(Range)
Initial number	N_0	7–145	10–25	
			26–50	
			51–75	
			76–100	
			>100	(101–125)
Environmental noise	s	0.25–1.25	Low	(0.15–0.45)
			Moderate	(0.45–0.75)
			High	(0.75–1.05)
Density dependence	g	0.53–4.07	Strong	(0.50–1.25)
			Moderate	(1.25–2.25)
			Weak	(2.25–3.50)
			Very weak	(3.50–5.00)
Expected trend	R	0.51–4.93	Rapid decline	(0.25–0.5)
			Slow decline	(0.5–1)
			No change	1
			Slow increase	(1–2)
			Rapid increase	(2–4)

[a]Based on parameter estimates from 18 streams in Idaho and Montana.

to 0.80 (mean = 0.46). A fourth parameter, R, was used to scale $\tilde{N}$ relative to N_0 (i.e., $\tilde{N} = R*N_0$). A value of R greater than 1 will lead to an expected increase in population size; a value less than 1 will lead to a decline.

Two hundred thousand combinations of N_0, γ, σ, and R were chosen randomly from the defined parameter space and used in 100-year simulations of the population. In each replication, the mean annual redd count (run), minimum redd count, and year of extinction (if applicable) were retained as output variables of interest. Although average run size provided an indication of the central tendency of the population, the cumulative frequency distribution of the minimum value recorded for each iteration for a given set of parameters determined the likelihood of dropping below a given value and is thus a measure of quasi-extinction risk. Input parameters and output variables were grouped into discrete levels to build contingency tables that would serve as link matrices between model parameters and outputs in the completed BBN. On average, 500 replications were used for each combination of discrete parameter ranges to estimate conditional probabilities.

Linking Land-Use Activities to Habitat

To predict population trends based on habitat, one must know something of the recent or current habitat conditions and have some expectation of the future conditions. In this application, I assumed that recent habitat conditions are determined solely by watershed history, in which watershed history is characterized by the magnitude and timing of disturbance (i.e., undisturbed, minor recent, major recent, or major old). Because recent habitat conditions are likely to have affected recent population trends, recent trend was included as an evidence node. If information on the recent population trend is available (decreasing, stable, or increasing), then this information together with watershed history determines my belief about recent habitat conditions using the dual flows of information discussed above. Habitat condition, both recent and future, is divided into four levels: submarginal, marginal, favorable, or superior—depending on the suitability of the habitat to support a self-sustaining population of Bull Trout. Future habitat condition results from a combination of recent habitat condition and proposed activity, in which future activity is classified according to its effect on the habitat (high or low positive, no effect, high or low negative).

The descriptive levels for watershed history and proposed activity are meant to be relative indicators of the intensity and magnitude of past anthropogenic or natural disturbance and proposed future management activities. Admittedly, these nodes may seem overly simplistic and imprecise. In part, this is unavoidable if one wishes to develop a model that may be generalized across a landscape covering 500,000 km^2 or more. This model is targeted at watersheds on the order of 10,000 ha, and there are thousands of such watersheds within the range of Bull Trout. Each watershed has its own unique combination of physiographic setting and disturbance history, including both anthropogenic and natural disturbances. The level of understanding of landscape impacts on fish is far too crude to develop tailormade specifications for each watershed. As understanding improves, we can begin to stratify watersheds into various types based on their biophysical setting and sensitivity to various types of disturbances. Belief networks will evolve to reflect this increased understanding and complexity. The current example, however, is limited to the type of relative, subjective judgments that are at least consistent with the everyday experience of biologists and land managers and can be applied universally.

Calibration of the conditional probability matrices linking habitat components and linking habitat to model parameters involved combining expert opinion with empirical evidence. A group of six colleagues from the Intermountain Research Station (see acknowledgments) with extensive experience in fisheries, habitat assessment, and watershed disturbance and recovery joined me in estimating the conditional probabilities needed to complete the network. Each individual independently filled out probability matrices such as the ones shown in Tables 9.1 and 9.2 and presented their estimates, with the rationale behind them, in an open discussion. Consensus values were developed during the open discussion. Finally, the consensus values were compared with the limited data available from Idaho and Montana and the matrix values updated to either be more consistent with the data or to resolve minor logical inconsistencies.

One consensus view reflected in the link matrices is that degraded habitat tends to recover or improve over time if left alone. Similarly, favorable or superior habitat is most likely to be found in undisturbed watersheds in which conditions remain relatively stable. Thus, when a proposed activity is judged to have no effect on an undisturbed watershed, the belief vectors for recent and future habitat conditions should be equivalent. For disturbed watersheds, activities having no effect lead to slight improvements in habitat conditions.

Linking Habitat to Viability Model Parameters

Two parameters of the viability model, the environmental noise component (σ) and the scalar determining expected trend (R), are directly linked to habitat. Future habitat interacts with natural variation to determine the environmental noise exhibited in the viability model. Natural variation refers to the level of unexplained variation in the time series of redd counts that might occur naturally, independent of habitat condition Factors that might contribute to natural variation include such things as the frequency of extreme hydrologic events, the stability of the parent geologic material within the watershed, or other interactions between climate, basin topography, and vegetation. Major disturbances such as stand-replacing wildfires or high-intensity floods were not included as part of the natural variation because they would be expected to have longer-lasting impacts; such risks are not addressed by this model. The link matrix between natural variation, future habitat, and environmental noise reflects a general perception that favorable and superior habitats tend to dampen or moderate the impacts of natural variation on environmental noise (Table 9.1).

The link between recent habitat, future habitat, and expected trend reflects the logical argument that if habitat conditions improve, the population increases; if conditions deteriorate, the population declines. The primary question concerns the amount of change. There was a general consensus among the group that rapid declines or increases are relatively rare and that the chances of decline under deteriorating or stable conditions are greater than the chances of increase under improving or stable conditions.

Results and Discussion

In the completed Bull Trout BBN, there are five root nodes (proposed activity, watershed history, natural variation, initial number, and density dependence) and four leaf nodes (recent trend, average run, minimum run, and years to extinction). The longest unidirectional path between a root node and a leaf node is between watershed history and minimum run via environmental noise, which is four arcs in length. All remaining paths are of length three or less. This produces a network where substantive changes in the belief vectors at the root nodes have readily observable impacts on the terminal leaf nodes. In networks with long paths between the root and leaf nodes, signals generally are attenuated such that major changes in belief at the root nodes may translate into only minor changes at the leaves. A general feature of BBNs is that the shorter the path between nodes, the greater the response in one to a change in the other, all things being equal.

Initialization of the BBN begins with specifying the belief vectors associated with the root nodes. In the Bull Trout network (Fig. 9.3), the root nodes have been initialized with uniform vectors, which indicates no prior information on the state of the system (i.e., if we know nothing about a node, all states are equally likely). There are numerous ways that one can use the Bull Trout network to explore various management scenarios; all involve entering findings at the appropriate nodes and examining the resultant changes in belief vectors of interest. For illustration purposes, two types of analyses are demonstrated that could be used to support management decisions.

Example 1: Analysis of a Specific Watershed

In the first example, a mix of activities is proposed for a hypothetical watershed for which there is limited information. The watershed has experienced some low-level logging activity in recent decades, and the forest plan calls for a series of small timber sales over the next 50 years, spread throughout the watershed. In light of evidence that the watershed supports a local population of Bull Trout, the forest supervisor asks the technical staff to identify a list of options and the respective risks and costs associated with each. The technical staff identifies five options covering the range of possible effects on future Bull Trout habitat conditions from high positive to high negative effects. These options include various levels of logging, riparian-area and hill-slope protection measures, and specific actions designed to enhance the habitat for Bull Trout.

No information is available on recent population trends, but a survey provides a rough estimate of the current number of spawning redds within the basin. Other analyses of populations in similar circumstances suggest that density dependence is most likely to be moderate, but weak or strong levels are not unreasonable. Based on the topography and climate of the basin, natural variation could fall anywhere but is most likely to be in the moderate range.

This information is entered into the Bull Trout BBN by entering findings for the respective root nodes. By sequentially entering findings for the proposed-activity node, a series of belief vectors for a minimum run are generated that can be used as relative indices of the risk to the population associated with each option. To facilitate comparison, the forest supervisor plots the biological risk, defined as the probability of a run of five or fewer (five redds might represent a lower threshold with unacceptable consequences) versus the net economic return (revenues minus costs) of each timber sale option (Fig. 9.4). From this graph, the supervisor decides that the increased economic return of the no-effect option is worth the additional risk over options with low or high positive benefits, but it would not be worth the additional risk to proceed with options that degrade the habitat.

Example 2: Search for Guidelines

In the second example, a fisheries biologist is asked to participate in developing a conservation strategy for Bull Trout within the federal lands of his region. Part of the strategy calls for setting aside selected watersheds that will act as population reserves. Potentially disruptive activities will be severely limited within water-sheds so designated. There are numerous candidate watersheds available; the biologist wants to be able to identify those most likely to support long-term populations. Because there are too many candidate watersheds to examine them all individually, the biologist looks to screen candidates by using the BBN.

The list of possible candidates ranges from relatively pristine streams in parks and wilderness areas to those heavily impacted by logging, mining, agriculture, and other activities. As a first step, the biologist hypothesizes that the areas with lowest risk will be undisturbed watersheds with high initial numbers and low natural population variation. Findings are entered to reflect these conditions, given that there is no other information about the streams. The biologist also minimizes the effect of management by setting the proposed activity node to "no effect." The resultant belief vectors (Fig. 9.5) provide a benchmark that can be used when searching for possible combinations of conditions that would be exceptions to this rule (i.e., streams not fitting this description that have equal or lower risks). The results suggest that even relatively pristine areas are not risk free, because there are factors other than habitat that threaten some populations (e.g., the spread of introduced competitors such as Brook Trout [*Salvelinus fontinalis*]).

To guide the search, the biologist uses the diagnostic features of the BBN and enters findings of ">60" for minimum run, ">100 yr" for years to extinction, and "no effect" for proposed activity. Propagation of the network reveals the most

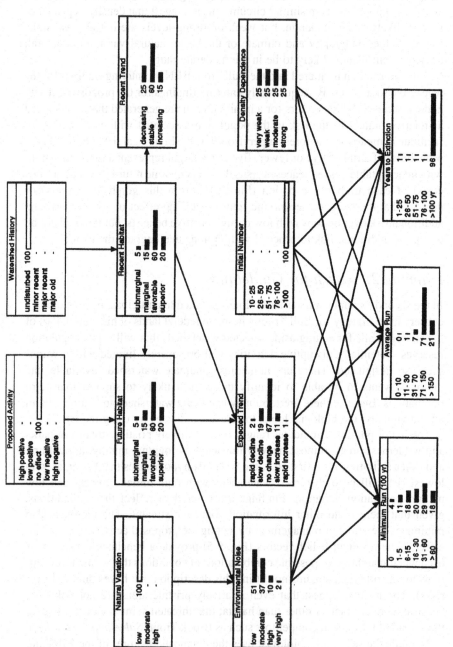

FIGURE 9.4. Bull Trout Bayesian belief network: example 1, no effect from proposed activity.

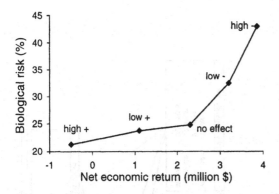

FIGURE 9.5. Biological risk (defined as the probability of falling to fewer than five redds during a 60-year period) versus net economic benefit for five management options.

likely conditions, given these constraints (Fig. 9.6). The results suggest that the most salubrious streams are those with low environmental noise, high initial numbers, and stable or increasing expected trends. To achieve such conditions requires low-to-moderate natural variation and improvements in habitat that increase the likelihood of favorable or superior conditions for both recent and future habitat.

Somewhat unexpected to the biologist is the level of uncertainty displayed for watershed history under this scenario. Such results are not surprising, however, given the length of the pathway and subsequent attenuation of signal between minimum run and watershed history, and the structure of the link matrices that make diagnostic reasoning inherently more equivocal than causal reasoning. The situation is not unlike that of the classical example of an unfair coin. If we know the coin is unfair, then we might be certain of observing three "heads" tossed in a row. Yet, observing three heads in a row is not convincing evidence that a coin is unfair, because such an event would be expected one out of eight times with a fair coin.

The biologist realizes that there are relatively few streams with large numbers of spawning adults within the region. Thus, the biologist would also like some guidance in choosing streams with smaller populations that are likely to persist. To explore this option, they set initial numbers to "51–75," minimum run to "31–60," and again hold proposed activity to "no effect". The picture that emerges from this scenario (Fig. 9.7) is remarkably similar to that from the previous scenario (i.e., look for steams with low environmental noise and improving conditions). The biologist has stumbled on a useful ecological finding—process is more important than numbers to population persistence. With this understanding, the biologists now redirects the search for emphasis areas, looking more at watershed potential than population numbers.

The biologist may use newly gained insights to identify a variety of combinations that produce a minimum run belief vector equal to or preferable to the belief vector generated under the initial hypothesis. For nearly any combination of watershed history and recent trend, there is a combination of proposed activity, natural variation, and initial number that produces the desired effect. Thus the

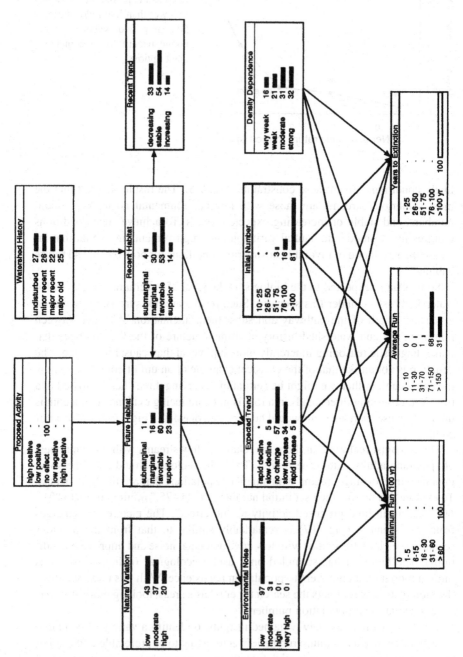

FIGURE 9.6. Bull Trout Bayesian belief network: example 2, benchmark.

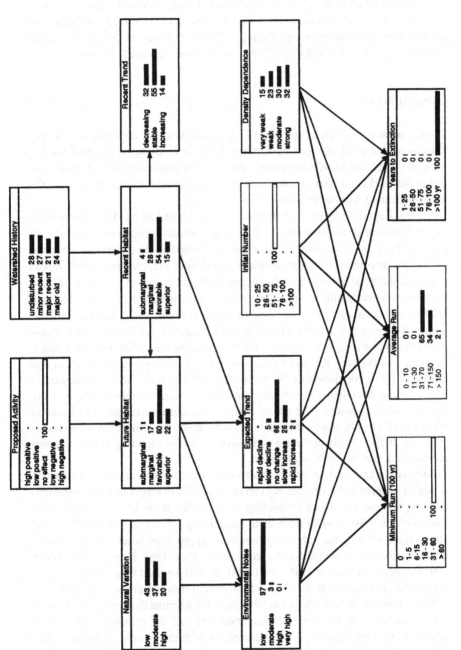

FIGURE 9.7. Bull Trout Bayesian belief network: example 2, minimum run greater than 60.

biologist is able to recommend a conservation strategy using a mix of watersheds in various stages of disturbance instead of being limited to only pristine areas. This increased flexibility allows a more robust strategy that incorporates a mosaic of conditions, which reduces overall risks from sources not directly addressed by the belief network (Rieman and McIntyre 1993). As expected, fewer options are available to reduce the risks sufficiently in the most degraded watersheds.

Conclusions

Conservation of Bull Trout and other TES species is a very complex and contentious issue that involves much uncertainty. Difficult land-use choices will have to be made to ensure population persistence. As this application shows, Bayesian Belief Networks can assist the decision process by quantifying and displaying the risks associated with alternative courses of action.

Several important features of BBNs are demonstrated above. First, the reasoning, beliefs, and assumptions that lead to an assessment are open to scrutiny by all parties affected by a decision. In this example, I relied on the expert knowledge of colleagues to construct the link matrices that project expected population trends and environmental noise from habitat conditions and proposed activities. The use of subjective probabilities from experts is problematic. Morgan and Henrion (1990) provide an excellent overview of potential biases and error that can arise with elicited expert judgment. Errors can arise because individuals lack the cognitive skills to assess probabilities accurately or because they might sense a strategic advantage in skewing the probabilities in one direction or another. Morgan and Henrion (1990) conclude that "one can only proceed with care, simultaneously remembering that elicited expert judgments may be seriously flawed, but often are the only game in town." The use of expert judgment in management of natural resources is not only unavoidable, it is desirable given the complexity of the issues involved. BBNs simply provide a convenient means of capturing expert opinions in ways that can be easily scrutinized.

Second, the approach recognizes that although the future outcome of an activity is always uncertain, this uncertainty can be quantified in ways that allow meaningful comparisons. The use of probability theory within the field of risk assessment and policy analysis is well established. As Ramsey (1926), de Finetti (1974), and others have shown, the rules of probability apply even when the probability merely reflects one's subjective degree of belief (Press 1989; Howson and Urbach 1991). Also, as Morgan and Henrion (1990) note, if we must trust management to experts, should we not at least know how certain they are of their answers?

Third, complex relationships can be captured in a format that promotes exploration of alternatives. The network described here and the viability model that it relies on are both relatively simple compared with the complex system models familiar to many ecological modelers. Yet, I contend that it is far easier to explore the dynamics of this system by using a BBN than it would be using a more traditional modeling approach, and the same can be said for more complex models. In my work, I have taken complex population models of trout and salmon dynamics (e.g., Lee and Hyman 1992) that generally are used only by technical experts and recast them as BBNs (Lee and Rieman, 1997). These networks were made accessible in spreadsheet formats and are being used by biologists throughout the Northwest (see Shepard et al. 1997). One can explore stochastic population dynamics far more efficiently with the BBNs than was ever possible by using the original format.

There is a loss in precision by using BBNs for viability analysis. BBNs are unlikely to replace more traditional analyses when there are sufficient data available to support them. In many cases, however, traditional approaches that rely solely on point estimates for parameters exaggerate the true precision of estimates and vastly understate the uncertainty in a particular course of action. In that regard, we might be better served by BBNs that highlight inherent uncertainties and keep us honest in our assessments.

Beyond their usefulness to practicing biologists and managers, BBNs also benefit researchers who strive to understand underlying mechanisms linking land-use activities to population viability. When individuals participate in developing a network and assigning values within the link matrices, they are forced to examine their assumptions about causal mechanisms and the available scientific evidence to support those assumptions. At times, the behavior of the network may challenge preconceived notions. Unexpected results prompt one to either reexamine one's preconception or examine the underlying causal linkages more carefully to ensure that the overall network behavior conforms to general experience. One of the more useful outcomes from developing this application was that individuals were forced to examine their research and experience from a novel perspective. This led to considerable discussion on the merits of various suppositions. Such discussions should lead to more focused research efforts that have direct policy implications.

Acknowledgments. I am grateful to my colleagues at the Rocky Mountain Research Station who assisted in developing the belief network demonstrated here and provided useful comments on the manuscript: John McIntyre, Bruce Rieman, Russ Thurow, Kerry Overton, James Clayton, and Gwynne Chandler. The use of trade or firm names in this publication is for reader information and does not imply endorsement by the U.S. Department of Agriculture of any product or service.

Literature Cited

Andersen SK, Olesen KG, Jensen FV, Jensen F (1989) Hugin—a shell for building Bayesian belief universes for expert systems. In: Proceedings of the Eleventh International Congress on Artificial Intelligence, pp 1080–1085. Reprinted in Shafer G, Pearl J (1990) Readings in uncertainty. Morgan Kaufman, San Mateo, CA

Bayes T (1763) An essay towards solving a problem in the doctrines of chances. Philosophical Transactions of the Royal Society of London 53:370–418. Reprinted (1958) in Biometrika 45:293–315

Behnke RJ (1992) Native trout of western North America. American Fisheries Society Monograph 6. American Fisheries Society, Bethesda, MD

Clemen RT (1996) Making hard decisions: an introduction to decision analysis. Duxbury Press, Pacific Grove, CA

de Finetti B (1974) Theory of probability, vols 1 and 2. John Wiley, New York

Dennis B, Taper M (1994) Density dependence in time series observations of natural populations: estimation and testing. Ecological Monographs 64:205–224

Dennis B, Munholland PL, Scott JM (1991) Estimation of growth and extinction parameters for endangered species. Ecological Monographs 61:115–143

Haas TC (1991) A Bayesian belief network advisory system for aspen regeneration. Forest Science 37:627–654

Henrion M (1986) Propagating uncertainty by logic sampling in Bayes' networks. Technical report. Department of Engineering and Public Policy, Carnegie-Mellon University, Pittsburgh, PA

Howell PJ, Buchanan DB (eds) (1992) Proceedings of the Gearhart Mountain Bull Trout Workshop, August 1992, Gearhart Mountain, OR. Oregon Chapter of the American Fisheries Society, Corvallis, OR

Howson C, Urbach P (1991) Bayesian reasoning in science. Nature 350:371–374

Lauritzen SL, Spiegelhalter DJ (1988) Local computations with probabilities on graphical structures and their applications to expert systems. Journal of the Royal Statistical Society B 50(2):154–227

Lee DC, Hyman JB (1992) The stochastic life-cycle model (SLCM): simulating the population dynamics of anadromous salmonids. Research paper INT-459. USDA Forest Service, Intermountain Research Station, Ogden, UT

Lee DC, Rieman BE (1997) Population viability assessment of salmonids by using probabilistic networks. North American Journal of Fisheries Management 17:1144–1157

Lee DC, Sedell JR, Rieman BE, Thurow RF, Williams JE, Burns D, Clayton J, Decker L, Gresswell R, House R, Howell P, Lee KM, MacDonald K, McIntyre J, McKinney S, Noel T, O'Connor JE, Overton CK, Perkinson D, Tu K, Van Eimen P (1997) Broadscale assessment of aquatic species and habitats. In: Quigley TM, Arbelbide SJ (eds) An assessment of ecosystem components in the interior Columbia Basin and portions of the Klamath and Great Basins. General Technical Report PNW-GTR-405. U.S. Department of Agriculture, Forest Service, Pacific Northwest Research Station, Portland, OR, pp 1057–1496

Meehan WR, Bjornn TC (1991) Salmonid distributions and life histories. In: Meehan WR (ed) Influence of forest and rangeland management on salmonid fishes and their habitats. Special Publication 19. American Fisheries Society, Bethesda, MD, pp 47–82

Morgan M, Henrion M (1990) Uncertainty: a guide to dealing with uncertainty in quantitative risk and policy analysis. Cambridge University Press, New York

Nehlsen W, Williams JE, Lichatowich JA (1991) Pacific Salmon at the crossroads: stocks at risk from California, Oregon, Idaho, and Washington. Fisheries 16:4–21

Nelson L, Davis J, Horner N (1992) Regional fishery investigation report, region 1. Job performance report F-71-R-17. Idaho Department of Fish and Game, Boise, ID

Olson RL, Willers JL, Wagner TL (1990) A framework for modeling uncertain reasoning in ecosystem management II Bayesian belief networks. AI Applications in Natural Resource Management 4:11–24

Pearl J (1988) Probabilistic reasoning in intelligent systems: networks of plausible inference. Morgan Kaufmann, San Mateo, CA

Press JS (1989) Bayesian statistics. John Wiley, New York

Ramsey FP (1926) Truth and probability. In: Braithwaite RB (ed) The foundations of mathematics and other logical essays (1931). Humanities Press, New York

Rieman BE, Apperson KA (1989) Status and analysis of salmonid fisheries: westslope Cutthroat Trout synopsis and analysis of fishery information. Job performance report, project F-73-R-1, subproject III, study I, job I. Idaho Department of Fish and Game, Boise, ID

Rieman BE, McIntyre JD (1993) Demographic and habitat requirements for conservation of Bull Trout. General technical report INT-302. USDA Forest Service, Intermountain Research Station, Ogden, UT

Shepard BB, Ulmer L, Sanborn B, Lee DC (1997) Status and risk of extinction for west-slope Cutthroat Trout in the upper Missouri River basin, Montana North. American Journal of Fisheries Management 17:1158–1172

Weaver TM (1992) Coal Creek fisheries monitoring study no X and forest-wide fisheries monitoring—1991. Montana Department of Fish, Wildlife, and Parks, Special Projects, Kalispell, MT

Whittaker J (1990) Graphical models in applied multivariate statistics. Wiley, Chichester

Williams JE, Johnson JE, Hedrickson DA, et al. (1989) Fishes of North America endangered, threatened, or of special concern: 1989. Fisheries 14:2–20

10

Using Matrix Models to Focus Research and Management Efforts in Conservation

Selina S. Heppell, Deborah T. Crouse, and Larry B. Crowder

Introduction

More than 1,000 North American species are listed or will soon be listed as endangered or threatened under the Endangered Species Act of 1973 (Glitzenstein 1993; U.S. Fish and Wildlife Service 1995). In addition, more than half the marine fish stocks used by U.S. commercial fisheries are depleted or declining (Sissenwine and Rosenberg 1993), with similar declines occurring in many freshwater fish species (Williams et al. 1989). The future of many of these species hinges on effective management and recovery plans that must be implemented in this decade (Sissenwine and Rosenberg 1993).

Unfortunately, our knowledge of a threatened species' life history and the potential conservation costs and benefits of various management alternatives is often extremely limited. Further, the research budgets of most resource management agencies are insufficient, and in many cases decisions must be made quickly. To enhance the effectiveness of conservation management under these constraints, it is important to evaluate the *relative* effectiveness of specific changes in vital rates (e.g., juvenile survival versus fecundity) on population responses (e.g., population growth rate, size, or structure) and the importance of uncertainty in our knowledge of each of these vital rates.

One set of tools for enhancing such decision making involves using deterministic matrix models to evaluate management alternatives as hypotheses. Rather than making quantitative predictions of population size through time or probability of extinction, deterministic model analyses focus on relative changes in population responses as certain parameters in the model are changed. One population response examined commonly is the intrinsic rate of increase (r) of a population. In a linear deterministic model, r (or ln [λ], where λ is the dominant eigenvector of a matrix) predicts exponential increase or decline in a population occupying a constant environment. Sensitivity analyses reveal how changes in stage-specific vital rates (e.g. survival, growth, or fecundity) affect λ.

We can apply this method to determine (1) which stage-specific vital rates contribute most to the asymptotic population growth rate; (2) how small perturbations of sensitive life history stages compare with similar perturbations of less sensitive stages; (3) which of an array of management alternatives is most (or least) likely to produce the desired result when the relative effects of each alternative can be estimated; and (4) where to focus limited research efforts to identify critical mortality sources and to refine parameters for future analyses (Schemske et al. 1994). In models that incorporate density dependence or stochasticity, the sensitivity of other response variables to changes in vital rates can also be examined. Matrix models can help managers make prudent management decisions currently, while simultaneously focusing limited research dollars on estimating those vital rates that most need refining to increase the accuracy of predictions. Sensitivity analysis of simple deterministic models can give insight on which parameters are likely to be most critical in more complex models that incorporate stochasticity. Additionally, these analyses have been used to compare population responses with perturbation across a range of life history strategies (Silvertown et al. 1993; Saether et al. 1996).

In this chapter, we emphasize the development and analysis of deterministic linear matrix models and briefly outline three case studies in which matrix models have provided useful insights for evaluating management and research alternatives in the field.

Matrix Modeling Approach

Matrix population models are surveyed exhaustively in the work of Caswell (1989). Since then, more specialized treatments have appeared, focusing on stochastic environments (Tuljapurkar 1990), optimal harvesting (Getz and Haight 1989), and density dependence (Ginzburg et al. 1990; Marschall and Crowder 1996; Grant 1998). Presentations of the basic theory and equations can be found in the work of Caswell (1986, 1989), van Groenendael and co-workers (1988), MacDonald and Caswell (1993), Noon and Sauer (1992), and Ferriere and associates (1996) and in the case studies reviewed in this chapter.

Deterministic linear models are relatively simple to produce and easy to interpret and provide analytical rather than simulation results. With a matrix model, we can calculate the proportion of individuals in an age or stage class each year, dependent on the survival and growth rates of individuals within each class and the fecundity of individuals that contribute to each class. A transition matrix (A) contains one row and column for each age or stage in the model, with each entry representing the probability of survival and transition to another stage or fecundity (Fig. 10.1). The asymptotic growth rate of a population (λ) is given by the dominant eigenvalue of the transition matrix. Because the models do not include variability in the matrix entries, populations converge to a constant proportion of individuals in each age or stage class (w, the right eigenvector of the matrix), so that

$$A(w) = \lambda(w) \tag{10.1}$$

The value of future reproduction by individuals in each age or stage class (v) is

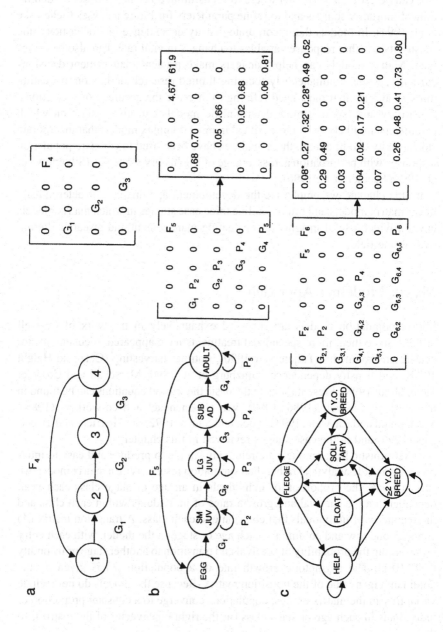

FIGURE 10.1. Life graphs and matrices for three linear deterministic matrix models. Circles denote an age or stage, represented by a row and column in each matrix. Arrows denote transition probabilities and fecundity: P = probability of surviving and remaining in a stage, G = probability of surviving and "growing" into another stage, F = number of same-sex offspring produced per reproductive individual. (a) A Leslie matrix, with each row and column representing a single time step. Individuals die after year 4; (b) stage-based model for Loggerhead Sea Turtles (Crowder et al. 1994). Individuals spend 1 year in stage 1 and multiple years in stages 2–5. This model has a postbreeding census (Caswell 1989); subadults that become adults in each time step reproduce, and a fecundity term appears in the fourth column (starred). (c) Male-only stage-based model for Red-cockaded Woodpeckers (Heppell et al. 1994). Multiple transitions are possible for fledglings, helpers, and floaters. Nonbreeding helpers increase fledgling survival rates (F_2). As in the Loggerhead Sea Turtle model, the postbreeding census requires the matrix to include fecundity terms for fledglings, floaters, and solitary males that become breeders just before the census (starred).

given by the left eigenvector, which can be obtained by transposing A (reversing its rows and columns). A number of mathematical software packages can calculate eigenvectors and eigenvalues, including MathCAD®, MATLAB®, Gauss™, and RAMAS®.

The elements of a deterministic matrix model are often based on means, calculated for several populations over one or more time periods. It is assumed that all individuals in an age class or stage are identical and that vital rates do not change over time (constant environment). Populations that are declining will always go extinct, whereas populations that are increasing will grow exponentially. Clearly, all populations violate these assumptions to some extent. Deterministic models cannot be used to estimate population size through time and are inappropriate for small or isolated populations that are subject to strong demographic stochasticity. However, these models can help us identify critical life stages and vital rates through a sensitivity analysis of the matrix. Also, the general model form of a transition matrix is flexible. Any number of life history stage classes can be modeled, and our case studies illustrate a number of useful modifications to the basic Leslie (1945) or Lefkovitch (1965) matrix.

Constructing a Deterministic Matrix Model

Compared with individual-based or stochastic models, relatively few data are necessary to construct a deterministic matrix model. This is an advantage, because data are limited for many species of concern to conservation biologists. In most stage-based models, mean annual survival, growth (or shrinkage), and fecundity probabilities are needed for each stage to be modeled. These rates are generally calculated for an annual time step, although any time step is acceptable so long as all the vital rates are calculated for the same time step (i.e., you cannot have eggs on a daily time step and adults on an annual time step in the same matrix). Generally, only females are modeled; however, some two-sex and male-only matrix models exist (Meagher 1982; Caswell 1989; Heppell et al. 1994).

One way to construct a matrix model is to start with a conceptual model, the life graph (Fig. 10.1) (Caswell 1982). Arrows between stages represent the probability of surviving and growing into the next stage (G) or creation of new individuals (F), whereas arrows that circle back into a stage represent the probability of surviving but remaining in a stage (P) (Fig. 10.1b, c). Any remaining individuals $[1 - (P + G)]$ are those that die or migrate out of the population; these are eliminated from the model population each time step. The simplest matrix is a Leslie model (Fig. 10.1a), in which surviving individuals grow into the next age class each time step, and all individuals die after a certain number of time steps. In a stage-based model, surviving individuals may remain in a stage for one or more time steps, as in our model for Loggerhead Sea Turtles (Crowder et al. 1994; Fig. 10.1b). Complex life histories may have transitions between several stages (Fig. 10.1c), as in our model for Red-cockaded Woodpeckers (RCW), which breed cooperatively and have many nonbreeding adult stages (Heppell et al. 1994).

Once the form of the model has been chosen, the transition probabilities and annual fecundities are calculated from available data. Because the parameters in a deterministic model are fixed, it is generally unwise to combine data from several populations that may experience different vital rates. Also, the modeler should consider when the "census" of the model takes place (e.g., just before or after breeding [birth-pulse models; Caswell 1989] or at an arbitrary "date" for populations with continuous reproduction [birth-flow populations; cf. Caswell 1989; Levin and Huggett 1990]). In birth-pulse models, transition matrices calculated for a prebreeding census do not have a row/column for newborns; instead, the fertilities are equal to fecundity × first-year survival. In models with a postbreeding census, adults that are counted at the beginning of the year must survive most of that year before reproducing, and any subadults that grow into adulthood that year must also reproduce. The fecundities in a postbreeding census are multiplied by the adult survival rate, and an additional fertility value appears in any subadult stages that have some probability of surviving and growing into adults (e.g., Fig. 10.1b, c).

Annual growth is straightforward in a Leslie model, in which individuals age each year and transition probability equals survival probability. But in a stage-based model, individuals may remain in a stage for one or more time steps, and the annual probability that an individual will survive and grow must be measured or calculated. In well-studied populations, average annual growth (γ) may be measured directly from transitions observed in marked individuals (e.g., Teasel; Werner and Caswell 1977). The matrix parameters then become (Caswell 1989)

$$P_i = \sigma_i (1 - \gamma_i) \tag{10.2}$$

$$G_i = \sigma_i \gamma_i \tag{10.3}$$

$$F_i = fec_i \sigma_1 \tag{10.4}$$

(prebreeding census), or

$$F_i = fec_i P_i + fec_{i+1} G_i \tag{10.5}$$

(postbreeding census), where fec_i is stage-specific fecundity and σ is newborn survival. In some organisms, such as sea turtles, a growth curve may provide the approximate number of years spent in each stage (Frazer 1983a; Crouse et al. 1987). A size-based model dependent on stage length is really an age-based model with blocks of ages grouped together. Caswell (1989) gives an equation for the annual growth probability (γ) of individuals in stage i:

$$\gamma_i = \frac{[\sigma_i/\lambda_{\text{init}}]^{T_i} - [\sigma_i/\lambda_{\text{init}}]^{T_{i-1}}}{[\sigma_i/\lambda_{\text{init}}]^{T_i} - 1} \tag{10.6}$$

where σ_i is annual survival probability, T_i is stage duration, and λ_{init} is an initial estimate of the population growth rate (Caswell 1989). The final λ of the matrix can be calculated through an iterative procedure in which λ_{init} is replaced by the calculated λ until $\lambda = \lambda_{\text{init}}$.

Are stage-based models "better" than age-based Leslie models? As with all model selection, what kind of model you choose depends on what data are available and what questions you wish to answer. Caswell (1989) discusses some statistical evaluations (e.g., log-linear models) that have been used to test the adequacy of age and stage as state variables. For some organisms, such as RCW, there are obvious life history stages that are independent of age. Sea turtles, fish, and other organisms with indeterminate growth may be more easily classified by size than age. In these cases, the sensitivity analysis is most useful to managers when individuals are grouped into stages that have identified mortality sources. But the population-level effects of a change in vital rates may not be immediately apparent in populations with long generation times. Leslie models can reveal transient effects on population dynamics that are caused by the time lag between birth and reproduction (Crowder et al. 1994; see Loggerhead case study below). These "waves" in the adult population are completely deterministic but may not be discernible in simulations based on stochastic population models. Both Leslie and stage-based models can be useful for population dynamics analysis.

Sensitivity and Elasticity Analyses of Linear Models

A sensitivity analysis is a quantitative comparison of the relative impact of model parameters on a population response that can be used to compare management alternatives qualitatively. The analysis calculates the change in the outcome of a model (e.g., the population growth rate or stage distribution) when a parameter in the model is altered. In linear models (i.e., those without density dependence) and models that do not include demographic stochasticity, the sensitivity of the asymptotic population growth rate (λ) can be used as an index to make predictions such as "increasing large juvenile [Loggerhead Sea Turtle] survival will have a relatively greater effect on population growth than saving eggs and hatchlings" (Crouse et al. 1987).

For simple linear models, the first step in a sensitivity analysis is to calculate the stable stage distribution (w) and the stage-specific reproductive values (v) of the model. These are the right and left eigenvectors associated with the dominant eigenvalue λ, which are usually scaled such that $\Sigma (w_i) = 1$ and the reproductive value of stage or age 1 individuals (v_1) = 1. Caswell (1978) used these eigenvectors to calculate the sensitivity of λ to changes in any matrix entry ($A_{i,j}$):

$$\frac{\partial \lambda}{\partial A_{i,j}} = \frac{v_i w_j}{\langle v | w \rangle} \tag{10.7}$$

where $\langle v | w \rangle$ is the inner product of the two vectors, $\{v_1 \times w_1 + v_2 \times w_2 \ldots \}$. Biologically speaking, the sensitivity analysis tells us how λ will change if a model parameter is increased or decreased by an infinitismal amount. An elasticity analysis (= proportional sensitivity; deKroon et al. 1986) calculates proportional changes in λ when matrix entries are changed by a small percentage. The elasticity of each matrix parameter is

$$\frac{A_{i,j}}{\lambda}\frac{\partial\lambda}{\partial A_{i,j}} = \frac{\partial\log\lambda}{\partial\log A_{i,j}} = \frac{A_{i,j}}{\lambda}\frac{v_i w_i}{\langle v|w\rangle} \tag{10.8}$$

The elasticities of the matrix elements sum to 1.0 (Caswell et al. 1984; deKroon et al. 1986), so elasticity analysis allows us to compare the effects of changes in parameters that are not on the same scale, such as fecundity and annual growth probabilities (Caswell 1989). For example, we use elasticities to compare the impact of a 10% increase in annual fecundity versus a 10% increase in the probability of surviving and remaining in a stage. Because the effect of a management proposal is often estimated as a proportional change in a vital rate, rather than an absolute change, elasticity analysis can be a highly useful comparative measure.

Often, we want to know the elasticity of a lower level variable that is used to calculate matrix entries, such as annual survival (σ) or annual mortality ($1 - \sigma$). The partial derivative may be calculated for any variable (x) in the model

$$\frac{x}{\lambda}\frac{\partial\lambda}{\partial x} = \frac{x}{\lambda}\sum_{i,j}\frac{\partial\lambda}{\partial A_{i,j}}\frac{\partial A_{i,j}}{\partial x} \tag{10.9}$$

or the sum of the partial derivatives with respect to x. This elasticity can be estimated by the average change in λ as x is increased and decreased by a small percentage:

$$S_{\text{prop}} = \frac{\lambda_{x+x(0.01)} - \lambda_{x-x(0.01)}}{\lambda\,x\,0.02} \tag{10.10}$$

where $\lambda_{x+x(0.01)}$ is the new λ calculated for the matrix as the parameter is increased or decreased by 1%. In the denominator, the λ from the original unperturbed matrix is multiplied by the total change in x (in this example, $0.01 + 0.01 = 0.02$, or 2%). This method for calculating proportional sensitivities is time-consuming but can be useful when the partial derivative in Equation (10.9) is difficult to compute. Also, this equation can be used to compare the relative effects of parameter changes on other response variables, such as stage distribution and population size in more complex nonlinear models.

Estimating the Effects of Management Alternatives

Elasticity analysis as outlined above can help managers decide which life stages are in most need of protection and which model parameters need additional research (Schemske et al. 1994). The next step is to determine which vital rates are most likely to be affected by a particular management proposal. By calculating λ for matrices that reflect the new vital rates imposed by a particular management plan, we can compare qualitatively the potential impacts of an array of proposals. To get an idea of the time scale needed to assess the effect of a management plan, we can use the growth rate and stable stage distribution from each new model to plot the equilibrium population size over time (e.g., Crowder et al. 1994; see

Loggerhead case study). Such trajectories are not exact but can be used to estimate how quickly a population could recover if vital rates remain fairly constant. In the Loggerhead Sea Turtle model, we produced a response surface to estimate the time needed for a 10-fold increase in population size, given that vital rates observed today continued for several decades (cf. Crowder et al. 1994; Fig. 10.2).

Additional considerations include the feasibility of particular management proposals and the potential magnitude of their effect on a particular life stage. Although adult survival rates may have the highest elasticities in many long-lived organisms, the natural survival rate of adults might already be so high that no management alternative is likely to improve it (Green and Hirons 1991). Or some sensitive life stages may be inaccessible to managers, such as the pelagic early life stages of sea turtles. The elasticity analysis can be used to compare potential effects in a quantitative manner. For instance, if the elasticity of a juvenile stage survival rate is one-fourth that of adults but a particular management proposal predicts a feasible survival increase roughly four times that possible for adults, focusing on the juvenile stage is warranted. We suggest that the best use for this type of analysis is to eliminate proposals that are unlikely to lead to population recovery, as in the captive-rearing program initiated for Kemp's Ridley Sea Turtles (*Lepidochelys kempi*) (Heppell et al. 1996). It is important to recognize that simply comparing the elasticities of a mean matrix does not reveal the causes of population decline, nor is it sufficient to warrant exclusion of research or manage-

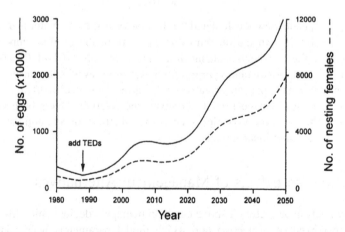

FIGURE 10.2. Transient "waves" in nesting Loggerhead female numbers and egg production resulting from shifts in the age distribution following management implementation. The population trajectory was produced by an age-based model with increasing large juvenile, subadult, and adult mortality for 30 years, followed by elimination of that mortality due to turtle excluder device (TED) implementation. These deterministic fluctuations in population size result from the extremely long time to maturity (21 years), and the large number of cohorts that were susceptible to trawling mortality while the population was declining. (Model formulation from Crowder et al. 1994.)

ment efforts directed at less sensitive life stages (Groom and Pascual 1998). Uncertainty in parameter estimates and annual variability or nonlinear changes in vital rates following a perturbation are just a few of the real-world factors that can alter the predicted effects of an elasticity analysis. However, it is a valuable tool for initial management recommendations for poorly known species and as a first step in population viability analysis (Ferriere et al. 1996).

Case Studies

We illustrate different aspects of the use of matrix projection models for aiding management decisions in three case studies, presented here in increasing order of complexity:

- a simple, stage-based matrix model and simulations to select between widely divergent management alternatives for threatened sea turtles, including using a modified age-based model to predict transient changes in population trajectories produced by shifts in the age structure of the population (Crouse et al. 1987; Crowder et al. 1994)
- a more complex stage-based matrix, in which an individual's life history stage is not dependent on age, designed to select from an array of management alternatives with multiple effects on endangered Red-cockaded Woodpeckers (Heppell et al. 1994)
- matrix models incorporating density dependence to depict and evaluate declining Brook Trout population dynamics more accurately with respect to a variety of management options (Marschall and Crowder 1995, 1996).

Case Study 1: Southeastern U.S. Loggerhead Sea Turtles

The evaluation of two divergent management strategies for federally listed, threatened Loggerhead Sea Turtles (*Caretta caretta*) in the southeastern United States provides a case study of the use of stage-based matrix models. By the mid-1980s, beach monitoring and nest protection projects existed on a number of beaches throughout the southeast, staffed primarily by local volunteers. In addition to increasing hatch success and learning about adult female fecundity and internesting behavior, these projects documented three things: (1) adult nesting females were declining (Richardson 1982; Frazer 1983b, 1987); (2) nest, egg, and hatchling losses to erosion and both natural and feral predators were high, sometimes approaching 95% (Hopkins and Murphy 1983); and (3) carcasses of drowned adult and juvenile turtles of both sexes often washed ashore (i.e., "strandings") (Schroeder 1987).

Incidental entrapment and drowning of sea turtles was documented in a variety of fishing gear, particularly shrimp trawls (Crouse 1984; Henwood and Stuntz 1987; Murphy and Hopkins-Murphy 1989). To address this problem, the National Marine Fisheries Service (NMFS) developed a turtle excluder device (TED) that released 97% of the turtles captured while allowing most shrimp to be retained in the net (Henwood and Stuntz 1987; National Research Council 1990). But TED development and requirements were highly controversial; members of the shrimping industry complained of potential shrimp losses, gear entanglement, and crew injuries from the new devices (Sea Turtle Conservation and the Shrimp Fishery 1990). They also noted that nest protection projects were yielding significant increases in hatch success and releasing large numbers of hatchlings at relatively low costs and with little socioeconomic impact. A classic management dichotomy was thus established: should managers require the disliked technology (TEDs) in an industry already in financial trouble or opt for more low-cost and widely popular nest protection projects?

To investigate the relative trade-offs between these disparate management alternatives, Crouse and co-workers (1987) developed a stage-based matrix model for southeastern Loggerheads, using a preliminary life table developed by Frazer (1983a). The model was based on stages identified by size classes because little was known about age in sea turtles, and most data were collected for turtles of different size classes. The model assumed a postbreeding census and included seven life stages: eggs/hatchlings, three successively larger juvenile stages, and three adult stages reflecting different fecundities as nesting females matured.

Both sensitivity and elasticity analyses of the matrix pointed to annual survival of the three juvenile size stages, particularly large juveniles (50–80 cm carapace length), as most important in determining future population growth rates (e.g., see Crouse et al. 1987 for assumptions). Most important, the model predicted that 100% survival of the egg/hatchling stage was unlikely to reverse current population declines. The model also indicated that future research should focus on refining our estimates of small juvenile survival and growth rates of all three juvenile stages. Refinement of fecundity and hatchling survival estimates would not change the qualitative predictions of the model. Due to the uncertainties in some of the model parameters and the obvious need for populations to reproduce, Crouse and associates (1987) did not advocate terminating nest protection projects but rather noted that nest protection projects without concurrent reductions in juvenile mortality (through the use of TEDs or some other mechanism) would likely be futile. Based in large part on this model, a National Academy of Sciences review panel recommended requiring TEDs "in most trawls at most times of the year" (National Research Council 1990). In December 1992, NMFS expanded seasonal TED regulations to require TEDs in all southeastern shrimp trawls by December 1994 (57 Fed. Reg. 57348-57359, Dec. 4, 1992).

Crowder and colleagues (1994) recently revised the original Loggerhead model (see Fig. 10.1b), reducing it to five stages by combining first-time breeders, remigrants, and the remaining adults into one adult stage. The authors used the population growth rates to compare the time it might take to see the results of TED regulations on nesting turtle numbers. Depending on how large a change in nesting numbers is required to produce an effect in the field (nesting turtle numbers fluctuate considerably from year to year anyway) and how much TEDs reduce annual mortality rates, it may be decades before we can be sure of the effect of TED requirements by counting adult females on the beach. However, the model results indicated that widespread use and enforcement of TEDs should lead to population increases of 3–6% per year.

The Loggerhead Turtle's extremely long time to maturity suggests that the population will take many years to reach a stable age distribution. The stage-based model does not tell us what the population size or stage distribution will look like in the next two or three decades as the population responds to the changes in vital rates associated with TED regulations. Crowder and co-workers (1994) expanded their five-stage model into a modified Leslie matrix to examine transient effects on the expected numbers of nesting females as the model population responded to new vital rates associated with the TED requirements. This 22 × 22 age-based matrix had all adults collapsed into a single stage, primarily because neither life span nor age-specific fecundity of sea turtles is known; thus, the model looked like a Leslie matrix (Fig. 10.1a) except for an adult survival entry in the lower right-hand corner of the matrix.

Because the population has been declining for several years and large juvenile turtles have been vulnerable to shrimp trawls, there is likely to be a "hole" in the population size distribution that will mature in the next decade or two. The age-based model revealed a transient dip in the nesting population in that time frame, completely independent of stochastic fluctuations (Fig. 10.2). To the inexperienced, this dip could be interpreted as a failure of the TEDs. By anticipating this transient decline in nesting females, such misinterpretations can be avoided.

Today, most monitored nesting beaches report that Loggerhead populations are stable or increasing (Turtle Expert Working Group 1998). A time-series analysis of strandings indicate that periods of TED use in South Carolina reduced stranding numbers by 40% (Crowder et al. 1995). Research continues to determine the population-level effects of TEDs and other management efforts for Loggerheads and other sea turtle species.

Case Study 2: Red-Cockaded Woodpeckers in North Carolina

Red-cockaded Woodpeckers (RCWs) are cooperatively breeding birds that inhabit long-leaf pine forests in the southeastern United States (Thompson 1971; Wood 1983; Walters 1990). RCW population dynamics are driven by territory availability, and males may spend several years as nest helpers before reproducing themselves. Because territories are only held by males and larger populations do not appear to be female limited (Walters 1990), Heppell and colleagues (1994) produced a male-only stage-based model for RCWs in the North Carolina sandhills. The object of this modeling exercise was to determine which life history stages had the greatest impact on population growth and to evaluate qualitatively several management proposals.

Unlike the sea turtle model, the woodpecker model was based on life history stages that are not dependent on age: fledglings, helpers, floaters (without a territory), solitary males (with a territory but no mate), 1-year-old breeders, and 2-year-old or older breeders (Fig. 10.1c). Multiple transitions between stages existed; for instance, fledglings could become helpers, floaters, solitary males, or breeders in their second year. An elasticity analysis indicated that fecundity, 2-year-old survival or longer, and helper-to-breeder transition probability were the most critical parameters in the model.

Heppell and co-workers (1994) hypothesized the effects of four different management proposals on RCW vital rates: (1) removal of nest cavity invaders, (2) translocation of female fledglings to solitary male territories, (3) drilling nest cavities into trees on existing territories, and (4) drilling nest cavities in trees in habitat unused by but suitable for RCWs. Each of these proposals affected one or more vital rates. The authors calculated the population growth rate for each plan based on increases in the affected vital rates of 5, 10, and 25%.

Because fledglings, nonbreeding helpers, and floaters cannot make the transition to breeder status unless territories are available, the authors held the transition to breeder probabilities constant for management options, which increased survival but did not increase the amount of suitable nesting habitat. Drilling cavities in unoccupied habitat, which increased fledgling survival and nonbreeder-to-breeder transition probabilities in the model, had the greatest impact on the intrinsic rate of increase of the population (Fig. 10.3).

The mortality sources of adult RCWs have not been studied extensively, and currently no management plan exists that might decrease annual mortality rates. But the elasticity analysis of the RCW model indicated that stage-specific survival

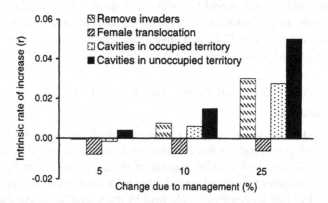

FIGURE 10.3. Intrinsic rates of increase predicted for four management scenarios applied to a male-only stage-based model for Red-cockaded Woodpeckers. Some management plans affect multiple stages or transition probabilities: removing cavity invaders increases fecundity; female translocation to solitary male territories increases solitary to breeder transition probability; drilling cavities in occupied (existing) territories increases fledgling survival, helper survival, and fledgling-to-helper transition probabilities; drilling cavities in unoccupied territories increases fledgling survival, fledgling-to-breeder, floater-to-breeder, and helper-to-breeder transition probabilities. Increases in matrix parameters (5, 10, or 25%) were assumed to be constant for all stages affected. (From Heppell et al. 1994; used with permission.)

had a large impact on the population growth rate. Heppell and associates (1994) examined the effects of decreasing the annual mortality of fledglings alone, birds that nest together on a territory (fledglings, helpers, and breeders), and all stages at once. When mortality was decreased, the population growth rate increased substantially, but the habitat restrictions led to an increase in the proportion of *nonbreeders* in the population. This analysis illustrates the need for critical thinking in model choice and parameterization and suggests that managers should think about the composition of their populations over time as well as population numbers.

Case Study 3: Southern Appalachian Brook Trout Populations

Brook Trout (*Salvelinus fontinalis*) populations in southern Appalachian mountain streams have declined in response to multiple anthropogenic effects including the introduction of an exotic salmonid species (Rainbow Trout, *Oncorhynchus mykiss*), a decrease in pH (through acid deposition), an increase in siltation and a decrease in shade and allochthonous nutrient input (from clear-cutting), and an increase in fishing pressure (Kelly et al. 1980; Larson and Moore 1985).

Marschall and Crowder (1996) developed a population model based on a simple size-classified projection matrix to address these multiple anthropogenic effects. They partitioned the Brook Trout population into 15 size classes based on total fish length; finer size divisions were used in the small size classes because changes in size in the model affect survival through the relation between size, number, and density dependence. The annual cycle of the model Brook Trout population consisted of a 6-month growth period (from spring to fall), fall spawning, overwintering without growth (from spawn until spring), and finally, emergence of larvae and beginning of growth again in the spring.

A matrix of monthly transition probabilities was calculated from available data on survival and growth rates. A distribution of growth rates was incorporated by calculating the proportion of juveniles that grew into each size class by using cohort-specific growth estimates from field data. Density-dependent survival was added to the model by multiplying the linear transition matrix by a second matrix with per capita survival probabilities on the diagonal entries; survival in the age 0 classes depended on body size and number of age 0 fish. Survival in the remaining size classes depended on size only (Marschall and Crowder 1995, 1996). Those fish that survived the growing season were then multiplied by a vector of size-specific fecundities and overwintering mortalities. The complexity of this model required the authors to calculate population size through time numerically and conduct a sensitivity analysis by using the approach described by Equation (10.10) (above).

Marschall and Crowder (1996) assessed the sensitivity of equilibrium population size and size class structure to parameter perturbations in several ways. First, they estimated elasticity to size-specific changes in growth and survival rates via simulation. Next, they assessed population response to a full factorial design of proportional parameter perturbations that would be associated with interactions with Rainbow Trout. Finally, the authors asked what magnitude of perturbations of suites of parameters associated with each ecological pressure were necessary to cause local extinctions.

Patterns in reproductive value among size classes and the damping effect of density-dependent survival in small fish led to high sensitivity values for survival of large juveniles and small adults and growth rates of small juveniles. Although equilibrium population size was relatively robust to changes in survival rates of most size classes, it was sensitive to changes in survival of fish in classes 60–100 mm and 101–140 mm (Fig. 10.4a). This roughly corresponds to Brook Trout late in their first growing season and early in their second growing season. An increase in survival of these fish resulted in an increase in equilibrium population size. Both measures of size distribution (number and proportion of large fish)

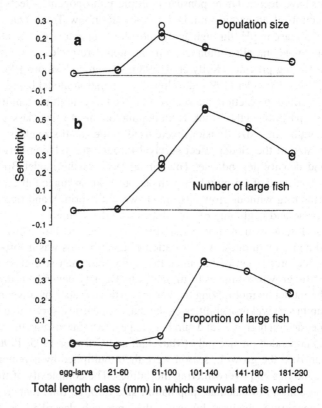

FIGURE 10.4. Sensitivity of population size (a), number of large fish (b), and proportion of large fish (c) to changes in stage-specific survival in a density-dependent matrix model of Brook Trout. Density-dependent survival of age 0 fish (larvae, 21–60 mm and 61–100 mm) was dependent on size and the number of age 0 fish. Points represent results from model runs with various levels of density dependence. (From Marschall and Crowder 1996; used with permission.)

were very sensitive to changes in survival rates of medium to large fish (Fig. 10.4b, c); increases in medium to large fish survival resulted in more large fish in the equilibrium population.

The factorial examination of population response to combinations of anthropogenic impacts revealed important information for management. Potential Brook Trout responses to Rainbow Trout competition and predation included a decrease in survival rate of small fish, a change in density dependence in survival of small fish, and a decrease in growth rates of all sizes. Population size tended to decrease with an increase in small-fish growth rate (producing a population with fewer but larger fish), but there was an interaction between the effect of small-fish growth rates and small-fish survival rate on ultimate population size. Brook Trout responded to decreases in pH, with decreased growth rate in all size classes, decreased survival rates of small fish, and decreased egg-to-larva survival rates. This combination of effects, at magnitudes documented in laboratory experiments, had severe negative impacts on the model population. Relatively small effects of changes in pH for Rainbow Trout resulted in local extinctions. By contrast, neither the increase in large fish mortality associated with sport harvesting nor the increase in egg-to-larva mortality associated with sedimentation caused local extinctions.

Discussion

Resources are limited to conduct research on population ecology of threatened, endangered, and other species of concern to conservation biologists. As a result, we need to invest our financial and human resources wisely, to maximize our ability to forecast population change. The age- and stage-based matrix modeling approaches we have outlined in this chapter and, in particular, the elasticity analyses of these matrices, provide guidance to focus research efforts. The analysis allows us to identify sensitive life stages and processes; these parameters have disproportionate effects on population responses and so need to be better understood to increase our confidence in making forecasts. For example, it is relatively inexpensive to estimate survival rates for sea turtle eggs or annual fecundity of female sea turtles, but the population growth rate is relatively insensitive to these components of the model (Crouse et al. 1987; Crowder et al. 1994). We would be much better off enhancing our understanding of survival rates in the small and large juvenile stages, although these data will be difficult to acquire. Similarly, in fish, it appears that we need to focus our efforts on making better estimates of survival in the juvenile stages. It is easier to work on adults (which can be tagged) or early larvae (for which estimates can be made using netting techniques), but the survival in the juvenile stage is often the most sensitive component in the model. If it is generally true that fish population growth rates are relatively sensitive to survival in the late larval and juvenile stages, this will be challenging to study because in many fishes these are difficult stages to assess quantitatively (Crowder et al. 1992). In the meantime, the elasticity analysis may help us focus interim management policies on those life history stages that are most likely to increase population growth.

The value of predicting the likely population-level effects of different management alternatives without expending either the time or money necessary to test them in the field is obvious, provided the input parameters realistically reflect the population's vital rates under the circumstances. Fully testing different management alternatives in the field could also be risky for endangered species. It is important to consider real management options, wherever possible, as alternative hypotheses. In the Loggerhead Sea Turtle model, it is theoretically possible to increase fecundity to a point at which population declines are arrested. But such a management alternative is meaningless, as we have no way to affect fecundity in sea turtles. Similarly, although small juvenile survival is an important vital rate to the model, to date, we do not have access to large numbers of small juvenile Loggerheads (it is believed they spend multiple years in a pelagic phase before returning to coastal waters at about 50-cm carapace length). So, when simulating the effects of TED requirements, we restricted the rate changes to large juveniles, subadults, and adults, a meaningful management alternative to consider.

By considering multiple management approaches on a relative basis, we can rank them by likely effect on population responses. For example, enhancing the transition from solitary males to breeders by translocation of females is unlikely to aid population growth based on the RCW model (because there are so few solitary males) but enhancing the transition from helper to breeder (via drilling artificial cavities, particularly in unoccupied habitat) should be successful. Similarly, with a reasonable size limit, it appears that it would be relatively difficult to reduce population sizes of Brook Trout via harvesting, but small changes in juvenile growth and survival related to Rainbow Trout competition or reduced pH could have dramatic effects. Again, the elasticity analysis approach should be useful to managers who must decide how to allocate limited resources to a suite of alternative management approaches.

The advantages to be gained from matrix model analysis come from the flexibility of the model format. In each case we presented above, the matrix modeling approach was modified to incorporate the particulars of the natural history of the organism to be modeled and the data available for calculating parameters. The complexity of each model reflects our current level of understanding of the population's vital rates. The sea turtle models are simple because the database to support a more complex model simply is not available at this point. By contrast, the RCW model is rather complex and reflects some of the natural complexity in the woodpecker life history. The Brook Trout model includes density dependence of vital rates when appropriate and examines both size and age classes. Matrix models allow ample flexibility to examine a variety of life histories, while permitting a number of well-developed mathematical procedures (elasticity calculations and others; cf. Caswell 1989) to aid analysis.

The deterministic linear modeling approach described in this chapter uses sensitivity analyses to qualitatively compare the impact of stage-specific survival, growth, and fecundity on population growth rates. There are a number of arguments against using models without variability in vital rates; however, in many situations the deterministic elasticity calculations are qualitatively robust. Dixon and associates (1997) examined the changes in elasticity through time in a stochastic population model for the endangered plant Northern Monkshood (*Aconitum noveboracense*). Four different matrices, calculated from 4 years of sampling, were run in a stochastic simulation with varying levels of autocorrelation. The authors calculated proportional sensitivities for the average matrix of each simulation. Although the transition probabilities and fertilities in each matrix were different, there were no qualitative changes in the elasticities; the most critical matrix parameters had the highest elasticities in every simulation. Further examination by using hypothetical population models revealed that except in cases in which life histories varied tremendously from one year to the next (e.g., a switch from iteroparity to semelparity), qualitative measures of elasticity were robust. However, uncertainty in parameter estimates can alter the relative "ranking" of elasticities, particularly when critical life history variables such as age at maturity and adult annual survival are unknown. In general, only large differences in elasticities should be regarded as useful indicators of the potential effects of management on a population.

The objective of matrix population models as we have used them is neither to produce a model for its own sake nor to make quantitative predictions of population growth rate or population size. Instead, we use matrix analysis as an heuristic device to synthesize available biological information regarding species of concern to conservation biologists. Elasticity analysis allows us to direct research efforts toward parameters or life stages that have the greatest effect on our ability to predict population responses. Sensitivity analysis and simulations altering vital rates appropriate to particular management options allow us to screen among arrays of management alternatives and to determine which are most (and least) likely to enhance threatened populations.

Acknowledgments. This research reported in this paper was supported by the following agencies: NOAA/NMFS (NA90AA-D-S6847); UNC Sea Grant (R/MER-21, R/MRD-27); NOAA Coastal Ocean Program SABRE Project R/SAB-4 (NA16RG0492-01); Center for Marine Conservation; and National Science Foundation Graduate Fellowship. Many individuals collaborated on these efforts including Hal Caswell, who provided an extremely helpful review of an earlier draft, and Libby Marschall, Tom Martin, John Quinlan, and Jeff Walters; we thank them for their advice and counsel. Two anonymous reviewers also provided valuable comments.

Literature Cited

Caswell H (1978) A general formula for the sensitivity of population growth rate to changes in life history parameters. Theoretical Population Biology 14:215–230

Caswell H (1982) Stable population structure and reproductive value for populations with complex life cycles. Ecology 63:1218–1222

Caswell H (1986) Life cycle models for plants. Lectures on Mathematics in the Life Sciences 18:171–233

Caswell H (1989) Matrix population modeling. Sinauer Associates, Inc, Sunderland, MA

Caswell H, Naiman RJ, Morin R (1984) Evaluating the consequences of reproduction in complex salmonid life cyles. Aquaculture 43:123–134

Crouse DT (1984) Incidental capture of sea turtles by commercial fisheries. Smithsonian Herpetological Information Service 62. Smithsonian Institution, Washington, DC

Crouse DT, Crowder LB, Caswell H (1987) A stage-based population model for Loggerhead Sea Turtles and implications for conservation. Ecology 68:1412–1423

Crowder LB, Rice JA, Miller TJ, Marschall EA (1992) Empirical and theoretical approaches to size-based interactions and recruitment variability in fishes. In: DeAngelis DL, Gross LJ (eds) Individual-based models and approaches in ecology. Chapman and Hall, New York

Crowder LB, Crouse DT, Heppell SS, Martin TH (1994) Predicting the impact of turtle excluder devices on Loggerhead Sea Turtle populations. Ecological Applications 4:437–445

Crowder LB, Hopkins-Murphy SR, Royle JA (1995) Effects of turtle excluder devices (TEDs) on Loggerhead Sea Turtle strandings with implications for conservation. Copeia 1995:773–779

deKroon H, Plaisier A, van Groenendael J, Caswell H (1986) Elasticity: the relative contribution of demographic parameters to population growth rate. Ecology 67:1427–1431

Dixon P, Friday N, Ang P, Heppell S, Kshatriya M, Lord C (1997) Sensitivity analysis of structured population models for management and conservation. In: Tuljapurkar S, Caswell H (eds) Structured population models in marine, terrestrial, and freshwater systems. Sinauer Associates, Sunderland, MA, pp 471–513

Ferriere R, Sarrazin F, Baron JP (1996) Matrix population models applied to viability analysis and conservation: theory and practice using the ULM software. Acta Oecologica 17:629–665

Frazer NB (1983a) Demography and life history evolution of the Atlantic Loggerhead Sea Turtle, *Caretta caretta*, nesting on Little Cumberland Island, Georgia. Dissertation, University of Georgia, Athens, GA

Frazer NB (1983b) Survivorship of adult female Loggerhead Sea Turtles, *Caretta caretta*, nesting on Little Cumberland Island, Georgia, USA. Herpetologica 39:436–447

Frazer NB (1987) Survival of large juvenile Loggerheads (*Caretta caretta*) in the wild. Journal of Herpetology 21:232–235

Getz WM, Haight RG (1989) Population harvesting: demographic models of fish, forest and animal resources. Princeton University Press, Princeton, NJ

Ginzburg LR, Ferson S, Akçakaya HR (1990) Reconstructability of density dependence and the conservative assessment of extinction risks. Conservation Biology 4:63–70

Glitzenstein ER (1993) On the USFWS settlement regarding federal listing of endangered species. Endangered Species Update 10(5):1–3

Grant A (1998) Population consequences of chronic toxicity: incorporating density-dependence into the analysis of life table response experiments. Ecological Modelling 105:325–335

Green RE, Hirons GJM (1991) The relevance of population studies to the conservation of threatened birds. In: Perrins CM, Lebreton JD, Hirons GJM (eds) Bird population studies: relevance to conservation and management. Oxford University Press, New York, pp 594–633

Groom MJ, Pascual MA (1998) The analysis of population persistence: an outlook on the practice of viability analysis. In: Fiedler PL, Kareiva PM (eds) Conservation biology for the coming decade, 2nd ed. Chapman and Hall, New York, pp 4–27

Henwood TA, Stuntz WE (1987) Analysis of sea turtle captures and mortalities during commercial shrimp trawling. Fish Bulletin 85:813–817

Heppell SS, Walters JR, Crowder LB (1994) Evaluating management alternatives for Red-cockaded Woodpeckers: a modeling approach. Journal of Wildlife Management 58:479–487

Heppell SS, Crowder LB, Crouse DT (1996) Models to evaluate headstarting as a management tool for long-lived turtles. Ecological Applications 6:556–565

Hopkins SR, Murphy TM Jr (1983) Management of Loggerhead Turtle nesting beaches in South Carolina. Study Completion Report E-1, Study VI-A-2. SC Wildlife and Marine Resources Department, Charleston, SC

Kelly GA, Griffith JS, Jones RD (1980) Changes in the distribution of trout in the Great Smoky Mountains National Park, 1900–1977. Technical paper 102. U.S. Fish and Wildlife Service, Washington, DC

Larson GL, Moore SE (1985) Encroachment of exotic rainbow trout into stream populations of native Brook Trout in the southern Appalachian Mountains. Transactions of the American Fisheries Society 114:195–203

Lefkovitch, LP (1965) The study of population growth in organisms grouped by stages. Biometrics 21:1–18

Leslie PH (1945) On the use of matrices in certain population mathematics. Biometrika 33:183–212

Levin LA, Huggett DV (1990) Implications of alternative life histories for the seasonal dynamics and demography of an estuarine polychaete. Ecology 71:2191–2208

MacDonald DB, Caswell H (1993) Matrix methods for avian demography. Current Ornithology 10:139–185

Marschall EA, Crowder LB (1995) Density-dependent survival as a function of size in juvenile salmonids in streams. Canadian Journal of Fisheries and Aquatic Sciences 52:136–140

Marschall EA, Crowder LB (1996) Assessing population responses to multiple anthropogenic effects: a case study with Brook Trout. Ecological Applications 6:152–167

Meagher TR (1982) The population biology of *Chamaelirium luteum*, a dioecious member of the lily family: two-sex population projections and stable population structure. Ecology 63:1701–1711

Murphy TM, Hopkins-Murphy SR (1989) Sea turtle and shrimp fishing interactions: a summary and critique of relevant information. Center for Marine Conservation, Washington, DC

National Research Council (1990) Decline of the sea turtles: causes and prevention. National Academy Press, Washington, DC

Noon BR, Sauer JR (1992) Population models for passerine birds: structure, parameterization, and analysis. In: McCullough DR, Barrett RH (eds) Wildlife 2001: populations. Elsevier Applied Science, New York, pp 441–464

Richardson JI (1982) A population model for adult female Loggerhead Sea Turtles (*Caretta caretta*) nesting in Georgia. Dissertation, University of Georgia, Athens, GA

Saether BE, Ringsby TH, Roskaft E (1996) Life history variation, population processes and priorities in species conservation: towards a reunion of research paradigms. Oikos 77:217–226

Schemske DW, Husband BC, Ruckelshaus MH, Goodwillie C, Parker IM, Bishop JG (1994) Evaluating approaches to the conservation of rare and endangered plants. Ecology 75:584–606

Schroeder BA (1987) 1986 Annual report of the sea turtle salvage and strandings network Atlantic and Gulf Coasts of the United States, January–December 1986. NOAA-NMFS-SEFC CRD-87/88–12. Coastal Resources Division, National Marine Fisheries Service, Miami, FL

Sea Turtle Conservation and the Shrimp Fishery (1990) Hearing before the Subcommittee on Fisheries and Wildlife Conservation and the Environment of the Committee on Merchant Marine and Fisheries. GPO Serial 101–83. U.S. House of Representatives, May 1, 1990, Washington, DC

Silvertown J, Franco M, Pisanty I, Mendoze A (1993) Comparative plant demography: relative importance of life-cycle components to the finite rate of increase in woody and herbaceous perennials. Journal of Ecology 81:465–476

Sissenwine MP, Rosenberg AA (1993) Marine fisheries at a critical juncture. Fisheries 18(10):6–14

Thompson RL (ed) (1971) The ecology and management of Red-cockaded Woodpecker. U.S. Bureau of Sports Fisheries and Wildlife, Tall Timbers Research Station, Tallahassee, FL

Tuljapurkar S (1990) Population dynamics in variable environments. Springer-Verlag, New York

Turtle Expert Working Group (1998) An assessment of the Kemp's Ridley (*Lepidochelys kempii*) and Loggerhead (*Caretta caretta*) Sea Turtle populations in the western North Atlantic. NOAA Technical Memorandum NMFS-SEFSC-409. National Marine Fisheries Service, Miami, FL

U.S. Fish and Wildlife Service (1995) Box score: listings and recovery plans for endangered species. Technical Bulletin. U.S. Fish and Wildlife Service, Washington, DC

van Groenendael J, de Kroon H, Caswell H (1988) Projection matrices in population biology. Trends in Ecology and Evolution 3:264–269

Walters JR (1990) Red-cockaded Woodpeckers: a "primitive" cooperative breeder. In: Stacey PB, Koenig WD (eds) Cooperative breeding in birds. Cambridge University Press, London, pp 67–101

Werner PA, Caswell H (1977) Population growth rates and age versus stage-distribution models for Teasel (*Dipsacus sylvestris* Huds). Ecology 58:1103–1111

Williams JE, Johnson JE, Hendrickson DA, Contreas-Balderas S, Williams JD, Navarro-Mendoza M, McAllister DE, Deacon JE (1989) Fishes of North America endangered, threatened, or of special concern: 1989. Fisheries 14:2–20

Wood DA (ed) (1983) Red-cockaded Woodpecker symposium II. Florida Game and Freshwater Fish Commission and U.S. Fish and Wildlife Service, Atlanta, GA

11
Variability and Measurement Error in Extinction Risk Analysis: The Northern Spotted Owl on the Olympic Peninsula

Lloyd Goldwasser, Scott Ferson, and Lev Ginzburg

Introduction

The degree of endangerment of a biological species can be described on a variety of different scales. For instance, U.S. federal law recognizes and protects a species as endangered if it is in danger of extinction and threatened if it is likely to become endangered within the foreseeable future (Fig. 11.1). The precise quantitative meanings of *danger, likely, foreseeable,* and even *extinction* were left undefined by Congress, except for the requirement that listing be based solely on the best available scientific and commercial data.

The classification suggested by the International Union for the Conservation of Nature and Natural Resources (IUCN 1994) recognizes a threatened species as vulnerable, endangered, critical, or extinct (Fig. 11.2). The criterion defining extinct status is that the species has not been sighted in the wild for an extended period of time. Mace and Lande (1991) proposed quantitative criteria for assigning species to the vulnerable, endangered, and critical categories. Each category is defined by a probability of extinction over some time scale: species having a 50% or greater probability of extinction over 5 years or two generations (whichever is longer) are designated as critical; a 20% probability of extinction over the longer of 20 years or 10 generations defines a species as endangered; and 10% probability of extinction over 100 years defines vulnerable status. Although they recognize that risk of extinction and its time scale are both relevant to the degree of endangerment, the extended time periods in their proposed criteria can impose high levels of uncertainty in the probability estimates when measurement errors are accounted for.

The analysis of extinction risk involves both the projection of population sizes into some point in the future and the assessment of the uncertainty associated with that projection. The choice of the time scale of the projection may strongly affect the strength of the conclusions because, although uncertainty inevitably increases with time, different processes contribute to uncertainty at different rates. The main sources of uncertainty that have been incorporated in most risk analyses to date have been the variability of the environment, which may raise or lower the vital rates of a population from year to year in an unpredictable fashion, and demo-

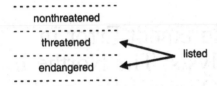

FIGURE 11.1. Categories recognized by the federal Endangered Species Act.

graphic stochasticity. The means of dealing quantitatively with these kinds of uncertainty are reasonably well established (Ginzburg et al. 1982; Lande and Orzack 1988; Ferson et al. 1989; Burgman et al. 1993). However, measurement error creates another crucial source of uncertainty that has not been investigated in most risk analyses, including our own. Errors in the estimates of the demographic parameters of the models can arise through sampling variation, and we suspect that the resulting measurement error can strongly affect both the interpretability of a risk analysis and any estimate of the uncertainty that it involves. With measurement error, it is not clear a priori how far into the future a population can be validly projected. The need for separate treatments of the two kinds of uncertainties has been stressed by Ferson and Ginzburg (1996).

In this chapter, we illustrate how the effects of measurement error can be incorporated into an extinction risk assessment for the Northern Spotted Owl (*Strix occidentalis caurina*) on the Olympic Peninsula. In analyzing the level of risk for this owl population under different scenarios, we examined the effects of different availabilities of suitable habitat and the effects of infrequent catastrophes, as well as the relative effects of environmental variability and measurement error. In incorporating measurement error into our analysis of extinction risk, we included the effects of uncertainty about the average demographic rates given the observed range of environmental variability. The Northern Spotted Owl provides a good case for developing these principles because it is fairly typical of applications of extinction risk assessment: despite years of attention and data collection, considerable uncertainty still persists about the demographic rates of each population and about the processes that determine those rates.

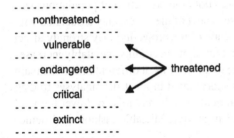

FIGURE 11.2. Categories recognized internationally by conservation biologists.

Model Structure

To simplify the bookkeeping, we follow standard practice and consider only females in the model. This simplification assumes that males are in sufficient abundance that they do not limit the dynamics of the population. Following Forsman and colleagues (1993), we recognize female owls in three categories:

- Juveniles, which includes all individuals between 0 and 1 year old (i.e., between hatching and 1 year after hatching)
- Subadults, which includes all individuals between 1 and 2 years old
- Adults which includes all individuals older than 2 years

Although the analysis by Burnham and co-workers (1994) of demographic data from all populations of the Northern Spotted Owl distinguished first-year adults into a second subadult stage, subadult2, the rates for this stage are statistically indistinguishable from those of the rest of the adult stage in the data from the Olympic Peninsula population (Forsman et al. 1993). Because we are primarily interested in the Olympic Peninsula population, we used only three stages.

Our model is a modification of a simple matrix model of the form

$$
\begin{pmatrix} \text{juveniles}_{t+1} \\ \text{subadults}_{t+1} \\ \text{adults}_{t+1} \end{pmatrix} = \begin{pmatrix} 0 & F_{\text{subadults}} & F_{\text{adults}} \\ S_{\text{juveniles}} & 0 & 0 \\ 0 & S_{\text{subadults}} & S_{\text{adults}} \end{pmatrix} \begin{pmatrix} \text{juveniles}_t \\ \text{subadults}_t \\ \text{adults}_t \end{pmatrix}
$$

where F represents fecundity and S represents survival rate over a yearly time step indexed by t. Our modification to this matrix model includes the effects of a limited number of territories, which imposes a ceiling on the number of pairs of owls that may breed in a given year. Given the total number of territories, we first fill as many of them as possible with adult owls and then fill any remaining ones with subadults. This procedure reflects the idea that the greater experience of older owls may give them a competitive advantage in establishing themselves in territories that have fallen vacant. Because of this ceiling, the number of breeding birds never exceeds the number of territories, although the number of floaters (nonbreeding adults without territories) theoretically can increase indefinitely if reproduction is high enough.

Demographic Parameters

The demographic rates for juvenile, subadult, and adult survival and fecundity that were used to parameterize the model are shown in Table 11.1. The values are based on those estimated for the Olympic Peninsula population as reported in the work of Forsman and co-workers (1993). The mean juvenile survival rate that they reported is probably an underestimate because it does not account for emigration by juvenile owls. Using data from two populations of the Northern Spotted Owl (Olympic Peninsula and Roseburg), Burnham and associates (1994)

TABLE 11.1. Demographic parameters used in this study.

Parameter	Juveniles	Subadults	Adults
Fecundity			
Mean value	0.0	0.206	0.380
Temporal variability (SD)	0.0	0.143	0.237
Measurement error (SE)	0.0	0.106	0.036
Survival rate			
Mean value	0.358	0.862	0.862
Temporal variability (SD)	0.033	0.033	0.033
Measurement error (SE)	0.064	0.017	0.017

computed an adjustment factor $E = 0.3158$ to compensate for this bias. This factor represents the proportion of juveniles that emigrate and survive but are not detected in subsequent sampling. Although comparable data are available for the Olympic Peninsula population alone, the sample size is probably too low for the resulting estimate to be reliable (E. Forsman, personal communication). Accordingly, we have used the correction $E = 1.4616$ to adjust our rate of juvenile survival from 0.245 reported by Forsman and colleagues (1993) to 0.358, which appears in the table. (More recently, Forsman et al. [1996] have estimated this rate to be 0.611.)

To perform a quantitative population viability analysis, it is necessary to have estimates for temporal variabilities of the vital rates caused by environmental stochasticity. We computed standard deviations from seven annual estimates of fecundity for the Olympic Peninsula population reported by Forsman and co-workers (1993). In this estimation, we used the respective sample sizes as weights for the data. Unfortunately, comparable information is not available to compute analogous standard deviations for survival rates. Previous investigators have found that the temporal variability of survival was not consistent but varied with other factors such as spatial location (Forsman et al. 1993; Burnham et al. 1994). This finding suggests that there exists no simple summary that completely characterizes the temporal variation in survival rate for the Northern Spotted Owl as a whole.

Eric Forsman (personal communication) kindly shared with us the estimates of year-by-year survival rates for the Olympic Peninsula population. Although there are statistical grounds to question a simple calculation of the standard deviation of the year-by-year estimates, for this modeling exercise, and in the absence of better estimates, we used these values to calculate the standard deviation in adult and subadult survival due to environmental variability. Because the sample sizes for juveniles are too low to provide a reliable estimate of their variability in survival (E. Forsman, personal communication), we used the same value, 0.033, as a ballpark estimate of juvenile survival variability as well. The values are shown in Table 11.1.

Forsman and associates (1993) also give the standard error for each estimated demographic rate, which reflects the uncertainty of the estimate due to sampling error (D. Anderson and K. Burnham, personal communications). We used these values as surrogates for measurement errors and, using a method described below, investigated their effect on our ability to estimate the population's viability. We should emphasize that our use of the reported standard errors as estimates of measurement errors is somewhat arbitrary and may be subject to controversy. However, the standard errors probably underestimate the overall measurement error (K. Burnham, personal communication); it is well known that there is a strong tendency to underestimate uncertainties in empirical data (Shlyakhter 1994; and references therein). A careful analysis of the data collection protocol or perhaps comprehensive comparison among the results from independent data collection teams would provide far better estimates, but neither has been conducted. The values we use are adequate for our primary purpose in this chapter, which is to explore the effects of measurement error on quantitative analyses of extinction risk rather than to make precise statements about the Spotted Owl.

Number of Territories

Although the number of breeding territories strongly affects the population's susceptibility to extinction, this number is not known with much precision for the Northern Spotted Owl on the Olympic Peninsula. Estimates range from more than 300 to about half this number (Hanson et al. 1993; Forsman 1994; Kenney 1994; Holthausen et al. 1994). Hanson and co-workers (1993, Table 1) report 163 known breeding pairs on the Olympic Peninsula. A recent study reports the number of confirmed breeding pair sites to be at least 155 but estimates the total number of sites with territorial pairs on the Olympic Peninsula to be between 282 and 321 (Holthausen et al. 1994, Table 2).

If the current number of territories includes sites that have already sustained some kind of environmental impact, it may overestimate the real number of territories on which owls can persist and breed over the long term. A criterion proposed by the U.S. Fish and Wildlife Service for a rule under section 4(d) of the Endangered Species Act suggests that territories must have more than 40% suitable habitat within a 2.7-mile radius of each owl site (USFWS 1993). Kenney (1994) estimated in a graphical analysis that less than 10% of extant territories are currently below this criterion, although there is still considerable uncertainty with respect to what the criterion actually implies about the number of territories (cf. Holthausen et al. 1994, Figure 8).

Because the number of available territories is both important and poorly known, we investigated a range of assumptions about its value so as to bracket the variation in risk of extinction that could result. For the high value, we chose 300 territories, which may be an optimistic estimate on the number of territories (cf. Holthausen et al. 1994). For the low value, we chose 200 territories, following an estimate in USDI (1992), as a pessimistic estimate of the number of territories. If this interval includes the true number of suitable breeding sites, then the actual probability of extinction is likely to fall somewhere between the values estimated for these bounds.

One might assume that the number of available territories is fixed during the foreseeable future, but such constancy is probably the least likely scenario. Because one of the primary effects of logging is to reduce the number of suitable breeding habitats for the Spotted Owl, one might want to subtract sites to account for future logging. Kenney (1994) estimated, however, that no additional owl sites could fall below the 40% suitable habitat criterion (USFWS 1993), even assuming that all privately held forest habitat was harvested. However, future growth of young forest (representing sites that have already been harvested) may, over time, acquire enough of the characteristics of old growth to provide breeding sites for the Spotted Owl. The extent of this potential increase in terms of the number of available breeding sites clearly depends on the future forest management practices, but it may be substantial. Calculations made by Robert Meier (personal communication) suggest that the temporal increase in the number of territories will be about 0.35% per year. This would amount to 35% over the next century. Some other estimates suggest as much as a 50% increase in potential owl habitat over this period from maturation of previously logged forests (e.g., USFS and BLM 1994). We included this increase in habitats in our simulations, but to be conservative, we used the lower value of 0.35% per year.

Demographic Stochasticity

Because populations consist of independent organisms, population sizes can take on only whole number values, not fractional ones. For any finite population size, survival rates cannot take on all values between 0 and 1 unless the rates are regarded as probabilities of survival that are faced by each individual independently, rather than ratios of whole numbers. If the rates represent probabilities, then the total number of survivors in the whole population is not a single deterministic value but is a random variate whose value comes from a distribution. It may be a binomial distribution because the total corresponds to the outcome of an independent "coin flip" by each member of the population. The term *demographic stochasticity* refers to the uncertainty in total population numbers due to the probabilistic nature of individual survival or reproduction. Even for a reasonably high probability of survival, mere chance could occasionally result in very few survivors in a given year. The smaller the population size, the larger the relative uncertainty, so demographic stochasticity is of particular concern in populations that are small enough already to be at risk of extinction (Burgman et al. 1993).

We could estimate the number of individuals who actually survive by drawing a random number for each living individual and comparing it with the survival rate and then tallying those whose random number is less than the rate as survivors. This method is computationally intensive, however, and instead we included demographic stochasticity in our simulations by using the methods described by Akçakaya (1991). To simulate the number of survivors at each time step, we generated a random variate from a binomial distribution with the survival rate as the mean and the number of individuals as the number of trials. The binomial variate is used as the number of survivors. To simulate the number of offspring, we generated a random variate from Poisson distribution with the product of the tabled fecundity and the number of potential mothers as its mean. The resulting Poisson variate is the total number of offspring. As a result of incorporating demographic stochasticity into the model, all the abundances in each of the stages are integer numbers.

Density Dependence

Density dependence in the Northern Spotted Owl is probably most strongly controlled by the existence of large and nonoverlapping territories held by breeding pairs. Because there is only a finite number of such territories available and because owls never reproduce unless they hold a territory, reproduction within the population is limited. We implemented this form of density dependence as a ceiling on the number of breeding pairs equal to the number of available territories. Above this number, any additional owls cannot produce offspring, even though they may be reproductively mature. However, if there are more territories than adults, then subadults may also form breeding pairs and reproduce at their own rate.

Allee effects (Allee 1931) may also influence the population dynamics of these owls; these effects are a kind of density dependence because they consist of changes in per capita rates with population size. These effects result in reductions in fecundity or survival rates at low (rather than high) population densities and can be due to inability to locate mates, insufficient group protection against competitors, inbreeding depression, and a variety of other ecological and genetic mechanisms. Such effects potentially lead to what has been called an extinction vortex (Gilpin and Soulé 1986) in which an already sparse population declines still further.

We acknowledge that Allee effects may be important in this system, but without specific empirical information on the topic, one cannot model them realistically and therefore cannot directly quantify their impact. Because we do not know the details of any Allee effects, we use a conservative method to explore the possible magnitude of their consequences. Instead of using zero or one as the population size at which we say a population is extinct, we raise this threshold to a higher number and suggest that if the population ever gets as low as the new threshold it should be considered extinct because of the possible influence of Allee effects. McKelvey and co-workers (1993) suggest that there may be a minimum size for a self-sustaining cluster of Spotted Owls. They suggest that this minimum size is as high as about 20 breeding pairs, although there is considerable uncertainty on this issue. Following a discussion of clustering by the Forest Service (Criterion 7, USFS 1992), we adopt the threshold of 15 breeding pairs as the "quasi-extinction" level (Ginzburg et al. 1982). If the population declines to this level, we will consider it as good as extinct because it may then no longer be large enough to maintain itself. A way to study different hypotheses about the strength of Allee effects in the population dynamics of the Northern Spotted Owl is to compare model results by using different threshold levels.

Windstorms

In the recent past, windstorms have destroyed large areas of potential owl habitat on the Olympic Peninsula. On a practical time scale, such destruction represents an effectively permanent loss of owl habitat relative to the slow rate of regrowth of the forest. To estimate the importance of these catastrophic windstorms, we simulated the population dynamics of the owl both with and without them.

According to Agee (1994; personal communication), approximatey three major blowdowns occur per century on average in the Olympic Peninsula. Based on this observation, we assumed that in any year there is a probability of 0.033 that a storm will destroy some territories. Agee (1994) estimated that the 1921 storm, which would cause moderate or greater damage to 13 of the current owl territories, was of approximately twice the severity of the average major storm, so an average storm might remove seven or eight territories. However, because some locations are more vulnerable to windstorms than are others, the effects of separate storms are likely to overlap rather than be distributed evenly over the landscape. For this reason, sites that have been struck by a windstorm are more likely to be struck again. We modeled this overlap by having each subsequent storm remove half as many territories than did the previous one. To maintain the mean effect suggested by Agee (1994; personal communication), in the simulation, the first storm removed 12 territories, the second 6, the third 3, the forth 1, and thereafter 0. We treat this destruction of habitat as permanent, because within the time scale of a century, this habitat does not return to the mature forest that Northern Spotted Owls prefer (Gutierrez 1996). Once a storm has occurred in any year over the course of a simulation, the affected territories remain unavailable as owl habitat, and the number of territories is decremented accordingly.

Windstorms constitute only one class of catastrophic events that might befall the Spotted Owl population. Historically, for instance, both storms and fires have destroyed large areas of potential owl habitat in the Olympic Peninsula. We neglected the impact of the latter on the persistence of the owl population for two reasons. First, changes in silvicultural practices make it unreasonable to use historical records to try to estimate the likelihood of occurrence of large fires. Second, modern methods of fire control may now be relatively effective in extinguishing fires after they begin. However, potential biotic catastrophes are harder to circumscribe by such reasoning. Disease outbreaks, for example, could directly affect the owl population or could indirectly affect owls by reducing the availability of their prey.

Of course, it is difficult to predict what biotic interactions will be important in the future, especially over the long term. For the purposes of this study, the analysis of catastrophic events was confined to windstorms.

Generation Time

The Mace and Lande (1991) criteria for classifying species uses multiples of the generation time as the horizon for estimating the risk of extinction, so it is necessary to calculate the generation time before applying the criterion. We calculate the generation time of the Northern Spotted Owl by using the standard definitions (Mertz 1970) and the demographic rates of Table 11.1. The resulting value is 8.88 years. Generation time represents a weighted mean of the product of fecundity and the number of survivors at each age, so an individual owl may live considerably longer than this span. Applying the Mace and Lande (1991) criteria to the Spotted Owl thus requires estimating their probabilities of extinction after 10 years, 89 years, and 100 years.

Population Trends

A traditional analysis of population change focuses on the long-term behavior of the deterministic component of population dynamics. From the mean values for fecundity and survival given in Table 11.1, we estimated the value of λ, the asymptotic annual rate of population increase (Caswell 1989; Ferson 1990), as the eigenvalue of the matrix

$$\begin{pmatrix} 0 & 0.206 & 0.380 \\ 0.358 & 0 & 0 \\ 0 & 0.862 & 0.862 \end{pmatrix}$$

The resulting value of λ was 0.9911, which suggests a slowly declining population under the assumption that the demographic rates do not change. This value is close to 1.0, which corresponds to a stationary population; estimates of its standard error depend on assumptions about covariances among the parameters. By raising 0.9911 to the hundredth power, we can compute a forecast based on λ that suggests a population decline of about 60% after 100 years.

There is, of course, little chance that the demographic rates of a real biological population will remain exactly unchanged for an extended period of time. Indeed, they are known to fluctuate with each new year. In our initial study of stochastic population trends, we simulated the effect of fluctuations in demographic rates from environmental variability as measured by interannual standard deviations listed in Table 11.1. We began each simulation with the population at stable stage distribution and consisting (somewhat arbitrarily) of twice as many adult and subadult owls as would fill the territories exactly. These extra adult and subadult owls constituted the initial pool of floaters. Although about a dozen time steps would have been sufficient, we estimated the stable stage structure by beginning the simulation with an even distribution and running it for 200 time steps to erase any transient behavior. We replicated each run 1,000 times, looking at the number of breeding pairs over a period of 100 years. All calculations were done by using RAMAS/stage (Ferson 1990).

Effects of Windstorms and Forest Maturation

The possibility of catastrophic windstorms does not appear to have a dramatic effect on the mean population size (Table 11.2) or on the extinction risks (Table 11.3) over 100 years. Considering that, on average, over the century-long simulation there will be three events that remove some of the available habitat, the influence of these catastrophes on the demography of owls is perhaps surprisingly low. The explanation for this small effect hinges on the fact that the impact only affects the ceiling carrying capacity. Because the population is declining anyway, the ceiling is reached only rarely, and so the impact on the actual owl population is negligible.

However, the maturation of young forests into habitat suitable for Spotted Owl breeding territories increases the mean population size (Table 11.2) by about 15%. Although we might expect it consequently to decrease the probability of extinction, we found negligible differences in the probability values as a result of including or excluding young forests (Table 11.3). In fact, our result is an overestimate, because we treated the removal of habitat as permanent, yet some regrowth may occur. Again, the reason for this is that a gradual addition of potential habitat would only be consequential if the population were a growing one. Because it is generally declining, it does not get an opportunity to take advantage of the improving situation.

TABLE 11.2. Mean number of breeding pairs over 1,000 replicates after 100 years to two significant digits, under various assumptions about habitat concerning the initial number of territories, the occurrence of windstorms, and the effect of maturation of young forests.[a]

	Initial territories	No windstorms	Windstorms
No regrowth	200	73	72
	300	110	110
With regrowth	200	85	82
	300	130	130

[a]"Initial territories" refers to the number of available breeding territories present at the beginning of the simulation. Both 200 and 300 territories were used for this value. "No regrowth" means there will be no increase in number of available owl territories resulting from maturation of young forests. "With regrowth" means breeding habitat increases at a rate of 0.35% per year as a result of maturation of young forests. "Windstorms" means that blowdowns occur with annual frequency 0.033. The first windstorm removes 12 territories, the second removes 6, the third removes 3, the fourth removes 1, the fifth and thereafter remove 0. "No windstorms" means that no territories are lost to blowdowns.

TABLE 11.3. Observed probabilities (as percentages) of quasi-extinction (falling below 15 breeding pairs) within 100 years, under various assumptions about habitat concerning the initial number of territories, the occurrence of windstorms, and the effect of maturation of young forests.[a]

	Initial territories	No windstorms	Windstorms
No regrowth	200	19.2	20.0
	300	9.4	11.7
With regrowth	200	20.0	19.2
	300	10.3	10.8

[a]The meanings of the labels are given in the legend for Table 11.2. These probabilities were estimated from simulations with 1,000 replications, so the tabled value ±2.8% forms 95% confidence intervals.

Review of Assumptions

We deliberately created the model with a number of conservatisms that are appropriate considering the status of the Northern Spotted Owl and the prevailing balance between scientific knowledge and ignorance about its life history and ecological fate. For instance, we assumed that no immigrants supplement the population on the Olympic Peninsula. Our use of quasi-extinction to account for possible influence of Allee effects was also conservative. We considered any population to be as good as extinct if it reached the quasi-extinction level of 15 breeding pairs. This will be conservative if Allee effects are not so severe, and there may be reason to suspect that this is the case. For instance, if Allee effects do not act as a strict threshold but instead become gradually stronger over a range of population densities, they would cause only an increased chance of actual extinction at such low levels rather than surety of an extinction vortex. Similarly, the model's strict ceiling on reproduction is conservative if the form of density dependence in the life history of owls is less exacting and more gradual in effect.

The strongest conservatism arises from the linear structure of the population model. Because the matrix model was parameterized with vital rates whose overall effect leads to a declining population, the model thereby guarantees that predicted population trajectories exhibit no trend for positive population growth from any level of abundance. Any increases in population size are the result of random fluctuations. As a consequence, the eventual fate of any population is a decline to extinction. However, if the observed vital rates are depressed only because recent habitat loss left the current population above its now lower carrying capacity, then the present decline in population may not continue in the future. Once the population declines to a level near its new carrying capacity, the vital rates may stabilize and the population may be capable of sustaining itself under these less crowded conditions. Our model does not assume that this will be the case and, indeed, assumes that the present gloomy life history characteristics will persist in the future. This assumption is surely very conservative because, if vital rates depend at all on habitat, then they should improve as the population comes into balance with a stable or even growing area of suitable habitat.

However, we also made a few modeling choices that may be considered optimistic. We assumed that there will be no further habitat loss due to human impacts from logging or other intrusions. This assumption may be unrealistic in the face of the continuing ecological pressure exerted by the human population, but investigating the consequences of general anthropogenic habitat destruction and other land-use effects is beyond the scope of the present study. We also assumed that there will be no loss of breeding birds to emigration out of the Olympic Peninsula population. Finally, we assumed that Allee effects do not become important until the population falls to 15 or fewer breeding pairs. This assumption could be relaxed by using an appropriately larger quasi-extinction level or perhaps by modeling Allee effects directly.

Some phenomena that have been included in the modeling exercise turned out not to have large effects on the demography of the owls. In particular, we observed that rare catastrophic windstorms probably affect owl extinction risk very little, unless they are of a magnitude or frequency much larger than has been seen in the historical record. Likewise, the increase in available breeding territories as a result of the maturation of young forests did not have a very large impact on extinction risk either. It did, however, introduce the potential that some trajectories actually increase over the century-long simulation rather than every single trajectory inexorably declining.

One of the most important assumptions made about the model was that measurement error is negligible. Although this assumption is commonly made, it represents the untenable belief that the numeric values used in the model are known precisely. In fact, because they are estimated from empirical studies, their measurement errors may be considerable. The next section considers this issue explicitly.

Measurement Error

Most population viability analyses estimate the risk of extinction by computing the probability of extinction under random environmental variability. This approach is sufficient only if the demographic rates and the magnitude of their year-to-year fluctuations are known exactly. However, such knowledge is rarely available in empirical situations, and even in fairly unchanging environments there can be considerable uncertainty associated with the estimate of each demographic rate. Indeed, Burnham and associates (1994; Burnham, personal communciation) found the variance in adult fecundity due to measurement error to be an order of magnitude greater than that due to year-to-year fluctuations. The effect of such uncertainty is to blur any projected population trajectory, even when the environmental variability is known completely. This blurring increases with time, rendering long-term projections particularly susceptible to uncertainties in the original estimates. More precise estimates of the demographic rates would result in narrower bundles of trajectories and a longer horizon for making meaningful population projections.

The distinction between measurement error and natural variability includes the difference between uncertainty about the value of the parameter at any given moment and changes in the parameter over time. If a parameter varies stochastically, then uncertainty about future fluctuations is inevitable and may often be substantial. For instance, a population that, on average, maintains itself adequately can be driven extinct by a series of bad years, and one job of risk assessment is to estimate the likelihood and probable impact of such a series. However, inaccuracy in the current estimates of vital rates can give projections that inevitably diverge from the actual course of the population, whatever the extent of environmental variability. Measurement error of the temporal variability of the vital rates also exists and may be fairly large. Considering it, however, would require a second-order approximation, which we did not attempt here.

Tossing a coin provides a good illustration of the effect of measurement error on risk analysis. If the probability of tossing heads is known to be p and the probability of tossing tails $(1 - p)$, well-known statistical results concerning the binomial distribution give both the expected number of heads after n tosses and the probability that the number of heads will exceed the number of tails by some quantity. We are able to perform a perfect risk analysis for any criterion of risk that we set for ourselves. The particular sequence of outcomes in n tosses is analogous to natural year-to-year variability in the environment. The solid curves in Fig. 11.3 show that the uncertainty about the result increases as the square root of the number of tosses.

Now suppose that the initial estimate of p was slightly inaccurate. For instance, suppose that we failed to detect that the coin was a little biased. At first, the

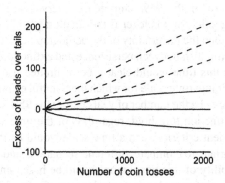

FIGURE 11.3. Comparison of the uncertainties in coin tossing due to measurement error and due to natural variability. Solid lines are for $p = .50$, dashed lines for $p = .54$. Each set shows the calculated probabilities of the expected excess of heads over tails and the ranges within which the excess is likely to lie 95% of the time. The ranges expand as the square root of the number of tosses, whereas the expected means diverge linearly. A small initial error in the estimate of p (.50 instead of .54) yields an excess that, on average, is no longer within the 95% probability range after about 600 tosses; after 2,500 tosses, the 95% ranges no longer overlap at all. (Figure redrawn from Ferson and Ginzburg 1996.)

difference between the actual and the expected tosses is negligible. Their divergence is linear, however, so that the eventual excess of heads over tails lies well outside the predicted range (Fig. 11.3). Although the analogy between coin tossing and population dynamics is only heuristic, it suggests that measurement error can play a significant role in limiting our ability to project population trajectories into the future. In general, for long periods of time, measurement error may be a dominant cause of our uncertainty about the future, whereas in the short run natural variability may dominate.

In assessing the effects of measurement error on our extinction risk analysis of owl population dynamics, we need to use the standard errors of the demographic rates (Table 11.1) to construct simulations reflecting the potential variation under the uncertainty in the estimates at the outset (as opposed to temporal variability that enters the picture during the simulation). There are several ways we might proceed. One way would be to let each mean vital rate be selected at random from a distribution (whose dispersion is characterized by its standard error) at the start of each trajectory and letting it be fixed at this random value for the 100 years of the simulation. Annual variation in the vital rate would be simulated as before with reference to the standard deviation for each vital rate. The resulting trajectories would therefore express uncertainty both from measurement error and from temporal variability. However, the output summaries would also confound both kinds of uncertainty.

Another way to assess measurement error's effect on our risk analysis would be to explore how it causes divergence under a fixed sequence of environmental variation. For instance, each trajectory would use the same sequence of environmental fluctuations, so that any spreading out of this trajectory over time shows the growing effects of the initial inaccuracy on the uncertainty of the projection. When the lower limit of the 95% confidence interval of the population size reaches 15 breeding pairs (or whatever is the smallest population size that could be self-sustaining), then the uncertainty of the demographic rates has rendered the model unable to distinguish between persistence and extinction of the population. Projections beyond this time would clearly be of limited use. Accordingly, we would take the maximum time scale for which meaningful estimates of population viability are possible to be the number of years before the 95% confidence interval drops below the extinction threshold, regardless of the time scales used by any arbitrary endangerment criteria. Such a time scale would be the longest one for which it is appropriate to compare different scenarios, and subsequent comparisons of the viability of this population should be based on it.

A third, more straightforward, way to explore the effects of measurement error is simply to bump the vital rates simultaneously up or down by magnitudes proportional to their respective measurement errors and compute the resulting variation in quasi-extinction risk. This technique is not generally possible in other applications of risk assessment because factors are generally related to the endpoint variable in complex ways. In this case, however, it yields an interpretable result because all the elements of the vital rate matrix contribute positively to population growth and contribute negatively to extinction risk.

TABLE 11.4. Probability (as a percentage) of quasi-extinction to 15 breeding pairs over three time horizons and two hypotheses about territory availability.[a]

Habitat	Time horizon	Low estimate	Point estimate	High estimate
200 territories	18 years	0	0	0
	89 years	0	13.7	99
	100 years	0	20.0	100
300 territories	18 years	0	0	0
	89 years	0	6.7	98
	100 years	0	9.4	99

[a]High and low estimates are given describing bounds on a point estimate that assumes measurement error has no effect. All estimates are based on 1,000 replicate simulations assuming windstorms and regrowth.

Table 11.4 presents the results of this investigation. The high estimates for risk were generated by decrementing the mean vital rates listed in Table 11.1 by their respective standard errors. The low estimates were derived by incrementing the vital rates by the same amounts. The risk estimates resulting from using the mean vital rates as listed in Table 11.1 are here labeled point estimates. We emphasize that these point estimates ignore the effect of measurement errors. Note that, despite the fact that plus or minus one standard error is a fairly modest range, the potential effect of ignorance due to measurement error is considerably larger than the risk consequences of the habitat assumption. For instance, if we make our simulations based on the best estimate of temporal variability, we compute a 13.7% quasi-extinction risk over 89 years for 200 territories. But this result is swamped by the uncertainty associated with measurement error: the true risk could be as large as 99% or as little as 0%. Given this range, it would be remarkable indeed if the best estimates happened to actually be, by chance, also accurate. Most prudent scientists or managers would be hesitant to rely on determinations having so little precision.

Discussion

It is clear from our analysis that the choice of time scale can strongly affect both the outcome and the reliability of a population viability analysis. Shorter time scales suffer less from measurement error but proportionally more from environmental variability. At very short time scales, divergent trajectories may still differ only negligibly. At very long time scales, measurement error may make it impossible to distinguish between even widely divergent trajectories. Although it is clear that some intermediate time scale can strike a balance between these different sources of uncertainty, there are probably no general rules that one can use to determine the ideal choice. The appropriate scale depends on characteristics of an organism's life history as well as on the particular model that one uses to study it and, possibly, on the question that one wishes to answer.

According to the criteria proposed by Mace and Lande (1991), the Olympic Peninsula population of the Northern Spotted Owl may merit classification as varied as nonthreatened, vulnerable, or endangered (Fig. 11.4). This conclusion depends on a number of factors. The most important factor is our ability to estimate the vital rates accurately. Other factors include Allee effects and the actual number of suitable breeding sites. Interestingly, it is only by taking measurement error into account that this population qualifies for endangered status; when the model takes environmental variability but not measurement error into account, the population appears to be "vulnerable" if there are initially 200 territories and "nonthreatened" if there are 300.

If one takes the Mace and Lande (1991) criteria as definitive and if one believes the point estimates, our analysis would suggest that estimating the number of territories is essential to determining the level of risk for the Olympic Peninsula population of Northern Spotted Owl. This conclusion would, in fact, be a rather naive one, however. The overwhelming influence of measurement error conclusively demonstrates that the number of territories is not really the burning question we might have thought it was. The width of uncertainty induced by measurement error in estimates of the demographic rates is as large as it is, primarily because the Mace and Lande (1991) criteria (among others) insist on such long-term extrapolations. Even using excellent data, one should hesitate to trust 100-year predictions based on only 6 years of information.

Given this problem, one may begin to doubt the practicality of criteria that require estimations of risks over long time horizons. Although we have to structure our protection efforts toward the longest term possible, purely technical

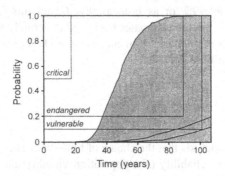

FIGURE 11.4. Quasi-extinction risk as a function of time for the Olympic Peninsula Spotted Owl population assuming 200 initial territories (higher bold curve) or 300 initial territories (lower bold curve). The shaded region represents the uncertainty of the risk estimates for the 200-territory curve. (The analogous region for the 300-territory curve is slightly smaller.) This uncertainty results from measurement error in the Spotted Owl vital rates. Also depicted are the quantitative criteria of endangerment proposed by Mace and Lande (1991) for threat categories recognized by the IUCN (1994). The smallest area in the upper left-hand corner represents critical status; the next larger enclosing area represents endangered status; and the largest area represents vulnerable status.

limitations induced by measurement errors may prevent us from being able to accurately estimate long-term risks. Because a large degree of uncertainty is unavoidable over long-time periods and it severely limits the value of any long-term analysis, one possible solution is to develop probabilistically equivalent criteria for shorter time periods. In particular, a criterion specifying a certain level of risk over a certain time horizon could be considered equivalent to a more stringent level of risk over a shorter time horizon. For instance, a 2% extinction probability for 35 years might be used in place of a 10% probability for 100 years. Although equivalent only under certain scaling assumptions (Akçakaya 1992), this approach may allow one to use shorter time horizons for which measurement error does not overwhelm our ability to forecast.

Akçakaya and Atwood (1997) described practical benefits of modeling based on time horizons of a few decades (or 50 years) in a recent analysis of the California gnatcatcher. Such benefits include use in sensitivity analyses, ranking management options according to predicted effects on a target species' viability and prioritizing conservation measures. The validity of such an approach probably depends on the particulars of the situation, because it may be possible for changes in some demographic rates to alter the relative ordering of some scenarios by their probabilities of extinction.

We believe that Mace and Lande (1991; Mace 1994) have initiated an important discussion on quantifying the degree of threat in conservation biology. However, the time scales at which these criteria are formulated may need to be reanalyzed and possibly defined differently, depending on relative accumulation of the two sources of uncertainty: variability and measurement error.

Acknowledgments. We thank David Anderson, Ken Burnham (National Biological Survey), Eric Forsman (U.S. Forest Service), James K. Agee (University of Washington), Daniel E. Varland, David P. Kenney, and Robert Meier (Rayonier) for sharing data and analyses with us, and H. Resit Akçakaya (Applied Biomathematics) for discussions and technical assistance. We also thank D. Anderson, K. Burnham, and E. Forsman for suggesting many improvements to the manuscript. This work was supported by a grant from Rayonier in 1994.

Literature Cited

Agee JK (1994) An analysis of catastrophic forest disturbance on the Olympic Peninsula. Unpublished report prepared for Rayonier, Inc., Olympia, WA

Akçakaya HR (1991) A method for simulating demographic stochasticity. Ecological Modelling 54:133–136

Akçakaya HR (1992) Population viability analysis and risk assessment. In: McCullough DR, Barrett RH (eds) Wildlife 2001: populations. Elsevier Applied Science, London, pp 148–157

Akçakaya HR, Atwood JL (1997) A habitat-based metapopulation model of the California Gnatcatcher. Conservation Biology 11:422–434

Allee WC (1931) Animal aggregations: a study in general sociology. University of Chicago Press, Chicago, IL

Burgman MA, Ferson S, Akçakaya HR (1993) Risk assessment in conservation biology. Chapman and Hall, London

Burnham KP, Anderson DR, White GC (1994) Estimation of vital rates of the Northern Spotted Owl. Appendix J, Final supplemental environmental impact statement on management of habitat for late-succession and old-growth forest related species within the range of the Northern Spotted Owl. U.S. Department of Agriculture, Forest Service, and U.S. Department of the Interior, Bureau of Land Management, Portland, OR

Caswell H (1989) Matrix population models: construction, analysis and interpretation. Sinauer Associates, Sunderland, MA

Ferson S (1990) RAMAS/stage: generalized stage-based modeling for population dynamics. Applied Biomathematics, Setauket, NY

Ferson S, Ginzburg LR (1996) Different methods are needed to propagate ignorance and variability. Reliability Engineering and System Safety 54:133–144

Ferson S, Ginzburg LR, Silvers A (1989) Extreme event risk analysis for age-structured populations. Ecological Modelling 47:175–187

Forsman E (1994) Letter to J Miles (Forest Practice Board member, Washington), April 19, 1994

Forsman ED, Giese A, Kelso D, Manson D, Maurice K, Swingle J, Townsend M, Zisa J (1993) Demographic characteristics of Spotted Owls on the Olympic Peninsula, 1987–1993. In: Jones J (ed) Annual report of the USDA Forest Service Wildlife Ecology Team. Ecosystem Processes Research Program, Pacific Northwest Research Station, Olympia, WA, pp 1–11

Forsman ED, Sovern SG, Seaman DE, Maurice KJ, Taylor M, Zisa JJ (1996) Demography of the Northern Spotted Owl on the Olympic Peninsula and east slope of the Cascade Range. Washington Studies in Avian Biology 17:21–30

Gilpin M, Soulé ME (1986) Minimum viable populations: processes of species extinction. In: Soulé ME (ed) Conservation biology: the science of scarcity and diversity. Sinauer Associates, Sunderland, MA, pp 19–34

Ginzburg LR, Slobodkin LB, Johnson K, Bindman AG (1982) Quasiextinction probabilities as a measure of impact on population growth. Risk Analysis 2:171–181

Gutierrez RJ (1996) Biology and distribution of the Northern Spotted Owl. Studies in Avian Biology 17:2–5

Hanson E, Hayes D, Hicks L, Young L, Buchanan J (1993) Spotted Owl Habitat in Washington: unpublished report by the Spotted Owl Scientific Advisory Group to the Washington Forest Practices Board, Washington Department of Natural Resources, Olympia, WA

Holthausen RS, Raphael MG, McKelvey KS, Forsman ED, Starkey EE, Seaman DE (1994) The contribution of federal and non-federal habitat to persistence of the Northern Spotted Owl on the Olympic Peninsula, Washington. Report of the Reanalysis Team, General Technical Report PNW-GTR-352. U.S. Department of Agriculture, Forest Service, Pacific Northwest Research Station, Portland, OR

IUCN (1994) International Union for the Conservation of Nature and Natural Resources, draft red list categories. World Conservation Union, Gland, Switzerland

Kenney D (1994) Habitat assessments for Olympic Peninsula Spotted Owl population. Unpublished report prepared for Rayonier, Inc, Olympia, WA

Lande R, Orzack SH (1988) Extinction dynamics of age-structured populations in a fluctuating environment. Proceedings of the National Academy of Science USA 85:7418–7421

Mace GM (1994) An investigation into methods for categorising the conservation status of species. In: Edwards PJ (ed) Large scale ecology and conservation biology: the 35th Symposium of the British Ecological Society with the Society for Conservation Biology, University of Southampton. Blackwell Scientific Publications, Boston, MA

Mace GM, Lande R (1991) Assessing extinction threats: toward a reevaluation of IUCN threatened species categories. Conservation Biology 5:148–157

McKelvey K, Noon BR, Lamberson RH (1993) Conservation planning for species occupying fragmented landscapes: the case of the Northern Spotted Owl. In: Kareiva P, Kingsolver JG, Huey RB (eds) Biotic interactions and global change, Sinauer Associates, Sunderland, MA, pp 424–450

Mertz DB (1970) Notes on methods used in life-history studies. In: Connell JH, Mertz DB, Murdoch WW (eds) Readings in ecology and ecological genetics. Harper and Row, New York, pp 4–17

Shlyakhter AI (1994) Uncertainty estimates in scientific models: lessons from trends in physical measurements, population and energy projections. In: Ayyub BM, Gupta MM (eds) Uncertainty modelling and analysis: theory and applications. Elsevier Science, BV, North Holland

USDI (1992) Recovery plan for the Northern Spotted Owl—Draft. U.S. Department of the Interior, Washington, DC

USFS (1992) Final environmental impact statement on management for the Northern Spotted Owl in the national forests. U.S. Department of Agriculture, Forest Service. U.S. Government Printing Office, Washington, DC

USFS and BLM (1994) Final supplemental environmental impact statement on management of habitat for late-succession and old-growth forest related species within the range of the Northern Spotted Owl. 1994–589–111/80005 Region 10. U.S. Department of Agriculture, Forest Service, and U.S. Department of the Interior, Bureau of Land Management. U.S. Government Printing Office, Washington, DC

USFWS (1993) Notice of intent to prepare an environmental impact statement on a proposed rule pursuant to section 4(d) of the Endangered Species Act for the conservation of the Northern Spotted Owl. Federal Register 58(248) 69132-69148

12

Can Individual-Based Models Yield a Better Assessment of Population Variability?

Yiannis G. Matsinos, Wilfried F. Wolff, and Donald L. DeAngelis

Introduction

The variability of a population is one of the key attributes influencing its ability to persist over a long period of time. Thus, population variability and the sources of that variability are of great interest to conservation biologists and to population ecologists in general. The central questions biologists want to answer are, can the factors controlling the variability in a population be identified, and, once identified, can they be included in a model of the population that can be used to predict changes in the population in response to those factors? We contrast two modeling approaches to this problem, the state variable approach and the individual-based approach.

Population variability is determined by the temporal variabilities of both reproduction and survival in a population. Part of this variability is a result of demographic stochasticity; the fact that, to some extent, births and deaths are influenced by factors that are effectively stochastic. The remainder of the variability is determined by temporal variability in the environment that changes the average survivorship and reproduction through time.

State Variable Models

The state variable approach, in which state variables are used to represent population sizes, is the most common approach in modeling populations. This approach is more appropriate for numerically large than small populations. In such populations, demographic stochasticity will generally not cause large excursions of the population size away from the mean value that the state variable represents.

Often, one state variable is used to represent the size of the total population, but in structured models, a set of state variables represents the numbers of individuals in various age, stage, or size classes within a population (e.g., the Leslie matrix model [Caswell 1989] or the McKendrick-von Foerster equation [Metz and Diekmann 1986]).

The temporal dynamics of the state variable models are described by differential, partial differential, or difference equations. Usually, the parameters of these models are assumed to be constant. However, noise can be introduced into the parameters, representing fluctuations in the environment, to produce stochastic difference or differential equation models (May and MacArthur 1972; Turelli 1977). Analysis of such models is difficult for all but the simplest models. Because a principal advantage of the state variable approach is that models may be analytically tractable, such stochastic state variable models are seldom used.

State variable models can be made spatially explicit to model populations in spatially heterogeneous regions by dividing the region into a number of subregions, connected by dispersal or migration, with variables representing the population sizes in the various subregions. This allows the model to incorporate some of the effects on variation in the total population that may result from the environmental conditions in different regions of space acting in phase or out of phase.

Thus, state variable models have the capability of representing complex population situations in which demographic stochasticity is not too important. However, such models become less mathematically tractable as they are made larger to deal with more complex situations. In small populations in which demographic stochasticity is expected to be important, birth-and-death models can be used (e.g., Pielou 1969), but these models are mathematically intractable in all but the simplest cases, and most of the advantages of an analytic formulation are lost.

In addition to the difficulties imposed by demographic stochasticity when using a state variable approach, we argue here that both the internal complexity of a population and the way in which environmental changes influence population dynamics often cannot be encompassed even by very complex state variable models. For example, the variability of a small population may depend critically on the detailed structure of the population through time, such as age and size structure, sex ratio, physiological conditions of the organisms in the population, genetic variability, and so forth. State variable models can take into account some of these detailed characteristics, but this again adds to their complexity, increasing the number of variables that must be followed.

Spatial heterogeneity further complicates the situation. In particular, resources, predators, refuge areas, disturbances that alter habitats, and other factors affecting populations, are often patchily distributed, and the relationships between the populations and these patchily distributed factors may be too complex to be represented by most conceivable spatially explicit state variable approaches. For example, important effects on population dynamics, and therefore variability, may result from (1) the details of the spatial relationships of patches (Cain 1991; Fahrig 1991), (2) the spatial pattern of disturbances that create new open patches available for colonization (Wu and Levin 1994; Moloney and Levin 1996), (3) the size distribution of patches (e.g., Hyman at al. 1991; Turner and Gardner 1991), and patterns of temporal changes in patches of resource availability (Fleming et al. 1994; Wolff 1994).

Using spatially explicit individual-based simulation models, the above authors have shown that such details of landscape configuration can have important consequences for the dynamics of a population. This holds not only for mobile animals, which have time and energy costs associated with their movement and the exploitation of resources, but also for plants, for which seed dispersal and survival may depend on details of the landscape.

From the above considerations, we conclude that traditional state variable models, although very useful in providing theoretical insights, may sometimes be too coarse to delineate the detailed mechanisms that can operate on complex landscapes. This may not make a difference in some cases, but in other cases it may.

Individual-Based Models

Spatially explicit individual-based modeling allows the effects of demographic stochasticity, the internal complexity within populations, and the subtle complexities of population interactions on a landscape to be taken into account in a fairly straightforward manner. In the individual-based approach, each individual in a population is modeled, differing in its own set of characteristics (e.g., age, sex, size, condition, social status) from all other individuals. This can reflect genetic differences as well as the different experiences of each organism (e.g., movements across a landscape, encounters with prey and predators, mating, and accidents), which are subject to stochasticity. In these models, each individual organism is also capable of making decisions (e.g., about their movements, foraging, predator avoidance, mating) that may or may not depend on temporally and spatially varying factors. In general, an individual and its interactions with the environment and other members of the population are modeled by decision rules based on the behavior and physiology of the organism.

In individual-based models, there are no state variables representing total population size or the sizes of various components of the population. Instead, these quantities are found by summing over the set(s) of relevant individuals. These models are almost by necessity computer simulation models, so there is seldom any hope of using analytic approaches. The emphasis is on developing rigorous and efficient computer codes.

The rapid increase in the execution speed and memory of computers during the past decade has played a major role in the development of individual-based models. Individual-based models are often complex, in the sense that they require extensive coding and usually considerable computer power and time. However, they are usually conceptually simple, at least much simpler than is generally realized. The basic features of an individual-based model are the following:

- A set of interacting individuals
- A set of rules that describe how local interactions are implemented
- A physical landscape, including resources

Each individual in such models is characterized by a set of variables regarding the different internal states (e.g., age, size). The spatial environment requires another variable, the location of the individual. By keeping track of the location of each individual, the model is thus able to identify which specific individuals might be interacting at a given moment (e.g., competing for food while sharing a particular spatial locale, usually represented as a unit cell on a spatial grid).

Individual-based models are mechanistic in approach. They use information at the level of the behavior and physiology of individual organisms. In principle, there is no limit to the amount of detail that can be put into the description of the interactions of individuals with their environment and with other individuals. This makes the individual-based modeling approach especially suitable for species for which individual organisms have been studied in some detail. Because complete physiological and behavioral information is rarely available for any given species, reasonable estimates must be made for those aspects that are not well known. Sensitivity to a complex array of environmental conditions can be incorporated into individual-based models. This includes stochasticity in environmental conditions on any time scale, because the time steps in the model are set at whatever size is appropriate to include all important environmental effects on individuals.

The fact that individual-based models consider not only the numbers of organisms in populations but also the condition of the individuals within the population confers another great advantage. There are conceivable cases in which population size remains relatively constant through time but in which conditions of the individuals are declining. Although the population may seem healthy from the point of view of numbers alone, it may actually be on the brink of collapse. The health of individuals could deteriorate long before this becomes manifest in a numerical collapse of the population. Thus the models allow one to follow not only the variability of numbers in a population, but the often equally important variability in the internal state of the individuals within the population.

Before discussing in more detail the uses and limitations of spatially explicit individual-based models, we describe a specific case study in some detail.

Case Study of an Individual-Based Model: Wading Birds in the Everglades

The purpose of this model is to predict one aspect of population variability, the variations in reproduction that can occur as a result of changing abiotic conditions and the resultant changes in patterns of prey availability. Although it is not a complete population model, it illustrates how this approach can be used to predict population variations.

An individual-based model was developed by Wolff (1994) (see also Fleming et al. 1994) to assist in answering questions about the decline of colonially nested wading birds in the Florida Everglades. Over the past decades, wading bird populations in the Everglades have experienced declines both in population numbers and in reproductive success concomitant with a series of anthropogenic changes in the Everglades. Several alternative hypotheses have been suggested in an effort to explain these declines, with two of them seemingly more consistent with observations: (1) development and/or drainage of former wetlands that has reduced the total size of the Everglades and resulted in general habitat loss and (2) an increase in the frequency of major drought events that has left insufficient time for standing stocks of fish and aquatic macroinvertebrate prey to replenish in some feeding areas.

The specific approach taken was to look at the endangered Wood Stork (*Mycteria americana*) by using a single-colony individual-based model. The principal parts of the model include (1) a submodel of the spatially heterogeneous landscape (i.e., spatially varying elevation), (2) a submodel for the resources (predominantly fish), which are variable both in time and space, (3) models for the behavior of the individual nesting adults, and (4) models for the energetics and growth of each nestling until it fledges (or dies).

The heterogeneity of the landscape was taken into account by subdividing a 40 × 40-km region around the colony into 25,600 square spatial cells of 250 m × 250 m each. Each cell had its own average elevation, so that the typical topography of the central part of the southern Everglades could be described. In the model, the transition from the wet to dry season was modeled by daily changes of water level that could also take into account rainfall events, which can cause reversals in water levels during the dry season. Long-legged waders such as Wood Storks usually require water depths within a certain range (for Wood Storks, 10–40 cm) to feed successfully on their prey of macroinvertebrates and small fish.

Simulation runs were started at the end of the wet season, with initial prey densities that are typical for the hydrological conditions simulated. Background competition for resources from other species of wading birds was taken into account by partitioning the resources based on local abundances and needs of the competitors.

The wading birds feeding in a cell could reduce the fish biomass of that cell; their foraging efficiency depended on resource levels and was reduced as resources became scarce. In the absence of foraging birds, cells previously depleted of fish could replenish, provided that they continued to be flooded.

A set of rules determined the behaviors of the birds from one time interval to the next. The time step for the simulation was taken to be 15 minutes because many discrete activities of the birds occur on such short time intervals. Changes in water depth were computed only once per day, however.

After nesting starts, the decisions made by the adults were guided by various constraints. Each adult attempted to meet its daily energy maintenance requirement. Wading birds could decide whether to forage alone or to join preexisting foraging groups (i.e., flocks). The location chosen by an individual in which to forage was based on partial information concerning the system. It was assumed that each individual had some knowledge, perhaps obtained by visual cues when flying or soaring, about the water depths of various locations in its foraging area. However, a bird was assumed not to know, a priori, if a cell also contained high concentrations of prey. If an adult stork did not capture prey during a 15-minute interval, it moved to another cell, possibly distant from the original cell, although it incurred higher travel costs from flying longer distances.

The growth of the nestlings was calculated on the basis of the energetic value of the food they received from their parents. Food requirements for growth and starvation thresholds were taken into account. For example, if a nestling did not attain a predetermined threshold of accumulated food before the rainy season started or if it received less than a specified threshold of food over 5 consecutive days, it died and was removed from the simulation. If the parents could not meet their own energetic needs, they abandoned their nest, causing their nestlings to perish.

Extensive simulations were performed in an effort to examine the causes of decline of the wading birds (Fleming et al. 1994; Wolff 1994). Instead of summarizing these results, we look here at the effects of a single mechanism on colony success. In particular, we demonstrate how the type of competition for food among nestlings can have an effect on numbers of successful fledglings at the population level. The relationship between success at the individual level and spatial heterogeneity depends on whether there is a scramble or contest competition for resources.

Altricial birds, such as storks, show total dependency on their parents for food. Egg laying (and hatching) is asynchronous, and it has been suggested (Fujioka 1985; Mock et al. 1987), especially in resource limited situations, that older nestlings have a competitive advantage over their younger siblings. We considered two different scenarios. In the first situation, the resource was unlimited, and as Figure 12.1 shows, all nestlings fledged successfully. In this scenario, all 150 nestlings fledged successfully, so the probability-of-success histogram (Fig. 12.2) was trivial. Both contest and scramble competition were simulated for the nestlings, but the results were indistinguishable.

In the second scenario, we decreased resources available initially by 30% in an effort to simulate the effects of competition. However, in contrast to nestlings, resource competition by adults is purely exploitative (i.e., there are no social dominance structures among them). The effects are shown in Figures 12.3 and 12.5. In the cases in which there was a 30% reduction of resources, assuming scramble competition, the chronology of hatching did not play any role. In the probability-of-success histogram (Fig. 12.4) the three hatching-order categories ended up with almost the same rates (9, 8, and 10%).

When the older nestling had a competitive advantage over its siblings, the probabilities to fledge successfully were very different (70%, 30%, 8%) (Fig. 12.6). In the case of contest competition, the parents in all but two nests (that were deserted) raised only one young. These results are intuitively clear. A modification of the rules for feeding among nestlings radically altered which of the nestlings fledged successfully.

We can also ask how the availability of resources affects the reproductive success of the colony under either competition scenario. The results show threshold effects for both cases (Figs. 12.7 and 12.8). For scramble competition and a reduction to about 80% of the initial resources, the probabilities to fledge

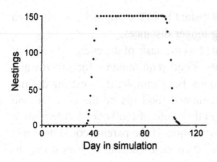

FIGURE 12.1. Simulation of reproduction when resources are not limiting for the Wood Storks. The nesting started early (on day 6 of the simulation). All nestlings survived and fledged.

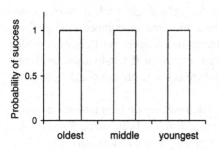

FIGURE 12.2. Probability of success for the oldest, middle, and youngest nestling for resource-unlimited conditions. All members of each category fledged.

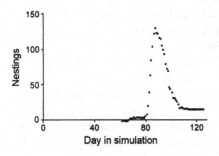

FIGURE 12.3. Simulation of reproduction when resources were reduced by 30% and when the nestlings compete by scramble competition for food brought back to the nest. The start of nesting was drastically delayed. Most nests were deserted. Only 12 nestlings fledged.

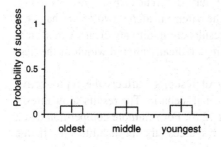

FIGURE 12.4. Probability of success of the oldest, middle, and youngest nestling for conditions of 30% reduction in resources. Scramble competition between the nestlings was assumed, and each type had the same probability of fledging.

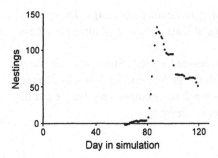

FIGURE 12.5. Simulation of reproduction when resources were reduced by 30% and when the nestlings compete by contest competition for food brought back to the nest. The start of nesting was again drastically delayed. However, this time all but two nests raised at least one offspring successfully. Fifty-four nestlings fledged.

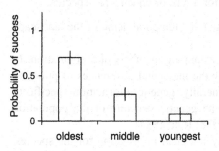

FIGURE 12.6. Probability of success for the oldest, middle, and youngest nestling for conditions of 30% reduction in resources. Contest competition between the nestlings was assumed. This time, the oldest nestlings had a much higher probability of fledging.

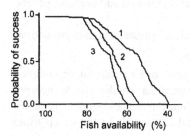

FIGURE 12.7. Probability of successful fledging of oldest, middle, and youngest nestling as a function of the fraction of fish removed from the resources available to the Wood Storks. The nestlings are assumed to compete by scramble competition.

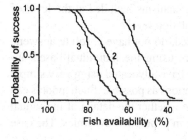

FIGURE 12.8. Probability of successful fledging of oldest, middle, and youngest nestling as a function of the fraction of fish removed from the resources available to the Wood Storks. The nestlings are assumed to compete by contest competition.

successfully decreased almost simultaneously for all three nestlings. Therefore, if one nestling fledges successfully, in general at least one of its siblings is success- ful as well.

For contest competition we still had threshold effects. However, the oldest nestling (i.e., the better competitor) may still fledge successfully when its younger siblings had little chance to do so. Only when food becomes very scarce did the older nestling have a decreased probability of fledging.

Conclusions

There are a number of advantages that an individual-based approach can offer when applied to assessing the prospects for a rare or endangered species.

1. It allows a characterization of the spatial and temporal detail of the landscape and resources.
2. It can incorporate individual differences among members of a population. In the above example, these were merely the incidental differences of pursuing different daily patterns. But more generally, genetic information specific to each organism can be included and the genetic structure of the population followed through time (e.g., Heuch 1978; Rice 1984).
3. It can use detailed behavioral and physiological information for the species.
4. It allows one to conduct detailed model management "experiments." As an illustration outside the scope of the example above, one could use such models to plan animal releases from a captive breeding program to determine if there are scenarios that provide consistently higher population survival and genetic diversity.
5. It provides output in the form of probability distributions, and thus predictions can be interpreted in a risk-analysis format.

In summary, the individual-based model allows one to calculate population variability that may come from a number of sources that are difficult to incorpo- rate in state variable models. These sources include landscape complexity, details of internal states of organisms, and behavioral rules such as the competition between sibling nestlings.

However, the amount of detail that can be incorporated in individual-based models leads critics to ask some questions: How much detail is it reasonable to include? Are there any guidelines to say at what stage the model is finished in some sense? How can one be sure that the mechanisms included at the level of individuals produce correct results at the population level?

These questions, and others that can be raised, do not have absolute answers. Concerning the first question, many modelers using the individual-based ap- proach (e.g., Fahrig 1991; Scheffer et al. 1994) have advocated using very simple models that have as few parameters and assumptions as possible. Such models can be used as research models, to try to determine whether the details of individual differences can, in principle, make a difference in population dynamics. The case study described above (Wolff 1994), however, attempts to gain insight into a particular environmental situation. Thus, it uses a level of detail appropriate to determining how certain types of behaviors and food availabilities for Wood Storks might affect population dynamics. If one were going to use Wolff's (1994) model to make specific management decisions, however, even greater detail, such as a more specific landscape, might be needed. Thus, the second question above is answered; the problem being addressed will determine the degree of detail needed.

The third question is more difficult. Of course, we need to improve our knowledge in identifying critical aspects of various behavioral and physiological traits before we can include them explicitly in models. Our ability to include the crucial mechanisms at the individual level depends on the understanding that physiological and behavioral ecologists and other life scientists have gained concerning the species in question. But the individual-based simulation approach at least provides a framework for inclusion of details at the individual level.

The great potential of individual-based modeling in no way obviates the need for state variable models. Because state variable models require far less information and are far easier to analyze, these models will continue to be the standard approach for many years.

Acknowledgments. This chapter was written with assistance from the National Park Service, U.S. Department of the Interior (Cooperative Agreement CA-5460-0-9001). The statements, findings, conclusions, recommendations, and other data in this chapter are solely those of the authors and do not necessarily reflect the views of the U.S. Department of the Interior, National Park Service.

Literature Cited

Cain ML (1991) When do treatment differences in movement behaviors produce observable differences in long-term displacements? Ecology 72:2137–2142

Caswell H (1989) Matrix population models: construction, analysis, and interpretation. Sinauer Associates, Sunderland, MA

Fahrig L (1991) Simulation methods for developing general landscape level-hypotheses for single-species dynamics. In: Turner MG, Gardner RH (eds) Quantitative methods in landscape ecology. Ecological Studies 82. Springer Verlag, Berlin, pp 417–442

Fleming DM, Wolff WF, DeAngelis DL (1994) The importance of landscape heterogeneity to Wood Storks in the Florida Everglades. Environmental Management 18:743–758

Fujioka M (1985) Food delivery and sibling competition in experimentally even-aged broods of the Cattle Egret. Behavioral Ecology and Sociobiology 17:67–74

Heuch I (1978) Maintenance of butterfly populations with all female broods under recurrent extinction and recolonization. Journal of Theoretical Biology 75:115–122

Hyman JB, McAninch JB, DeAngelis DL (1991) An individual-based simulation model of herbivory in a heterogeneous landscape. In: Turner MG, Gardner RH (eds) Quantitative methods in landscape ecology. Ecological Studies 82. Springer Verlag, Berlin, pp 443–475

May RM, MacArthur RH (1972) Niche overlap as a function of environmental variability. Proceedings of the National Academy of Sciences USA 69:1109–1113

Metz JAJ, Diekmann O (eds) (1986) The dynamics of physiologically structured populations. Lecture Notes in Biomathematics 68. Springer Verlag, Berlin

Mock DW, Lamey TC, Ploger BJ (1987) Proximate and ultimate roles of food amount in regulating egret sibling aggression. Ecology 68:1760–1772

Moloney KA, Levin SA (1996) The effects of disturbance architecture on landscape-level population dynamics. Ecology 77:375–394

Pielou EC (1969) An introduction to mathematical ecology. Wiley-Interscience, New York

Rice WT (1984) Disruptive selection of habitat preference and the solution of reproductive isolation: a simulation study. Evolution 38:1251–1260

Scheffer M, Baveco JM, DeAngelis DL, Lammens EHRR, Shuter BJ (1994) Stunted growth and stepwise die-off in animal cohorts. The American Naturalist 145:376–388

Turelli M (1977) Random environments and stochastic calculus. Theoretical Population Biology 12:140–178

Turner MG, Gardner RH (eds) (1991) Quantitative methods in landscape ecology. Ecological Studies 82. Springer Verlag, Berlin

Wolff WF (1994) An individual-oriented model of a wading bird nesting colony. Ecological Modeling 72:75–114

Wu J, Levin SA (1994) A spatial patch dynamic modeling approach to pattern and process in an annual grassland. Ecological Monographs 64:447–464

13

Potential of Branching Processes as a Modeling Tool for Conservation Biology

Frédéric Gosselin and Jean-Dominique Lebreton

Deterministic theory in population ecology thus seems to be of little help in providing a framework for probability theory. We had better not adhere too much to our deterministic concepts and ideas, but start afresh.

—J. Reddingius (1971)

Introduction

Reaching some predictive ability is a long-term purpose in many applied scientific fields. Conservation biology is no exception to the rule, because land managers dealing with endangered species frequently expect predictive evaluations of alternative management plans. Such predictions, often developed in the framework of population viability analysis (PVA), generally rely on some kind of modeling (Boyce 1992). In such a multidisciplinary endeavor and in the context of the strong social pressure presently typical of many environmental problems, it is not surprising that a variety of modeling tools has been used with, in general, an emphasis on biological relevance. Simulation is frequently used without any mathematical analysis, and many models exist only as computer programs (e.g., Woolfenden and Fitzpatrick 1991; Stacey and Taper 1992). This state of the art is summarized by Boyce (1992), who says of PVA, "Any attempt is qualified that involves some simulation or analysis with the intent of projecting future populations, or estimating some extinction or persistence parameter."

The priority afforded to extinction implies the need to account for stochastic events in the models to represent adequately the possibility of extinction. For instance, within the more restrictive framework of minimum viable population (MVP) estimation (Gilpin and Soulé 1986), frequent use is made of density-dependent stochastic models. Even within this narrower focus, the variety of model structures in the literature and the absence of a backbone theory for extinction models are striking. There is therefore a clear need for a class of models with, simultaneously, enough biological relevance and enough mathematical tractability to be useful in solving problems (Reddingius 1971; Chesson 1978).

What are the features that should be considered in such canonical extinction models? Despite the diversity of modeling approaches used until now, there is a clear consensus on three structural characteristics:

1. Extinction models must consider several distinct types of variability in population processes (e.g., Chesson 1978; Shaffer 1981; Gilpin 1987), namely, demographic stochasticity (or within-individual variation), environmental stochasticity over time, density dependence, dispersal over spatial units, and environmental variability over space and time.
2. Extinction models should emphasize, at least in a first step, demographic aspects prior to genetical ones (Lande 1988).
3. To be useful in practice, extinction models should be written in terms of demographic parameters (Boyce 1992). In this context, discrete-time models are more realistic to account for the seasonality and for the age structure of most species, in particular for vertebrates, the subject of many PVA (Eberhardt 1985; Lebreton and Clobert 1991).

The aim of this chapter is to introduce a class of extinction models, called discrete-time branching processes (BP) and to present mathematical results about them that are useful in the context of population extinction. In particular, we emphasize a paradoxical form of stability when ultimate extinction is certain, called quasi-stationarity, which provides a clear conceptual background to the interplay of persistence and extinction. Quasi-stationarity is often implicit in many PVAs, especially in relation to a geometric probability distribution of time to extinction (Goodman 1987; Woolfenden and Fitzpatrick 1991; Gabriel and Bürger 1992). Although quasi-stationarity has already been explicitly used in some stochastic finite-state population models (e.g., Verboom et al. 1991; Day and Possingham 1995), BPs are among the simplest individual-based infinite-state models in which quasi-stationarity can be studied formally. We hope, in turn, to convince the reader that BPs are suitable for playing a theoretical and practical role in the study of population extinction similar to that of matrix models (e.g., Caswell 1989; Heppell et al., this volume) in the study of population growth.

Our chapter is organized as follows: after having first recalled the general features of BPs, we consider the simplest case of density-independent growth and introduce the key notion of quasi-stationarity. Then we investigate BPs that account for an age structure and apply such a BP to a population of White Storks. We then introduce density dependence and random environment to a BP, first separately, then simultaneously, together with an age structure. Finally, we discuss the relevance of BPs as extinction models.

The notation and abbreviations used are given in Appendix 13.1. In particular, we denote by Pr(A) the probability of the event A and by E(X) the expectation of the random variable X. Furthermore, we denote by $N = \{0, 1, 2, \ldots\}$ and $N^* = \{1, 2, \ldots\}$ the sets of non-negative and positive integers, respectively. In this chapter, time t takes discrete values, in N.

General Features of BPs

Characteristics Relevant to PVA

BPs are stochastic processes built to model simultaneously the multiplicative nature of population growth and random differences between individual demographic performances. They have four characteristics particularly relevant to PVA:

1. The population size in the model takes only non-negative integer values. As a consequence, contrary to models with real-valued population sizes, extinction is defined unambiguously as reaching a population size equal to 0.
2. BPs are fundamentally stochastic: the population size at time t, Z_t, is obtained stochastically from that at time $t - 1$, Z_{t-1}, by transition probabilities. As a consequence, even conditional on a particular value of Z_{t-1}, Z_t is an integer valued random variable. Thus, the process cannot be reduced to a deterministic process plus white noise.
3. Individuals are considered explicitly in a BP: the transition probabilities depend on the particular demographic rules retained when building the model, and these demographic rules are defined at the individual's level. The core property of a BP in this respect is the branching property, according to which individuals reproduce and die independently of each other;
4. Although the properties above are presented for the simple case of a single type of individual, BPs can be generalized to account for several types of individuals (e.g., age classes, spatial cells), environmental variability, and density dependence.

BPs as Infinite Markov Chains with an Absorbing State

Technically, for the simple case of a single type of individuals, a BP is a Markov chain (i.e., Z_t depends on the past only through the previous time $t - 1$, indeed only through the value of Z_{t-1}). The state space of the Markov chain is made up of all possible values of Z_t (i.e., all non-negative integers). We call such a chain, which has an infinite denumerable number of states, an infinite Markov chain. Markov chains that are commonly used in ecology (e.g., for modeling succession) (e.g., Facelli and Pickett 1990) or for simple population models (e.g., Verboom et al. 1991; Day and Possingham 1995) have only a finite number of states and, as such, are called finite Markov chains.

In accordance with the above definition of extinction, in a BP, 0 is an absorbing state (i.e., once the population is extinct, it remains so). BPs are thus specific infinite Markov chains with an absorbing state. In a finite Markov chain with an absorbing state, under mild conditions, absorption is certain ultimately:

$$\Pr\left(\lim_{t \to \infty} Z_t = 0\right) = \lim_{t \to \infty} \Pr(Z_t = 0) = 1 \tag{13.1}$$

A striking difference is that this is no more the case for an infinite Markov chain with an absorbing state. In the case of BP, divergence to infinity can occur as well as extinction; indeed, under mild conditions,

$$\Pr\left(\lim_{t \to \infty} Z_t = 0\right) + \Pr\left(\lim_{t \to \infty} Z_t = \infty\right) = 1 \tag{13.2}$$

This is so because, when there is one absorbing state, the ultimate probability of presence in any other state tends toward 0 (i.e., all other states are transient). In a finite Markov chain, absorption is then certain. In an infinite Markov chain, the divergence to infinity, moving through infinitely many transient states during the course of divergence, remains possible. This is one of the many counterintuitive intricacies of stochastic processes, here as a consequence of the infinite number of states.

Individuals' Replacement and the Branching Property

The transition from $t - 1$ to t can be represented as usual for Markov chains by a transition matrix, which, in the case of BPs, has infinitely many rows and columns. Fortunately, in BPs, the transition from Z_{t-1} to Z_t can alternatively be efficiently represented as the sum of individuals' contributions. According to this point of view, from time $t - 1$ to t, each individual j is replaced by a random number of individuals $X_{t-1,j}$, taking non-negative integer values 0, 1, 2, ... The probability distribution of $X_{t-1,j}$ is based on probabilities of death and reproduction. When the individual j dies without reproducing, it is replaced by zero individuals (i.e., $X_{t-1,j}$ takes the value 0). This is the case for individual 2 at time $t - 1$ in Figure 13.1. When the individual j survives and/or reproduces, it is replaced by one or more individuals (i.e., $X_{t-1,j}$ takes a positive value). This is illustrated by individuals 1 and Z_{t-1} at time $t - 1$ in Figure 13.1. In a process with a single type of individuals, no difference is made between the case in which the individual is still

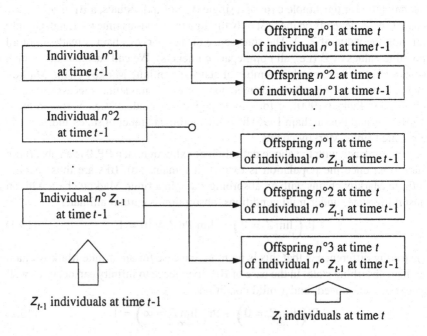

FIGURE 13.1. Structure of a discrete time branching process, with its key feature, the independence of individuals, or branching property.

present or the case in which it is replaced by a new one. More sophisticated rules, accounting for the age of individuals, will be developed with multitype BPs.

The stochastic replacement of individuals can be summarized in a single equation giving Z_t, conditional on Z_{t-1}:

$$Z_t = \sum_{j=1}^{Z_{t-1}} X_{t-1,j} \tag{13.3}$$

The branching property states moreover that the individuals' contributions [i.e., the $(X_{t-1,j})$ $(j = 1, \ldots, Z_{t-1})$] are independent of each other. An example of realization of this transition from time $t - 1$ to time t is given in Figure 13.1.

Density-Independent BP: The Bienaymé-Galton-Watson BP

Definition

The Bienaymé-Galton-Watson BP (BGW BP) is the simplest BP, because the individuals' performances are identical and constant in time (i.e., the $X_{t-1,j}$ are identically distributed over time $t - 1 = 0, 1, \ldots$ and over individuals $j = 1, \ldots, Z_{t-1}$). In particular, there is no age structure in this model, and the expected individual performance, denoted by $m = E(X_{t-1,j})$, is constant. This definition as well as the properties that we will state about BGW BPs are well known and can be found in, for example, the work of Athreya and Ney (1972), Jagers (1975), or Asmussen and Hering (1983).

Demographic Background: BGW BP as a Single Age Class Model

The BGW BP can be expressed easily in terms of demographic parameters, as we now show by a simple example. Let us consider the females of a sexually reproducing population such that a female survives with probability s until next year and, independently, gives birth to $0, 1, 2, \ldots$ 1-year-old females with respective probabilities $\pi_0, \pi_1, \pi_2, \ldots$. For a female to be replaced by k individuals, where k is in N, she has to give birth to $k - 1$ individuals and survive (with probability $s \, \pi_{k-1}$), or, exclusively, give birth to k individuals and die (with probability $(1 - s)$ π_k). Hence, the common probability distribution of the random variables $(X_{t-1,j})$ is given by

$$\text{PR}(X = k) = s \, \pi_{k-1} + (1 - s) \, \pi_k \tag{13.4}$$

One can then show that $m = s + h$, where $h = \sum_{k=1}^{\infty} k \pi_k$ is the average net fecundity, expressed in females aged 1 per female. This model has a single age class, because the survivors of newborn individuals are considered as adults at age 1. It could be easily modified to split the net fecundity, with expected value h, into two stochastic components: the fecundity in newborn females per female, with expected value f, and the survival probability from birth until age 1, p. One would then have $m = s + fp$.

General Behavior of BGW BP

Each individual is replaced, on the average, by m individuals. From the conditional expectation formula $E(Z_t \mid Z_{t-1}) = m\, Z_{t-1}$, one gets

$$E(Z_t) = mE(Z_{t-1}) \tag{13.5}$$

Hence, expected population size varies exponentially:

$$E(Z_t) = m^t E(Z_0) \tag{13.6}$$

If $m > 1$, the expected population size diverges over time, but Equation (13.5) tells us that extinction is still possible. Indeed, if $m > 1$, the process goes extinct with a non-null probability r and diverges toward infinity with probability $1 - r > 0$; this case is called supercritical.

However, when $m \le 1$, ultimate extinction is certain, because, from Equation (13.6), the population size cannot diverge to infinity with a positive probability. The so-called critical case, $m = 1$, is unrealistic in practice (Lebreton 1981). The case $m < 1$ is called subcritical.

At first sight, the supercritical case is the most attractive, as a natural stochastic counterpart of exponential growth, the oldest model of population growth (Malthus 1798). However, the emphasis in PVA is on regulation, which tends to keep population size away from infinity. We see below that realistic forms of density dependence imply the certainty of (ultimate) extinction. This qualitative reasoning, which we also discuss below, gives a special interest to the subcritical case.

Quasi-Stationarity of Subcritical BGW BP

Because in subcritical BGW BPs extinction is certain, we are naturally led to study the behavior of the process before extinction. Let us consider a small initial population size ($Z_0 > 0$). If at least some of the Z_0 individuals have survived and/or have given birth to new individuals, the population is not extinct at the next time step ($Z_1 > 0$). As long as this goes on (i.e., as long as extinction has not taken place), we obtain a series of positive random variables $Z_1, Z_2, \ldots$. The key result is that, in subcritical BGW BPs, the probability distribution at time t of a population size conditioned on nonextinction converges, when t tends to infinity, to a probability distribution $(b_k)_{k \in N^*}$ (Fig. 13.2), irrespective of the initial population size (e.g., Seneta and Vere-Jones 1966; Jagers 1975).

The probability distribution $(b_k)_{k \in N^*}$ is called the quasi-stationary distribution (QSD) because

1. It is stationary in the sense that the probability distribution of Z_t converges to it irrespective of the initial value (i.e., the value of $Z_0 > 0$).

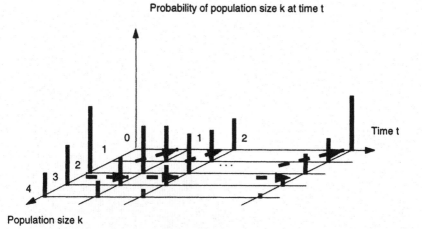

FIGURE 13.2. Distribution of Z_{n+1} as a function of the distribution of Z_n, using the transition matrix of the Markov chain representation of a BP.

2. It is only "quasi"-stationary, and not stationary in the usual sense, because the BP converges to it only conditional on nonextinction.

Indeed, the stationary distribution sensu stricto is the long-term probability distribution of Z_t. In practice, it can be approached by simulation, by looking at the probability distribution of Z_t for large t over a large number of replicates. In a subcritical process, the only stationary distribution is extinction, which is confirmed by simulation because, for t sufficiently large, most of the Z_t values equal 0 (Fig. 13.3a). The QSD can be approached by keeping only the fraction of the replicates with $Z_t > 0$) (Fig. 13.3b).

Let us stress that the convergence to the QSD is a convergence in distribution (i.e., a convergence of probability distributions) and that the QSD $(b_k)_{k \in N^*}$ is a probability distribution, not a number. As for every probability distribution, we can associate with the QSD some numbers, such as its expectation or its variance.

Under mild conditions, the expectation, $\sum_{k=1}^{\infty} kb_k$, toward which $E(Z_t \mid Z_t > 0)$ converges when t tends to infinity, and further moments of the QSD are finite (Heathcote et al. 1967; Bagley 1982).

Even if the convergence to the QSD is conditional on nonextinction, the existence of a QSD has implications for the way the BGW BP goes extinct. Indeed, in BGW BPs, the convergence to the QSD is accompanied by the convergence, as time tends to infinity, of the probability of nonextinction at the next time step conditional on nonextinction at present. The limit of this convergence is denoted by λ. In the particular case of BGW BPs, $\lambda = m$ (Fig. 13.2). This entails, under mild conditions, that the expected population size at time t is asymptotically proportional to $\lambda^t = m^t$, as we already knew from Equation (13.6). We therefore

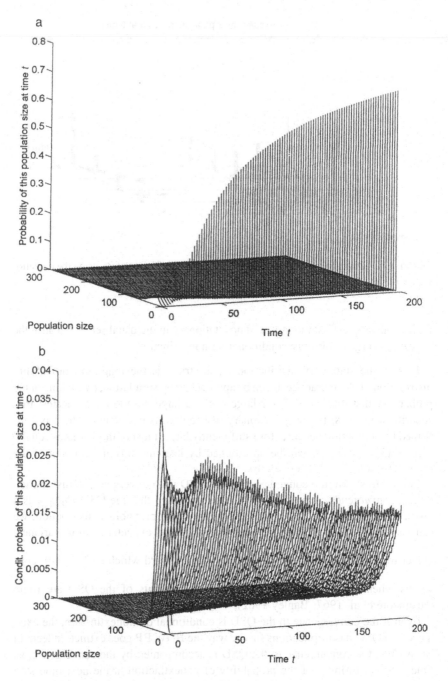

FIGURE 13.3. Convergence toward a stationary distribution conditional on nonextinction (a) or quasi-stationary distribution (b).

call the limit λ the asymptotic growth rate of the population. Technically, under mild conditions, the asymptotic growth rate is the dominant eigenvalue of the substochastic matrix Q (Appendix 13.2).

We define quasi-stationarity of a BP as the simultaneous existence of a QSD and of an asymptotic growth rate λ. We then say that the BP is quasi-stationary. Because quasi-stationarity implies that the probability of immediate extinction conditioned on nonextinction at present converges to $1 - \lambda$, quasi-stationarity further entails that the probability distribution of the time to extinction will be asymptotically geometric, with reason $1 - \lambda$.

Multitype BGW BP and a Case Study: The Alsace White Stork

Despite the theoretical interest of the previous results, more general models are needed for reaching some realism and applicability. We propose in this section a first such generalization, which consists of considering different kinds of individuals. This will allow us to apply these theoretical results to an age-structured declining White Stork (*Ciconia ciconia*) population.

Multitype BGW BP

We now distinguish individuals according to characteristics such as their age, so that individuals are classified into d types ($d \in N^*$). The possible states of the stochastic random variable Z_t representing the population at time t are no longer single integers, but d-uplets of non-negative integer numbers, representing the number of individuals in the d different types at time t: Z_t is now a random vector with values in N^d. Each type then has its own offspring probability distribution (i.e., the probability distribution of the X variables introduced above now depends on the type of the individual considered). Moreover, the X variables are multivariate with d integer-valued components, because an individual may produce offspring individuals of several types (e.g., an adult aged i may survive as an adult aged $i + 1$ and give birth to young, that is, to individuals aged 0).

A BP that relies on the same assumptions as a BGW BP except that it allows for different types of individuals is called a multitype BGW BP (MT BP). The scalar m is then replaced by a non-negative $d \times d$ matrix $M = (m_{\alpha,\beta})_{1 \leq \alpha, \beta \leq d}$ so that

$$E(Z_t \mid Z_{t-1}) = MZ_{t-1} \text{ and } E(Z_t) = ME(Z_{t-1}) \qquad (13.7)$$

The entry $m_{\alpha,\beta}$ of the matrix M is the average number of individuals of type α produced by an individual of type β at the next time step. In particular, with age intervals whose length equals the time step, M is a Leslie matrix. As a consequence, "the exact stochastic analogue of Leslie's theory . . . can be regarded as a special case of the general theory of the multi-type Galton-Watson process" (Pollard 1966).

The supercritical, critical, or subcritical behavior of a MT BP is determined by μ, the dominant eigenvalue of M (i.e., the asymptotic multiplication rate of the deterministic counterpart of the model) (Joffe and Spitzer 1967). In particular, the process is subcritical if and only if μ is less than 1. Then, ultimate extinction (i.e., reaching the state, [0, 0, . . ., 0]) is certain, the MT BP has a QSD associated with the asymptotic growth rate $\lambda = \mu$ (Joffe and Spitzer 1967). Therefore, when μ < 1, the MT BP is quasi-stationary. Furthermore, provided the QSD has a finite first moment and M is a Leslie matrix, we can prove that the QSD is then a stochastic extension of the stable age distribution (e.g., Caswell 1989) of matrix population models.

Case Study: The Alsace White Stork

The White Stork population in the Alsace area (eastern France) has been decreasing rapidly since 1960 (Fig. 13.4), with a recent recovery as a consequence of a reintroduction program. All other western European populations, migrating through Gibraltar and wintering in western Africa, have similarly decreased (Bairlein 1991).

The dynamics of the Alsace White Stork population thus seem amenable to some modeling by a subcritical BP. It can be analyzed, as a first approximation, by a density-independent, fixed-environment, Leslie matrix (deterministic) model (Table 13.1).

Survival probabilities have the strongest influence on the population multiplication rate, as a result of the relatively high generation time (Lebreton and Clobert 1991), which is about 6 years (Lebreton 1978). Yet, the adult survival probability varies significantly with rainfall in the wintering area in the Sahel zone, and rainfall causes a negative trend over time (Kanyamibwa et al. 1990, 1993). This explains the strong decrease in numbers (more than 15% a year),

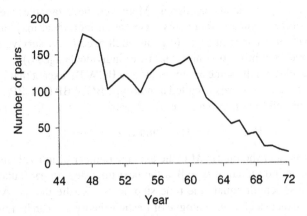

FIGURE 13.4. Size (number of breeding pairs) of the Alsace White Stork population from 1944 to 1973 (before reintroduction of captive birds).

TABLE 13.1. Leslie matrix model for the Alsace White Stork population.[a]

$$
E(Z_t) = \begin{pmatrix} N_1 \\ N_2 \\ N_3 \\ N_4 \end{pmatrix}_t = \begin{pmatrix} 0 & 0 & U_3 R\,pa & R\,pa \\ q_1 & 0 & 0 & 0 \\ 0 & q_2 & 0 & 0 \\ 0 & 0 & q_3 & q_4 \end{pmatrix} \begin{pmatrix} N_1 \\ N_2 \\ N_3 \\ N_4 \end{pmatrix}_{t-1} = M\,E(Z_{t-1})
$$

Z_t: random vector of population size in each type (i.e., age class) at time t
N_α: expected number of individuals in age class α (4: age 4 and more)
q_α: survival probability from age class α to age class $\alpha + 1$, if $1 \le \alpha \le 3$, or from age 4 to age 4, if
  $\alpha = 4$
p: survival probability from birth to age 1
U_3: probability of reproduction of 3-year-old individuals
R: probability of successful reproduction
a: expected number of young females successfully raised per female

[a]From Lebreton (1978).

which was leading the population to rapid extinction, prior to the first reintroductions in the early 1970s.

We can associate with the Leslie matrix M in Table 13.1 a MT BP that makes it possible to study the approach to extinction. Each parameter in the Leslie matrix M is then replaced by a random variable whose expectation equals this parameter, as shown in Table 13.2.

The dominant eigenvalue of M, μ (i.e., the asymptotic multiplication rate of the model population) is smaller than 1. Then, there is a QSD. On the basis of empirical data, the QSD for the Alsace White Stork during its regime of decrease seems to have only a few individuals (Table 13.3; Lebreton 1978). Once the QSD is reached, the extinction time follows a geometric probability distribution with annual probability of extinction $1 - \lambda = 1 - \mu$. The cumulative risk of extinction reaches 0.95 within less than 20 years (Lebreton 1978).

Extinction actually has not occurred so far, partly because the reintroduced individuals are partially sedentary due to being kept captive during a few months and are not affected by the Sahel drought. Indeed, their estimated annual probability of survival is 0.91 (Kanyamibwa 1991), well above the threshold for population stability (0.75).

More realistic models would take account of random environmental variation, in particular of the aforementioned relationship between the Sahel rainfall and adult survival. A projection based on a varying environment Leslie matrix model (Fig. 13.5) confirmed that nearly all variation in population numbers may be explained in that way. This reinforces the relevance of a BP in a random environment for studying the approach to extinction when demographic stochasticity becomes predominant over environmental stochasticity.

For MT BGW BPs in random environments, the existence of a QSD is only conjectural (however, if some density dependence is introduced the case may be different; see below). The difficulty is that there is no simple equivalent of the criticality parameter $E(\ln m_t)$ of (monotype) BGW BPs in random environments

TABLE 13.2. Random variables corresponding to the various transitions for each individual in one time step between age classes in a branching model of the Alsace White Stork population.[a,b]

Survival:

Age transition	Probability distribution for transition
1 to 2	Bernoulli probability distribution (probability q_1)
2 to 3	Bernoulli probability distribution (probability q_2)
3 to 4	Bernoulli probability distribution (probability q_3)
4 to 4	Bernoulli probability distribution (probability q_4)

Reproduction:

Age transition	Probability distribution for transition
3 to 1	Bernoulli probability distribution to be a breeder (probability U_3)
	If breeder, reproduction according to the probability distribution L
	If young, Bernoulli probability distribution (probability 0.5) of being a female
	If young and male, not taken into account
	If young and female, survival according to a Bernoulli probability distribution (probability p)
4 to 1	Reproduction according to the probability distribution L
	If young, Bernoulli probability distribution (probability 0.5) of being a female
	If young and male, not taken into account
	If young and female, survival according to a Bernoulli probability distribution (probability p)

[a]From Lebreton (1978).
[b]A Bernoulli probability distribution with parameter q yields zero individual with probability $1 - q$ and one individual with probability q. The probability distribution L yields two young with probability 0.25, three young with probability 0.5, and four young with probability 0.25. In this setting, for instance an individual aged 3 will be replaced by at most one individual aged 4 if it survives, and by possibly several individuals of age 1 if it gives birth to young that survive until age 1.

TABLE 13.3. Summary of results from a branching model of the Alsace White Stork population.[a,b]

Expected values of parameters	Asymptotic behavior of the branching model
q_α: 0.600	Existence of a QSD and of an asymptotic growth rate $\lambda = 0.818$
p: 0.393	Asymptotic probability of immediate extinction:
U_3: 0.450	$\Pr(Z_t = 0 | Z_{t-1} = \text{QSD}) = 1 - \lambda = 0.182$
R: 0.800	Stable age structure: (0.38, 0.18, 0.13, 0.31)
a: 3.000	Expected number of female storks aged 3 and more in the QSD: 1.643 females
	Expected number of individuals in the QSD: 7.27 individuals

[a]From Lebreton (1978).
[b]Parameters are as in Tables 13.1 and 13.2.

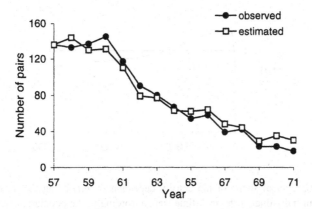

FIGURE 13.5. A varying environment Leslie matrix model for the Alsace White Stork: model projection and observed population sizes. Adult survival varies with rainfall in the Sahel. Subadult survival was estimated to be three of five of the adult survival. Fecundity was observed once each year. (From Kanyamibwa and Lebreton 1992.)

(see below). This problem is well known in the context of random environment models based on random products of matrices (i.e., without demographic stochasticity) (Tuljapurkar 1990). A rough numerical approach (Lebreton 1978), in which random environment is introduced via a series of independent and identically distributed random variables, tended to confirm the existence of a QSD. The average number of individuals in the QSD was moderately sensitive to various levels of environmental variability in parameters, with a maximum of twice as many individuals as in the fixed environment case, as would be expected for a small population that would be more sensitive to demographic stochasticity.

More General BPs

After having considered the above BP that accounts for different types of individuals, in this section we investigate BPs that incorporate density dependence or a random environment. We then consider more complex models that include simultaneously different types of individuals, density dependence, and a random environment. For all these models, we give sufficient conditions for the certainty of ultimate extinction and quasi-stationarity of the BP.

Density-Dependent BGW BP

A major drawback of BGW BP is that demographic parameters are independent of population size, although as noted by Nunney and Campbell (1993), "Most (extinction models) include some form of population regulation."

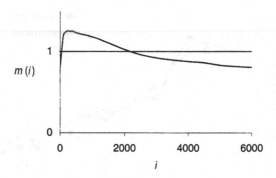

FIGURE 13.6. Typical shape of the density dependence: $m(i)$ being the mean offspring number of an individual at the next time step, according to the population size i.

Such a regulation usually leads to consideration, as in Figure 13.6, of a model with $m > 1$ for "medium" population sizes and $m < 1$ for large population sizes (as, e.g., Caswell 1989) where m is still the mean number of offspring per individual. One feels intuitively that the introduction of a regulation such that $m < 1$ for large population sizes leads to a strong probability of the population becoming extinct. We will indeed see that such regulation, by itself, entails certain ultimate extinction.

A BP that fulfils the assumptions of a BGW BP, except the one concerning independence of demographic parameters relative to population size, is called a density-dependent BGW BP (DD BP). It is characterized by a specific offspring number probability distribution for each population size. We therefore introduce the random variable $_{(i)}X_{t-1,j}$, representing the (random) offspring number of the jth individual among i at time $t - 1$, whose probability distribution may now depend (solely) on population size i. The random variables $(_{(i)}X_{t,j})_{t\in N,(i,j)\in N^{*2}}$ are however still supposed to be independent (branching property). We then set $_{(i)}m = E(_{(i)}X)$.

A DD BP $(Z_t)_{t\in N}$ is still a Markov chain with an infinite state space and an absorbing state (as in 13.2). Thus, under mild conditions, $\Pr\left(\lim_{t\to\infty} Z_t = 0 \ or \ \lim_{t\to\infty} Z_t = \infty\right) = 1$. Furthermore, a sufficient condition for certainty of (ultimate) extinction is that for all population sizes i over some threshold i_0, the expected number of offspring per individual $_{(i)}m$ in a population of size i satisfies $_{(i)}m \le m < 1$ (Gosselin, 1998d). This condition implies that when the population is big enough each individual is replaced, on the average, by less than m individuals, with $m < 1$. This sufficient condition for certain extinction seems realistic and comes up to our previous expectations about regulation (e.g., Fig. 13.6).

Under further technical conditions (e.g., the existence of a maximum offspring number; cf. Gosselin, 1998d), the DD BP has a QSD and an asymptotic growth rate λ (Gosselin, 1998d).

BGW BPs in Random Environments

Another major limitation of BGW BPs is the absence of environmental stochasticity (i.e., of random variation of demographic rates over time), despite the key role it plays in extinction processes (e.g., Shaffer 1987; Lande 1993).

Such environmental variation can, however, be introduced to the BP via a sequence of random variables, called the environment process, that accounts for the random variability of the environment. Then the offspring number probability distribution varies randomly over time t, depending on the environment variable at time t. In particular, the mean of the offspring probability distribution at time t, m_t, is now a random variable. BPs that are like BGW BPs except that they include such random environmental variability are called BGW BPs in random environments (BP RE).

The asymptotic results presented above still hold when the environment process is independent and identically distributed over time. The new criteria for supercriticality, criticality, and subcriticality are, respectively, $E(\ln m_t) > 0$, $E(\ln m_t) = 0$, and $E(\ln m_t) < 0$, under additional assumptions (Smith and Wilkinson 1969; Athreya and Karlin 1971). Thus, provided $E(\ln m_t) < 0$, the BP RE certainly ultimately goes extinct and has a QSD and an asymptotic growth rate. Results about more general environmental processes were obtained by Athreya and Karlin (1971) but under somewhat more stringent conditions that make these extensions of little help in biology. For instance, the time reversibility (or exchangeability in the terms of Athreya and Karlin 1971) required is not met when the environment process is an autoregressive moving average (ARMA; e.g., Box and Jenkins 1970) process.

Closer to Realism

Mathematical results are now available for BPs that combine density dependence, environmental variability, and several kinds of individuals (Gosselin 1998b). For instance, sufficient conditions for certainty of ultimate extinction and quasistationarity to hold are that (1) the density dependence and the random environment considered are such that, in some way, the mean number of offspring per individual tends to zero when population size tends to infinity; and (2) the random environment does not have any temporal autocorrelation (for more details and further results, see Gosselin 1998b). However, spatial autocorrelation of the random environment (as found in, e.g., Hanski and Woiwod 1993) can be incorporated.

Discussion

We discuss the relevance of BPs as extinction models, first in terms of structure and second in terms of theoretical results available. We then discuss their use as PVA extinction models and the further work needed in this direction.

Relevant Structure for Extinction Models . . .

BPs have several features that make them attractive as extinction models. The first is the representation of population sizes as integers, which induces a clear definition of extinction, as reaching population size 0.

The second is the branching property, according to which population size at any time is obtained as the sum of independent integer random variables representing the individuals' contributions. With a cost associated to the restrictive assumption of independence (discussed later), the use of a sum of random individual contributions has at least two practical advantages: first, the individual contributions can be written easily in terms of demographic parameters (see further details below); second, demographic stochasticity is represented in a canonical way via the randomness of the individual contributions, which is particularly relevant at least for small or spatially subdivided populations in which demographic stochasticity cannot be neglected (Chesson 1978; Mode and Pickens 1986; Gabriel and Bürger 1992).

Although the branching property may seem restrictive, the independence stated is conditional on the state of the full system at a given time. This makes it possible to consider the effect on the individual contributions of the abiotic environment, of population size, and even the population size of interacting species (see an example in Lebreton 1990). If the independence hypothesis seems yet not acceptable, a splitting of the time scale may make it acceptable (Gosselin 1998b): compensatory mortality in a quarry species during the hunting season could, for instance, be represented by a series of density-dependent monthly mortalities.

A third feature of BPs that makes them suitable in practice is their flexibility. Density dependence, environmental variability, and different types of individuals (according to their age, sex, or discrete spatial location) can be considered, simultaneously if needed. It may, however, be difficult to distinguish whether a model combining these different demographic features is a BP. Actually, a necessary and sufficient condition for a model to be a BP is that it is in discrete time, relies on integer numbers, and satisfies the following statement: it is conditional on the state of the population (i.e., the number of individuals in each type) at time $t - 1$, conditional on the random environment at time $t - 1$, and conditional on any event concerning the further past, the number of offspring in the different types at the next time step t of each individual in the population at time $t - 1$ must be drawn at random, independently from offspring random numbers of other individuals, according to a probability distribution that may depend solely on the type of the individual, the state of the population, and the random environment at time $t - 1$.

Altogether BPs appear as fully stochastic extinction models, contrary to models obtained by adding real valued white or pink noise to a deterministic equation.

. . . with Relevant Results Available . . .

On the basis of the relevant structure of BPs, one should expect the theoretical results to be relevant, in particular when realism is enhanced by considering density dependence. The two main such results are the certainty of ultimate extinction and quasi-stationarity.

Certainty of ultimate extinction is a general feature of BPs, as soon as population size is prevented from divergence to infinity (Equation [13.2]). This feature seems realistic to us, although it has been open to discussion. Indeed for Chesson (1978), "If population sizes are not allowed to become arbitrarily large it seems that, in the absence of evolution, eventual extinction is a fact of life," whereas the mathematicians Athreya and Ney (1972) state that, because of certain ultimate extinction, "an unmodified branching process is thus not a satisfactory model for most biological situations."

In our opinion, the biological realism of certain extinction comes simply from the fact that it may be very likely or be unlikely over, for example, hundreds of time steps. Actually, quasi-stationarity, defined as the existence of a QSD and of an asymptotic growth rate, tells us how, asymptotically, the population goes extinct. Because eventual extinction is certain, the asymptotic growth rate λ is less than 1. Then, the asymptotic slow or rapid occurrence of extinction will depend on whether λ is close to 1 or not.

Quasi-stationarity further implies the asymptotic geometric probability distribution of time to extinction, with a geometric rate of $1 - \lambda$. It is satisfying that this result, general in ad hoc extinction models studied by simulation (e.g., Goodman 1987; Woolfenden and Fitzpatrick 1991), holds for a reasonably wide class of models. Indeed, certain extinction and quasi-stationarity are common features of BPs that incorporate very different demographic features, as outlined in the sections above.

For instance, "realistic" DD BP share these two properties with subcritical (density-independent) BGW BP. The key difference between decreasing density-independent populations, modeled by subcritical BGW BPs, and density-dependent populations, modeled by DD BPs, may, however, lie in the asymptotic risk of immediate extinction $1 - \lambda$, which may become negligible when a "strong" density dependence is introduced in a subcritical BGW BP. The sharp borderline between stable populations and decreasing ones doomed to extinction, which characterizes deterministic models, therefore vanishes in BPs. It is replaced by a continuum of situations spreading (e.g., from a value of $1 - \lambda$ equal to 0.15 for the decreasing density-independent multitype White Stork population through $1 - \lambda \approx 1.5 \ 10^{-3}$ for a weakly DD BP) (Gosselin 1996, model with no dispersal in scenario I), to $1 - \lambda \approx 10^{-43}$ in the strongly DD BP treated by Lebreton (1981).

Two additional remarks widen the applicability of our results. First, if the branching property were to be left, for example, because of dependencies between mated individuals (e.g., McCarthy et al. 1994), certainty of extinction and quasi-stationarity persist in the more general framework of absorbing Markov chains (Gosselin, 1998a, b,d). This renders the independence condition in the branching property less stringent and makes our results apply to models of territorial species, such as those of Woolfenden and Fitzpatrick (1991),[1] Schneider and Yodzis

---

[1]Provided their reproduction number probability distribution fulfils certain conditions.

(1994), Bart (1995), and McKelvey and co-workers (unpublished data) (Gosselin 1998b).

Second, if we want to investigate quasi-extinction (Ginzburg et al. 1982) of a quasi-stationary BP, we can use a modified version of quasi-stationarity relative to quasi-extinction by making the states below the quasi-extinction threshold absorbing. Although the quasi-extinction QSD and asymptotic growth rate are generally not linked in a simple way to the QSD and asymptotic growth rate relative to extinction, they account for the same qualitative phenomena, such as the asymptotically geometric probability distribution of the time to quasi-extinction.

Finally, note that further results about BPs are available (see Gosselin 1997, 1998a,b,d).

. . . That Are Underused in PVA

BPs have a long history among applied probabilists (e.g., Galton 1873; Yule 1924), and some of their features have been used implicitly in many models (e.g., North et al. 1988; Burkey 1989; Gabriel and Bürger 1992; Beissinger 1995). However, BPs have seldom been used explicitly to model extinction (Lebreton 1978, 1982, 1990; Mode and Jacobson 1987a, b; Gosselin 1996). This current state of affairs results partly from the complexity of the mathematics required, partly from the difficulty in the transfer of knowledge on such multidisciplinary matters, but also from the lack of results until recently for the most realistic BPs, those with density-dependent regulation. The two main advantages of an explicit use of BPs as extinction models, as opposed to ad hoc models, lie first in their clearly defined structure, which simplifies the building of models, and second in the availability of theoretical results.

The branching property, one of the specificities of BPs, makes it easy to build the overall random transition from one time step to the next as the sum of independent random individual contributions. These individual contributions can be easily formulated, based on a life cycle graph of the species on which we superimpose probability distributions (as in the White Stork example). BPs can therefore be easily used to model specific population-environment systems.

Besides, the clearly defined structure of BPs and the independence assumption in the branching property simplify computer programming of Monte Carlo simulations based on BPs.

Another advantage of BPs is the availability of theoretical results that can be used to interpret PVA simulation results in the same way as, for instance, the stable age theory enlightens the use of Leslie matrix models (Caswell 1989).

Consider as an illustration the White Stork BP model. Any other PVA of the White Stork would have likely obtained a similar declining trend. Even a simple deterministic Leslie matrix model (such as that in Table 13.1) would have found exactly the same asymptotic growth rate λ. However, the Leslie matrix theory does not tell anything about the meaning of λ relative to extinction. To the contrary, the theoretical results presented in this chapter allow us to interpret this declining trend as the interplay between extinction (at a rate that asymptotically tends to $1 - \lambda$) and the stabilization conditional on nonextinction in the QSD. This interplay between extinction and asymptotic stabilization conditional on non-extinction makes such a population behave asymptotically as a single individual whose survival probability from one time step to the next would be λ, the asymptotic growth rate of the population. In the White Stork case, the high value of $1 - \lambda$ tells us that asymptotically the population will become extinct relatively quickly.

The aforementioned interplay between extinction and "quasi-stabilization" is valid as soon as quasi-stationarity occurs, including for DD BPs. The above parallel between deterministic models and BPs is, however, only valid in density-independent, constant-environment situations, in which the expected population sizes of the BPs are given by a Leslie matrix model. For DD BPs, we are not aware of any simple deterministic model that would yield the successive expected population sizes of the BPs as functions of only the previous expected population size. Indeed, the interplay between stochasticity and density dependence makes $E(Z_t)$ depend not only on $E(Z_{t-1})$ but on the whole probability distribution of Z_{t-1}.

Theoretical results also appear useful to frame simulations. We give three illustrations of this statement:

1. Theoretical results emphasize certain quantities, with known asymptotic behavior, such as the expected population size conditional on nonextinction (e.g., Lebreton 1978; Gosselin 1996).
2. They lead to meaningful procedures for estimating some parameters. For example, once quasi-stationarity is reached, mean time to extinction is best estimated by fitting a truncated geometric distribution to the empirical probability distribution of time to extinction. In simulations over a finite time interval, one may account in this way for the nonobserved tail of the probability distribution of time to extinction and estimate properly the mean time to extinction.
3. Last, theoretical results may help to distinguish the transient part of the simulation results (i.e., the part that is influenced by the initial conditions) from the asymptotic one (e.g., Goodman 1987; Gabriel and Bürger 1992). The relative role of the two components, and thus the relevance of the asymptotic results, likely depend on a ratio $\frac{|\lambda_2|}{\lambda}$, where λ, the asymptotic rate of growth, is the largest eigenvalue, in modulus, of an equivalent of the matrix Q introduced in Appendix 13.2. The second largest eigenvalue is λ_2. If $1 - \frac{|\lambda_2|}{\lambda}$ is much bigger than $1 - \lambda$, quasi-stationarity will be reached rapidly, before many extinctions have occurred, in which case the results presented in this chapter are very relevant. On the contrary, if $1 - \frac{|\lambda_2|}{\lambda}$ is not much bigger than $1 - \lambda$, many extinctions are likely to take place before quasi-stationarity is reached. In this last case, quasi-stationarity is less interesting (cf. Day and Possingham 1995; Gosselin 1998b, c).

Thus, far from invalidating or challenging past PVA simulations, our theoretical results will help (1) to render explicit some underlying qualitative behaviors at work in simulation models (cf. White Stork example); (2) to frame the simulation (e.g., by illuminating which quantities should be studied, which estimation procedures should be used, . . .); and (3) to compare different PVA, as a result of the similar theoretical framework.

Further Steps Required to Use BPs More Efficiently in PVA

Any further progress will first require that some conjectures are solved. The most urgent one concerns the situation in which there is both a finite number of types and a random environment, the existence of a QSD under more general conditions than above. Results about a random environment with temporal autocorrelation, as studied by simulation by Mode and Jacobson (1987a, b), are particularly critical.

Additional technical tools would also be needed for the use of BPs in PVA simulations. For instance, a procedure that would reflect when the asymptotic behavior (see above) is reached with a given precision, would be very useful, as well as an accurate procedure for the estimation of the asymptotic growth rate.

More globally, the user may need to have a clearer picture of the assumptions, typical simulations, and theoretical results of different kinds of stochastic extinction models. Other kinds of stochastic population models can be clustered roughly into stochastic difference equation models (DeAngelis 1976), birth and death processes (Cohen 1969; Wissel 1989; Wissel and Stöcker 1991; Wissel and Zaschke 1994), diffusion approximations (Lande and Orzack 1988; Dennis et al. 1991), and stochastic patch-occupancy metapopulation models (Gyllenberg and Silvestrov 1994; Hanski 1994; Day and Possingham 1995; Gosselin 1998c) or other finite state Markov chains (Verboom et al. 1991). A first comparison between different continuous time models was performed by Durrett and Levin (1994).

We complement this analysis by three remarks. First, BPs and diffusion approximations lead to qualitatively different results, because diffusion approximations do not induce an asymptotically geometric probability distribution of the time to extinction (Lande and Orzack 1988; Dennis et al. 1991). Second, continuous time and discrete state homogeneous Markov chains are, when considered at regular discrete times, discrete-time Markov chains (e.g., Reddingius 1971). Thus, the framework presented here and that studied, for example, in the work of Gosselin (1998d) embrace many continuous-time stochastic frameworks (e.g., Wissel 1989; Wissel and Stöcker 1991; Mangle and Tier 1993; Wissel and Zaschke 1994). Third, stochastic patch-occupancy models and other finite absorbing Markov chain population models have the same theoretical properties as subcritical BPs (i.e., certain extinction and quasi-stationarity; cf. Verboom et al. 1991; Day and Possingham 1995; Gosselin 1998c).

Conclusion

The canonical structure of BPs and the theoretical results available allow us to handle, within the same conceptual framework, PVA models with a variety of demographic features. The reader interested in applying BPs structures and results to a specific PVA has the choice between developing his or her own model and using a generic extinction simulation tool (for a review of some of these models, see Lindenmayer et al. 1995). Among generic simulation tools, the latest version of the software ULM (Legendre and Clobert 1995) takes explicitly into account the theoretical results presented here.

Last, the tuning of a BP model to a specific population-environment system based on empirical data will meet problems of parameter estimation in small populations and of detection and assessment of a specific functional form of density dependence. These questions, which depend critically on the quality of the data available, are, however, not specific to BPs. Classically, the range of strategies thus spreads from a detailed PVA model, relying on extensive data (as in, e.g., Woolfenden and Fitzpatrick 1991), to a series of different scenarios corresponding to varying assumptions and parameter values, when data are not sufficient.

Acknowledgments. This chapter is partly based on a talk at the Second European Congress of Mathematics Applied to Biology and Medicine (Lyon, France, December 1993). We thank Claude Millier (Scientific Director, ENGREF) and Jean Jacod (Probability Lab, Paris VI University) for their cooperation in this project and for comments on a former version of the manuscript.

Literature Cited

Asmussen S, Hering H (1983) Branching processes. Birkhäuser, Boston, MA

Athreya KB, Karlin S (1971) Branching processes with random environments. I Annals of Mathematical Statistics 42:1499–1520; II Annals of Mathematical Statistics 42:1843–1858

Athreya KB, Ney PE (1972) Branching processes. Springer Verlag, New York

Bagley JH (1982) Asymptotic properties of subcritical Galton-Watson processes. Journal of Applied Probability 19:510–517

Bairlein F (1991) Population studies of White Storks *Ciconia ciconia* in Europe. In: Perrins CM, Lebreton J-D, Hirons GJM (eds) Bird population studies: relevance to conservation and management. Oxford University Press, Oxford, UK, pp 207–229

Bart J (1995) Evaluation of population trend estimates calculated using capture recapture and population projection methods. Ecological Applications 5:662–671

Beissinger SR (1995) Modeling extinction in periodic environments: Everglades water levels and Snail Kite population viability. Ecological Applications 5:618–631

Box GEP, Jenkins GM (1970) Time series analysis forecasting and control. Holden-Day, San Francisco, CA

Boyce MS (1992) Population viability analysis. Annual Review of Ecology and Systematics 23:481–506

Burkey TV (1989) Extinction in nature reserves: the effect of fragmentation and the importance of migration between reserve fragments. Oikos 55:75–81

Caswell H (1989) Matrix population models. Sinauer, Sunderland, MA

Chesson P (1978) Predator-prey theory and variability. Annual Review of Ecology and Systematics 9:323–347

Cohen JE (1969) Natural primate troops and a stochastic population model. American Naturalist 103:455–477

Day JR, Possingham HP (1995) A stochastic metapopulation model with variability in patch size and position. Theoretical Population Biology 48:333–360

DeAngelis DL (1976) Application of stochastic models to a wildlife population. Mathematical Biosciences 31:227–236

Dennis B, Munholland PL, Scott JM (1991) Estimation of growth and extinction parameters for endangered species. Ecological Monographs 61:115–143

Durrett R, Levin S (1994) The importance of being discrete (and spatial). Theoretical Population Biology 46:363–394

Eberhardt LL (1985) Assessing the dynamics of wild populations. Journal of Wildlife Management 49:997–1012

Facelli JM, Pickett STA (1990) Markovian chains and the role of history in succession. TREE 5:27–30

Gabriel W, Bürger R (1992) Survival of small populations under demographic stochasticity. Theoretical Population Biology 41:44–71

Galton F (1873) Problem 4001. Educational Times 17

Gilpin ME (1987) Spatial structure and population variability. In: Soulé ME (ed) Viable populations for conservation. Cambridge University Press, Cambridge, UK, pp 125–140

Gilpin ME, Soulé ME (1986) Minimum viable populations: the processes of species extinctions. In: Soulé ME (ed) Conservation biology: science of scarcity and diversity. Sinauer, Sunderland, MA, pp 13–34

Ginzburg LR, Slobodkin LB, Johnson K, Bindman AG (1982) Quasiextinction probabilities as a measure of impact on population growth. Risk Analysis 2:171–181

Goodman D (1987) The demography of chance extinction. In: Soulé ME (ed) Viable populations for conservation. Cambridge University Press, Cambridge, UK, pp 11–34

Gosselin F (1996) Extinction in a simple source/sink system: application of new mathematical results. Acta Oecologica 17:563–584

Gosselin F (1997) Modèles stochastiques d'extinction de population: propriétés mathématiques et leurs applications. Unpublished PhD thesis, Paris 6 University, Paris

Gosselin F (1998a) Asymptotic behaviour of some discrete-time Markov chains conditional on non-extinction. I-Theory. Mimeographed research report 98–04. Biometrics Unit, INRA/ENSAM/University Montpellier II, France

Gosselin F (1998b) Asymptotic behaviour of some discrete-time Markov chains conditional on non-extinction. II-Applications. Mimeographed research report 98–05, Biometrics Unit, INRA/ENSAM/University Montpellier II, France

Gosselin F (1998c) Reconciling theoretical approaches to stochastic patch-occupancy metapopulation models. Bulletin of Mathematical Biology 60:955–971

Gosselin F (1998d) Asymptotic behavior of some discrete time Markov chains conditional on non-extinction. I-Theory; II-Applications. Technical Reports 98-04, 98-05. Groupe de Biostatistique et d'Analyse des Systèms. Université de Montpellier II, France

Gyllenberg M, Silvestrov DS (1994) Quasi-stationary distributions of a stochastic metapopulation model. Journal of Mathematical Biology 33:35–70

Hanski I (1994) A practical model of metapopulation dynamics. Journal of Animal Ecology 63:151–162

Hanski I, Woiwod IP (1993) Spatial synchrony in the dynamics of moth and aphid populations. Journal of Animal Ecology 62:656–668

Heathcote CR, Seneta E, Vere-Jones D (1967) A refinement of two theorems in the theory
of branching processes. Theory of Probability and Its Applications 12:342–346

Jagers P (1975) Branching processes with biological applications. Wiley, London

Joffe A, Spitzer F (1967) On multitype branching processes with $\rho \leq 1$. Journal of Mathematical Analysis and Applications 19:409–430

Kanyamibwa S (1991) Dynamique des populations de Cigogne Blanche (*Ciconia Ciconia
L*) en Europe Occidentale: contribution à la conservation des populations naturelles.
Unpublished thesis, Montpellier II University, Montpellier, France

Kanyamibwa S, Lebreton JD (1992) Variation des effectifs de Cigogne Blanche et facteurs
du milieu: un modèle démographique. In: Mériaux JL, Schierer A, Tombal C, Tombal JC
(eds) Les cigognes d'Europe. Institut Européen d'Ecologie, Metz, France, pp 259–264

Kanyamibwa S, Schierer A, Pradel R, Lebreton J-D (1990) Changes in adult survival rates
in a western European population of the White Stork *Ciconia ciconia*. Ibis 132:27–35

Kanyamibwa S, Bairlein F, Schierer A (1993) Comparison of survival rates between
populations of the White Stork *Ciconia ciconia* in central Europe. Ornis Scandinavica
24:297–302

Lande R (1988) Genetics and demography in biological conservation. Science 241:1455–
1460

Lande R (1993) Risks of population extinction from demographic and environmental
stochasticity and random catastrophes. American Naturalist 142:911–927

Lande R, Orzack SH (1988) Extinction dynamics of age-structured populations in a fluctuating environment. Proceedings of the National Academy of Sciences of the USA
85:7418–7421

Lebreton J-D (1978) Un modèle probabiliste de la dynamique des populations de la
Cigogne Blanche (*Ciconia ciconia L*) en Europe Occidentale. In: Legay JM, Tomassone
R (eds) Biométrie et Ecologie. Société de Biométrie, Paris, pp 277–343

Lebreton J-D (1981) Contribution á la dynamique des populations d'oiseaux. Modèles
mathématiques en temps discret. Unpublished thesis, Lyon I University, Villeurbanne,
France

Lebreton J-D (1982) Applications of discrete time branching processes to bird population
dynamics modelling. In: ANAIS da 10 ª conferência Internacional de Biometria.
EMBRAPA-DID/DMQ/Sociedade Internacional de Biometria, Brasil, pp 115–133

Lebreton J-D (1990) Modelling density dependence environmental variability and
demographic stochasticity from population counts: an example using Wytham Wood
Great Tits. In: Blondel J, Gosler A, Lebreton J-D, McCleery R (eds) Population biology
of passerine birds: an integrated approach. NATO ASI series. Series G: Ecological
sciences, vol 24. Springer Verlag, Berlin, pp 89–102

Lebreton J-D, Clobert J (1991) Bird population dynamics management and conservation:
the role of mathematical modeling. In: Perrins CM, Lebreton J-D, Hirons GJM (eds)
Bird population studies: relevance to conservation and management. Oxford University
Press, Oxford, UK, pp 105–125

Legendre S, Clobert J (1995) ULM: a software for conservation and evolutionary biologists. Journal of Applied Statistics 22:817–834

Lindenmayer DB, Burgman MA, Akçakaya HR, Lacy RC, Possingham HR (1995) A
review of the generic computer programs ALEX RAMAS/space and VORTEX for
modelling the viability of wildlife metapopulations. Ecological Modelling 82:161–174

Malthus TR (1798) An essay on the principle of population, as it affects the future improve-
    ments of society, with remarks on the speculations of Mr. Godwin, M. Condorcet, and
    other writers. John Murray, London
Mangle M, Tier C (1993) Dynamics of metapopulations with demographic stochasticity
    and environmental catastrophes. Theoretical Population Biology 44:1–31
McCarthy MA, Franklin DC, Burgman MA (1994) The importance of demographic uncer-
    tainty: an example from the Helmeted Honeyeater. Biological Conservation 67:135–142
Mode CJ, Jacobson ME (1987a) A study of the impact of environmental stochasticity on
    extinction probabilities by Monte Carlo integration. Mathematical Biosciences 83:105–
    125
Mode CJ, Jacobson ME (1987b) On estimating population size for an endangered species in
    the presence of environmental stochasticity. Mathematical Biosciences 85:185–209
Mode CJ, Pickens GT (1986) Demographic stochasticity and uncertainty in population
    projections—a study by computer simulation. Mathematical Biosciences 79:55–72
North PM, Boddy AW, Forrester DR (1988) A computer simulation study of stochastic
    models to investigate the population dynamics of the Screech Owl (*Otus asio*) under
    increased mortality. Ecological Modelling 40:233–263
Nunney L, Campbell KA (1993) Assessing minimum viable population size: demography
    meets population genetics. TREE 8:234–239
Pollard JH (1966) On the use of the direct matrix product in analysing certain stochastic
    population models. Biometrika 53:397–415
Reddingius J (1971) Gambling for existence. Acta Biotheoretica 20 Suppl:1–208
Schneider RR, Yodzis P (1994) Extinction dynamics in the American Marten (*Martes
    americana*). Conservation Biology 8:1058–1068
Seneta E, Vere-Jones D (1966) On quasi-stationary distributions in discrete-time Markov
    chains with a denumerable infinity of states. Journal of Applied Probability 3:403–434
Shaffer M (1987) Minimum viable populations: coping with uncertainty. In: Soulé ME (ed)
    Viable populations for conservation. Cambridge University Press, Cambridge, UK, pp
    69–86
Shaffer ML (1981) Minimum population sizes for species conservation. Bioscience
    31:131–134
Smith WL, Wilkinson WE (1969) On branching processes in random environments. Annals
    of Mathematical Statistics 40:814–827
Stacey PB, Taper M (1992) Environmental variation and the persistence of small popula-
    tions. Ecological Applications 2:18–29
Tuljapurkar S (1990) Population dynamics in variable environments. Lecture Notes in
    Biomathematics 85. Springer-Verlag, New York
Verboom J, Lankester K, Metz JAJ (1991) Linking local and regional dynamics in stochas-
    tic metapopulation models. Biological Journal of the Linnean Society 42:39–55
Wissel C (1989) Metastability a consequence of stochastics in multiple stable population
    dynamics. Theoretical Population Biology 36:296–310
Wissel C, Stöcker S (1991) Extinction of populations by random influences. Theoretical
    Population Biology 39:315–328
Wissel C, Zaschke S-H (1994) Stochastic birth and death processes describing minimum
    viable populations. Ecological Modelling 75/76:193–201
Woolfenden GE, Fitzpatrick JW (1991) Florida Scrub Jay ecology and conservation. In:
    Perrins CM, Lebreton J-D, Hirons GJM (eds) Bird population studies: relevance to
    conservation and management. Oxford University Press, Oxford, UK, pp 542–565

Yule GU (1924) A mathematical theory of evolution based on the conclusions of Dr JC Willis, FRS. Philosophical Transactions of the Royal Society of London B 213:21–87

Appendix 1. Notation and Abbreviations

In this chapter, we denote by $N = \{0, 1, 2, \ldots\}$ and $N^* = \{1, 2, \ldots\}$ the set of non-negative and positive integers, respectively. Time t takes discrete values, in N.

We denote by $\Pr(A)$ the probability of the event A and by $E(X)$ the expectation of the random variable X (if it is defined). If $\Pr(B) > 0$, we denote by $\Pr(A \mid B) = \dfrac{\Pr(A \cap B)}{\Pr(B)}$ the conditional probability of A with respect to B. When both X and Y are random variables, we denote by $E(X \mid Y)$ the conditional expectation of X relative to Y, a classical notion in probability theory.

Z_t denotes the random value of population size at time t. Note that either the branching process (BP) has only one type of individuals and Z_t takes non-negative integer values or the BP has at least two types of individuals (i.e., the BP is multitype [MT]) and Z_t is a random vector, whose components are non-negative integers representing the number of individuals in each type. We also used $X_{t,j}$ to denote the offspring number at the next time $t + 1$ of the jth individual in the population at time t. When the BP is density dependent, we denote by $_{(i)}X_{t,j}$ the offspring number at the next time $t + 1$ of the jth individual in a population of size i at time t. We then denote by $m = E(X_{t,j})$ and $_{(i)}m = E(_{(i)}X_{t,j})$ the associated expectations. In a BP in random environment, m_t is the (random) mean offspring number at time t conditional on the value of the environmental process ζ_t at time t. The equivalent quantity in a MT BP is a matrix, that we denoted by $M = (m_{\alpha,\beta})_{1 \leq \alpha, \beta \leq d}$, whose dominant eigenvalue, μ, describes the asymptotic behavior of the BP. We denote by $(b_k)_{k \in N^*}$ the quasi-stationary distribution (QSD) of a BGW BP and by μ the asymptotic growth rate of a BP.

Finally, we used the following abbreviations in this chapter:

- PVA, population viability analysis
- BP, (discrete-time) branching process
- BGW BP, Bienaymé-Galton-Watson branching process
- QSD, quasi-stationary distribution
- MT BP, multitype Bienaymé-Galton-Watson branching process
- DD BP, is density-dependent Bienaymé-Galton-Watson branching process
- BP RE, is Bienaymé-Galton-Watson branching process in random environment

Appendix 2. Basic Results on Markov Chains with a Denumerable Number of States

Let $(Y_t)_{t \in N}$ be a sequence of random variables with values in N. This sequence of random variables is an homogeneous Markov chain if for every t in N and (i_{t+1}, i_t) in N,

$$P_{i_t, i_{t+1}} = \Pr(Y_{t+1} = i_{t+1} \mid Y_t = i_t) = \Pr(Y_{t+1} = i_{t+1} \mid Y_t = i_t, Y_{t-1}, \ldots, Y_0)$$

Thus, for every t and j in N, we have

$$\Pr(Y_{t+1} = j) = \sum_{i=0}^{\infty} \Pr(Y_t = i)\, p_{i,j}$$

that is, in matrix terms

$$\begin{pmatrix} \Pr(Y_{t+1} = 0) \\ \Pr(Y_{t+1} = 1) \\ \vdots \\ \Pr(Y_{t+1} = j) \\ \vdots \end{pmatrix} = \begin{pmatrix} p_{0,0} & p_{1,0} & \cdots & p_{i,0} & \cdots \\ p_{0,1} & p_{1,1} & \cdots & p_{i,1} & \cdots \\ \vdots & \vdots & \vdots & \vdots & \vdots \\ p_{0,j} & p_{1,j} & \cdots & p_{i,j} & \cdots \\ \vdots & \vdots & \vdots & \vdots & \vdots \end{pmatrix} \begin{pmatrix} \Pr(Y_t = 0) \\ \Pr(Y_t = 1) \\ \vdots \\ \Pr(Y_t = i) \\ \vdots \end{pmatrix} = P \begin{pmatrix} \Pr(Y_t = 0) \\ \Pr(Y_t = 1) \\ \vdots \\ \Pr(Y_t = i) \\ \vdots \end{pmatrix} = P^{t+1} \begin{pmatrix} \Pr(Y_0 = 0) \\ \Pr(Y_0 = 1) \\ \vdots \\ \Pr(Y_0 = i) \\ \vdots \end{pmatrix}$$

where P is the following infinite matrix

$$P = \begin{pmatrix} p_{0,0} & p_{1,0} & \cdots & p_{i,0} & \cdots \\ p_{0,1} & p_{1,1} & \cdots & p_{i,1} & \cdots \\ \vdots & \vdots & \vdots & \vdots & \vdots \\ p_{0,j} & p_{1,j} & \cdots & p_{i,j} & \cdots \\ \vdots & \vdots & \vdots & \vdots & \vdots \end{pmatrix}$$

and $P^{t+1} = P P \ldots P$, where there are $t + 1$ terms in the product. P is called the transition matrix of the Markov chain $(Y_t)_{t \in N}$.

In a monotype branching process (BP), the sequence of random population sizes $(Z_t)_{t \in N}$ is a Markov chain with transition matrix

$$P = \begin{pmatrix} 1 & p_{1,0} & \cdots & p_{i,0} & \cdots \\ 0 & p_{1,1} & \cdots & p_{i,1} & \cdots \\ \vdots & \vdots & \vdots & \vdots & \vdots \\ 0 & p_{1,j} & \cdots & p_{i,j} & \cdots \\ \vdots & \vdots & \vdots & \vdots & \vdots \end{pmatrix}$$

where (1) 0 is an absorbing state; (2) as a consequence of the branching property, the $(i + 1)$th column of P represents the ith convoluate of the probability distribution of the random variables X in the density-independent case and $_{(i)}X$ in the density-dependent case.

We then denote by Q the substochastic matrix, which is the restriction of P to the nonextinction state space: $Q = (p_{i,j})_{i \geq 1, j \geq 1}$:

$$Q = \begin{pmatrix} p_{1,1} & \cdots & p_{i,1} & \cdots \\ \vdots & \vdots & \vdots & \vdots \\ p_{1,j} & \cdots & p_{i,j} & \cdots \\ \vdots & \vdots & \vdots & \cdots \end{pmatrix}$$

Then, in the case of the subcritical Bienaymé-Galton-Watson BP, if we skip the technicalities of eigenvalues and vectors of infinite matrices, $(1, 0, 0, \dots)$ is an eigenvector of the matrix P, associated with the dominant eigenvalue 1, and the quasi-stationary distribution $(b_k)_{k \in N}*$ is an eigenvector of Q associated with the dominant eigenvalue of Q, m. The same holds, for example, for density-dependent BP, under further conditions, by replacing m by the asymptotic growth rate λ.

14

Role of Genetics in Conservation Biology

Sabine S. Loew

Introduction

Many species are represented solely by populations that are highly fragmented, isolated, or captive and therefore lack the number of individuals considered necessary for healthy sustainable populations. These species run an increased risk of local and global extinction due to environmental, demographic, and genetic stochasticity (Shaffer 1981) and may benefit from special management attention. Two major goals of conservation genetics are to minimize loss of genetic variation in managed populations and define the taxonomic units (i.e., rare populations, subspecies, species) we ought to conserve. In this chapter, I outline the rationale for managing genetic diversity, introduce molecular tools available to assess genetic variation and taxonomic relationships, and provide examples illustrating applications of population genetics to conservation biology. This chapter is not an exhaustive review. The discussion applies to many captive and natural threatened populations ranging from insects to primates. The apparent bias among the examples toward mammals and birds is not a value statement but simply a reflection of the general interest in conservation biology toward higher vertebrates as well as my own background in mammalian biology.

Concept of Genetic Diversity

The genetic composition of a population can be described by the alleles present at different loci for a representative sample of individuals within a population. Genetic diversity can be quantified by the number and distribution of alleles within and between individuals and populations. Diploid organisms have two alleles per locus. Individuals are considered homozygous at this locus if both alleles are the same and heterozygous if they are different from one another. Genetic variation can be described at a single locus or multiple loci by using several methods (Lande and Barrowclough 1987; Lacy et al. 1995; Hamrick and Godt 1996). For example, a polymorphic locus is defined as having several alleles with the population frequency of the most common allele being smaller than 0.99. The proportion of polymorphic loci, in turn, is a measure of the number of variable loci among all sampled loci within a population. Genetic variation may also be described by quantitative variation in traits derived from the actions (and interactions) of many genes, termed *quantitative genetic variation*.

The most frequently used measure of populationwide genetic diversity is the amount of heterozygosity. Individual heterozygosity describes the observed proportion of heterozygous loci in an individual (Mitton and Pierce 1980), and average heterozygosity reflects the proportion of heterozygous individuals in a population measured across several loci (Hartl and Clark 1989). The theoretical predictions based on the Hardy Weinberg Principle provide the framework within which to evaluate the amount and distribution of genetic variation documented in natural populations. Deviations from the expected genotype frequencies within a population, for example, can be indicative of past bouts of strong selection or inbreeding.

To evaluate how loss of genetic diversity affects population survival, it is crucial to distinguish the significance of different types of genetic variation. For example, single-locus diversity is measured by individual heterozygosity, whereas diversity associated with polygenic quantitative traits is measured by phenotypic variation. Empirical studies suggest that most phenotypic changes in a population are the result of small alterations at numerous loci rather than a consequence of major mutations at a single locus (Lande 1981; Lande and Barrowclough 1987; Lande 1995). Consequently, the adaptive potential of a population may depend more on variation of quantitative traits determined by multiple loci than on single-locus polymorphisms (Lynch 1996).

Lande and Barrowclough (1987) contrast the adaptive importance of these types of genetic variation and discuss their maintenance in the context of neutral and stabilizing selection. They suggest that a population of several hundred individuals is necessary to maintain amounts of quantitative genetic variation necessary for evolution. Similarly, Lynch (1996) emphasizes that quantitative genetics focuses on the evolutionary properties of morphological and behavioral traits and therefore can provide insights into the effects of small population size on fitness and ultimately extinction risks. He goes further than Lande and Barrowclough (1987) in his suggestion that maintenance of the adaptive genetic variation of populations requires more than 10,000 reproductive individuals and that current conservation policies leave most endangered species at risk of losing genetic integrity.

A general prediction from population genetics theory is that existing heterozygosity erodes by 50% within $1.39N_e$ generations, where N_e is the effective size of a randomly mating population (Wright 1931; Hartl and Clark 1989; see definition below). Small populations run a greater risk of becoming genetically depauperate and, therefore, are at the center of attention of conservation genetics. In the following sections, I discuss the merits of genetic diversity, tools to assess genetic variation and phylogenetic uniqueness, factors affecting genetic diversity, and the question of what to preserve.

Merits of Genetic Diversity

Genetic Diversity and Adaptation

Long-term survival of a species depends on its adaptation to current and future biotic and abiotic aspects of its environment. Natural selection results in the survival and propagation of those individuals that are best adapted to prevailing conditions. In a genetically diverse population, individual fitnesses vary, and differential survival affects gene frequencies within and between populations. On changes in the environment, natural selection may favor different genotypes and hence alter the distribution of gene frequencies. However, a population depauperate of genetic variation may not carry any individuals that are genetically pre-adapted to the new environment and hence runs an increased risk of extinction. To maximize the probability of long-term survival of species, especially in changing environments, conservation geneticists seek to maintain high genetic diversity, although this variation may be represented by neutral single locus variation or by variation in fitness-related quantitative traits.

Genetic Diversity and Inbreeding Depression

Inbreeding (nonrandom mating with respect to relatedness) skews genotype frequencies within populations toward increased frequencies of homozygotes. Inbreeding in combination with selection against homozygotes can reduce the reproductive performance of naturally outbreeding populations (Wright 1977; Falconer and Mackay 1996) and therefore may decrease long-term survival (Ralls et al. 1988; Thornhill 1993; Frankham 1995c, 1998). In general, inbred offspring are expected to be less fit than offspring produced by random matings (Ralls et al. 1979, 1988; Thornhill 1993; Falconer and Mackay 1996). This reduction of fitness with inbreeding (inbreeding depression) is manifested by reduced growth rate, fertility, fecundity, survival, developmental stability, or changed mating behavior among inbred offspring (Lerner 1954; Wright 1977; Ralls and Ballou 1982; Miller et al. 1993; Keller et al. 1994; Rave et al. 1994; Falconer and Mackay 1996). Likely mechanisms for inbreeding depression have been debated for decades (Charlesworth and Charlesworth 1987; Shields 1993; Thornhill 1993), and the two competing explanations (among others such as partial and associative overdominance), the overdominance and dominance hypotheses, are compatible with many theoretical and empirical results. Proponents of overdominance argue that heterozygous genotypes are on average fitter than homozygotes and as inbreeding reduces heterozygote frequencies, it depresses population performance (Mitton 1993). The dominance hypothesis contends that inbreeding depression is due to the increased expression of recessive deleterious alleles resulting from an increase in the number of homozygous loci.

Distinction of these hypotheses is not merely of academic value but could have implications for conservation management, especially for small inbred populations. If inbreeding depression is a consequence of the reduction of heterozygous individuals, mating among relatives will result in a population with less fit individuals, irrespective of its mating history. By contrast, inbreeding should no longer result in depression of individual fitness if recessive deleterious alleles underlie inbreeding depression and can be purged from populations with regular inbreeding (Lande 1988; Hedrick 1994; Fu et al. 1998). Byers and Waller (1999) reviewed evidence for the purging hypothesis among plant populations and concluded that "purging appears neither consistent nor effective enough to reliably reduce inbreeding depression in small and inbred populations."

Lande (1995) suggested that inbreeding depression is the result of segregation of deleterious alleles and that there is little evidence to support the overdominance hypothesis. He concluded that gradual inbreeding will only reduce the risk of inbreeding depression due to lethal and sublethal mutations but will not succeed in purging mildly deleterious alleles (Hedrick 1994; Lande 1995). This prediction is consistent with Ballou's (1995) results that a long history of inbreeding in 25 captive populations resulted in a slight reduction but not complete removal of inbreeding depression.

Experimental studies of inbreeding effects in wild populations are rare and have produced mixed results. Brewer and co-workers (1990) tested the prediction of the dominance hypothesis that small isolated populations of White-footed Mice with low genetic diversity would show less inbreeding depression than large central populations. The severity of fitness depression in inbred litters did not correlate with genetic diversity, and deleterious effects did not diminish through several generations of inbreeding (support for overdominance hypothesis). However, maternal care was only depressed in inbred mothers from genetically diverse populations (support for dominance hypothesis). Brewer and associates (1990) concluded that "overdominance of fitness traits probably contributed as much to the genetic load as did deleterious recessive alleles."

Keane (1990a, b) bred wild-caught mice from a single population. Whereas significant inbreeding depression was documented for full-sibling and half-sibling matings, mating among cousins resulted in high reproductive success, and there was potential for outbreeding depression (fitness reduction due to matings among genetically distant individuals). Jimenez and colleagues (1994) documented severe inbreeding depression in White-footed Mice when they reintroduced inbred offspring of wild-caught mice into natural habitat. Inbred mice showed continual weight loss and suffered higher mortality than noninbred mice after release. These deleterious effects of inbreeding were much more severe in the natural environment than in captivity. In a "natural experiment" involving a Song Sparrow population, outbred individuals had a higher probability of survival during severe population crashes. Keller and associates (1994) concluded that environmental and genetic effects on survival interact and that inbreeding depression among the Song Sparrow population was expressed when the population experienced environmental stress (i.e., severe winter weather). These studies reiterate a point made by Hedrick and Miller (1992) who caution that inbreeding depression may be greater in natural populations than in laboratory animals due to more severe environmental conditions.

Genetic Diversity at Selected Sites: The Case of the Major Histocompatability Complex

Proponents of the neutral theory argue that most genetic variability at the genomic or molecular level is selectively neutral and allele frequencies within populations are merely a function of mutation rate and the effective population size (Kimura 1968; Kimura and Ohta 1971). Neutralists nevertheless agree that deleterious alleles are eliminated through directional selection and that adaptive evolution is mediated through fitness differences among alleles (for further discussion, see Li 1997).

The selectionist-neutralist debate has spilled into the conservation biology arena in discussions about the kind of genetic variation that ought to be preserved in endangered taxa (Hughes 1991; Vrijenhoek and Leberg 1991; Vrijenhoek 1994; Avise 1995; Miller 1995; Lynch 1996). Unfortunately, the precise relationships between most fitness-related traits and genetic diversity at specific allozyme loci, DNA, or quantitative markers continue to be elusive. Nevertheless, a recent debate focused on conservation management for maximizing heterozygosity at fitness-related traits, such as the major histocompatibility complex (MHC) (Hughes 1991; Vrijenhoek and Leberg 1991; Hedrick and Miller 1994).

MHC molecules play a key role in the immune response of mammals and birds (possibly all vertebrates) and have been linked to kin recognition based on individual odor profiles in mice (Yamazaki et al. 1979, 1983; Klein 1986; Egid and Brown 1989; Potts et al. 1991; Brown and Eklund 1994) and possibly humans (Wedekind et al. 1995). MHC genes encode cell-surface proteins that bind foreign molecules and aid in recognition and elimination of these potentially harmful antigens. Extensive allelic diversity at MHC loci has been documented for most populations studied (Klein 1986; Nei and Hughes 1991). Notable exceptions are the virtually monomorphic Syrian Hamster, mouse populations on North Sea islands, and the Cheetah, all of which probably lost overall genetic diversity due to severe population crashes or small population sizes for many generations (Streilein et al. 1984; McGuire et al. 1985; O'Brien et al. 1985). A variety of mechanisms for maintaining MHC polymorphism have been proposed in the past few decades (for reviews see Potts and Wakeland 1990; Nei and Hughes 1991; Alberts and Ober 1993; Klein et al. 1993; Parham and Ohta 1996). The major explanations include (1) maternal–fetal interactions (Clarke and Kirby 1966; Hedrick and Thomson 1988), (2) disassortative mating preference based on MHC genotypes (Yamazaki et al. 1976; Hedrick 1992a; Manning et al. 1992; Wedekind et al. 1995), (3) disease-based overdominance leading to an increased chance of survival in environments with infectious diseases (Doherty and Zinkernagel 1975; Hughes and Nei 1988; Nei and Hughes 1991), and (4) disease-based frequency dependence based on the selective advantage of new mutant alleles (Snell 1968; Bodmer 1972).

The role of pathogen-mediated mechanisms in maintaining MHC variability is of particular importance to conservation biology. If current high levels of MHC variability are a consequence of disease-based overdominance selection in the past, then new pathogens may pose a significant threat to the future survival of populations with low MHC variability. Host-parasite coevolution as a selection mechanism has been supported by studies that have documented a relationship between MHC haplotypes and disease resistance such as Marek's disease and fowl cholera in chicken and malaria in humans (Briles et al. 1977; Lamont et al. 1987; O'Brien and Everman 1988; Hill et al. 1991). Slade (1992) attributed low MHC variability in Southern Elephant Seals to their evolution in a pathogen-free marine environment. Reduced genetic diversity at MHC loci has also been linked to disease susceptibility in Cheetahs (O'Brien et al. 1985).

This evidence and the apparent positive selection for a balanced MHC polymorphism led Hughes (1991) to believe that MHC diversity is paramount to the survival of small endangered populations. He then made the controversial argument that "all captive breeding programs for endangered vertebrate species should be designed with the preservation of MHC allelic diversity as their goal" (Hughes 1991). However, Vrijenhoek and Leberg (1991) cautioned that Hughes' (1991) management recommendations are based on assumptions that need experimental evaluation before their general validity is established. Selective breeding for maintenance of MHC diversity could, for example, increase loss of whole genomic diversity and hence bear the risk of inbreeding depression (Haig et al. 1990; Falconer and Mackay 1996). In addition, disease protection is not only a consequence of the adaptive immune response of organisms and hence the diversity of MHC molecules but involves numerous other genes associated with the humoral immune system (Janeway and Travers 1994). Vrijenhoek and Leberg (1991) recommended preservation of genetic diversity at the level of the whole genome instead of selected loci while intensifying efforts to monitor and understand the relationship between disease susceptibility and MHC variability.

Critics point out that several populations with low MHC variability are viable and healthy, such as northern European Beaver populations (Ellegren et al. 1993). However, the apparent well-being of a population with low genetic variability at either neutral or selected loci may be (1) the result of selection in the past and survivors represent the fittest genotypes; (2) because low variability is the result of drift or inbreeding in combination with selection, and although the population is healthy now, it lacks the potential for future adaptations; or (3) because there is actually no correlation between fitness and genetic variability. Similarly, a positive correlation between low fitness and low genetic diversity does not necessarily mean a causal relationship exists. Instead low reproductive success may be explained by nongenetic factors, such as predation, low densities, or disrupted behavior.

Tools to Assess Genetic Diversity and Phylogenetic Uniqueness

Molecular tools can contribute to conservation biology by revealing the genetic structure of the population, evolutionary history, and evolutionary potential of taxa (Ashley 1999). This information can serve to detect past, present, and possibly future population declines, besides establishing evolutionary uniqueness. Natural populations of threatened species are managed to conserve genetic diversity to enhance individual fitness and maintain the evolutionary potential for future adaptations. Such genetic management generally requires information on demography (e.g., migration rates, population size). The amount and distribution of genetic variation within and between populations can provide relatively quick indirect estimates of migration, population subdivisions, and isolation, which can serve to define the appropriate scale for short- and long-term management (Moritz 1994b). Furthermore, characterization of the genetic make-up (i.e., the pedigree) of captive individuals can help to determine appropriate breeders to maximize offspring survival through inbreeding avoidance and genetic compatibility (Ryder 1986; Garner and Ryder 1992).

The proliferation of genetic markers in recent years has facilitated reliable estimates of some forms of genetic diversity and phylogenetic analysis at several taxonomic levels. The genetic variation detected by these molecular markers differs quantitatively and qualitatively as a result of the kind and number of genomic sites they assay. Protein electrophoresis uncovers putative genetic variation associated with protein coding regions, whereas DNA-DNA hybridization, restriction fragment length polymorphism (RFLP), and DNA sequencing analysis reveal differences at the DNA level for coding as well as noncoding regions. DNA sequencing data generally provide genetic distance estimates among taxa based on single locus comparisons whereas the other methods reveal variation at several loci. Resolution power of any molecular marker depends on the number of independent linkage groups (sites) it assays and the evolutionary rate of change associated with those groups. Animal and plant mitochondrial DNA (mtDNA), for example, usually is maternally inherited as one nonrecombining linkage group. Although the rate of sequence evolution varies along the mtDNA molecule, it is generally higher than the rate of single-copy nuclear DNA (scnDNA) (Moritz et al. 1987). The intramolecular variability of the rate of evolution makes mtDNA ideally suited to resolve taxonomic differences at various levels of divergence. In addition maternal inheritance uniquely qualifies mitochondrial markers for tracing of maternal genealogies. Lack of recombination, however, renders all mitochondrial genes part of one linkage group that are subjected to the same stochastic (e.g., random lineage extinction) and deterministic events (e.g., selective sweeps). Therefore they do not provide independent estimates of evolutionary change.

Choice of the appropriate marker for a particular task has to be based on the resolution power necessary to determine genetic differences and similarities among the individuals, populations, or other taxonomic units of interest. In addition, quality and quantity of the DNA source as well as to the required sample size have to be taken into consideration as these factors affect the applicability and cost-efficiency of a particular molecular tool. In general, rapidly evolving DNA (e.g., mini- and microsatellite DNA) generates genetic differences among individuals and populations that provide information on individual identity, paternity determination, and intraspecific genetic isolation among populations. Genetic markers with moderate evolutionary rates (e.g., allozymes, mtDNA control region) are used to resolve genetic distances at intermediate taxonomic levels. Resolution of deep branching patterns or divergence among distantly related taxa requires highly conserved regions (e.g., ribosomal RNA genes) within which evolutionary changes accumulate slowly (for reviews, see Moritz et al. 1987; Avise 1994; Simon et al. 1994; Avise et al. 1995). Following is a brief introduction to commonly used genetic tools and some examples of application in conservation biology (see Avise and Hamrick [1996] for an excellent recent collection of case studies in conservation genetics).

Protein Electrophoresis (Lewontin and Hubby 1966)

Electrochemical differences among proteins allow separation in either starch or acrylamide gels. The net charge, size, and shape of the protein determine the speed and direction of movement through a gel matrix in an electric field, and the position of each protein is visualized with specific histochemical stains. Homozygous and heterozygous genotypes are detected as single and double bands for monomeric proteins; polymeric proteins show more than two bands for heterozygotes. The quick and inexpensive resolution of genetic variation associated with unlinked loci and the Mendelian inheritance of allozyme polymorphisms made protein electrophoresis a good tool to assay heterozygosity within and between populations and estimate gene flow (Koehn and Eanes 1978; Selander and Whittam 1983; Philipp and Gross 1994).

Among vertebrates, however, the level of protein polymorphism is frequently too low to resolve specific genetic issues such as parentage. Molecular analysis of nuclear or mtDNA markers replaces protein electrophoresis in such cases (Gilbert et al. 1991; Packer et al. 1991; Sherwin et al. 1991; Martin et al. 1992a, b). There are numerous applications of this technique in conservation biology, including the identification of gene pools to guide conservation priorities, establishment of restocking programs, and evaluation of causes of decline in endangered species (e.g., Vrijenhoek et al. 1985; Quattro and Vrijenhoek 1989; Wayne et al. 1991; Petit et al. 1998).

DNA-DNA Hybridization

The complementary strands of the DNA duplex are connected by hydrogen bonds that are unstable at high temperatures and become single stranded when boiled. With cooling temperature, the complementary nucleotide sequences reanneal (i.e., become double stranded). In general, the greater the similarity or homology between DNA sequences, the more hydrogen bonds exist between matched nucleotides and the greater the thermal stability of the DNA duplex. To determine the homology of DNA sequences, single-stranded DNA of two individuals is allowed to anneal. DNA-DNA hybridization simultaneously samples genetic differences across most of the genome and hence provides valuable genetic distance estimates to resolve phylogenies, especially at intermediate taxonomic levels (Sibley and Ahlquist 1981). However, it does not provide qualitative data on character states crucial for many phylogenetic analyses. Few laboratories are set up for routine DNA-DNA hybridizations. This tool is rarely applied in conservation biology.

Restriction Fragment Length Polymorphism (RFLP)

Restriction enzymes cut double-stranded DNA at specific recognition sites that consist of short oligonucleotide sequences (typically four, five, and six base pairs). Restriction of DNA with one or several enzymes results in a range of fragment sizes that are electrophoretically separated in either agarose or polyacrylamide gels. Complex banding patterns can be detected when either all or selected fragments are visualized by means of a chemical stain (e.g., ethidium bromide) or radioactive DNA probes (i.e., short labeled DNA fragments that bind and hence mark complementary sequences). Furthermore, the development of the polymerase chain reaction (PCR) allows amplification of homologous locus-specific DNA fragments that can be subjected to restriction fragment analysis. PCR-based RFLP analysis has the advantages that very little tissue provides sufficient amounts of DNA and that large copy numbers of each fragment facilitate visualization of the banding patterns. As restriction sites are distributed throughout nuclear and cytoplasmic DNA, RFLP analysis can be applied to a variety of DNA regions by using a number of DNA sources, as discussed below.

Single-Copy Nuclear DNA (scnDNA) (Quinn and White 1987; Karl and Avise 1993)

In contrast to repetitive DNA, single copy nuclear DNA (scnDNA) is represented only once or possibly a few times in a haploid genome. As scnDNA is found in both coding and noncoding regions, evolutionary rates of change associated with these sites vary considerably. Genetic variation is detected by hybridizing a locus-specific scnDNA probe to digested whole genomic DNA. Alternatively, scnDNA can be PCR-amplified at specific loci and digested with restriction enzymes, and electrophoretically separated fragments are visualized with ethidium bromide staining. As scnDNA polymorphisms are numerous and their Mendelian inheritance can be established through pedigree analysis, scnDNA provides a wealth of genetic markers to estimate genetic diversity. Karl and colleagues (1992) showed the usefulness of scnDNA in a study on the genetic population structure of the endangered Green Turtle. Previous mtDNA analysis had shown high nest site fidelity among females and suggested severe limitation of gene flow between breeding populations (Bowen et al. 1992). Analysis of scnDNA, however, revealed only a moderate degree of genetic substructure, suggesting moderate levels of male-mediated gene flow between rookeries (Karl et al. 1992).

Random Amplified Polymorphic DNA (RAPD) (Williams et al. 1990; Welsh and McClelland 1990)

Random amplified polymorphic DNA (RAPD) analysis uses primers of a randomized oligonucleotide sequence to amplify DNA from various anonymous sites. The detected polymorphism can be applied to estimate genetic distances among closely related species or recent hybrids and in some cases to perform paternity exclusions (Lewis and Snow 1992; Levitan and Grosberg 1993; Milligan and McMurray 1993; Avise 1994). However, RAPD fragments do not always amplify reliably or show Mendelian inheritance, which renders some polymorphisms unsuitable for population genetics and parentage analyses (Riedy et al. 1992).

Ribosomal RNA Genes (rRNA)

Ribosomal RNA (rRNA) plays a crucial role in protein assembly, and therefore most regions of the molecule are structurally and functionally highly constrained (Mindell and Honeycutt 1990; Hillis and Dixon 1991). Nuclear rRNA is considered middle-repetitive DNA and generally consists of repeat units that contain highly conserved coding regions and less-conserved noncoding spacer regions (Avise 1994). Probes for rRNA genes are readily available and can detect intra- and interspecific restriction fragment length differences, which are either the consequence of heterogeneity in the length of noncoding DNA regions or the position of restriction sites. Several studies have found the hypervariability of the noncoded part of the nuclear rRNA gene families useful to differentiate populations and closely related species (Williams et al. 1985; Davis et al. 1990).

Major Histocompatability Complex (MHC) Genes (Klein 1986)

MHC genes code for cell surface proteins crucial to the immune response of animals as discussed above. The MHC is a family of tightly linked loci, many with scores of alleles, making it one of the most variable regions in the genome. Allelic variation associated with those regions can be detected through RFLP analysis using DNA probes (Klein 1986; Nei and Hughes 1991). MHC variability has been used to study inbreeding and fitness, genealogies, and population histories. Specific MHC haplotypes are thought to be associated with kin recognition and mate choice in mice and humans and disease resistance in birds and humans (Briles et al. 1977; Tiwari and Terasaki 1985; Hedrick et al. 1991; Howard 1991; Hughes 1991; Potts et al. 1991; Vrijenhoek and Leberg 1991; Hedrick 1992a; Ellegren et al. 1993; Klein et al. 1993; Brown and Eklund 1994; Wedekind et al. 1995).

Minisatellite DNA (Jeffreys et al. 1985a, b)

Minisatellite DNA consists of highly conserved core areas of short oligonucleo-tide sequences (10–65 base pairs rich in GC nucleotides) that are strung together in long arrays. These arrays of repetitive DNA vary considerably in size within and between genomes due to differences in the number of tandem repeat units (core areas) across 10–25 loci. Minisatellite DNA is inherited in a Mendelian fashion. Consequently, RFLP analysis of minisatellite DNA reveals unique inher-ited banding patterns for each individual ("DNA fingerprints") ideally suited for assigning individual genetic profiles and parentage (Burke and Bruford 1987; Burke et al. 1989; Rabenold et al. 1990; Westneat 1990; Morin and Ryder 1991; Martin et al. 1992b).

Genetic variation associated with minisatellite DNA has been revealed by means of multilocus and single-locus DNA fingerprinting (Bruford et al. 1992). Multilocus DNA fingerprinting simultaneously assays genetic variation associ-ated with numerous minisatellite loci through hybridization with a labeled mini-satellite probe (Loew and Fleischer 1996). This relatively quick assay of multi-locus variation is one of the major advantages of minisatellite DNA fingerprinting over microsatellite DNA, especially when the study population is suspected to be inbred. The detected polymorphism, can provide a snapshot of genetic variability within populations and, in some cases, relative estimates of genetic similarity between populations (Gilbert et al. 1990; Reeve et al. 1990). Useful applications in conservation biology have been developed by Fleischer and co-workers (1994) on the Palila, an endangered Hawaiian Honeycreeper, and by Fleischer and asso-ciates (1995) on endangered Clapper Rails, leading to recommendations for trans-locations between populations.

The complex banding pattern of multilocus DNA fingerprinting makes it less suitable for population genetics studies for determining exact degree of related-ness beyond full sibs (Lynch 1988, 1990, 1991; Burke et al. 1991; Jin and Chak-raborty 1993, 1994). Coancestry coefficients are generally correlated with band-sharing coefficients (Lynch 1988, 1990; Kuhnlein et al. 1990; Reeve et al. 1990; Rave et al. 1994) and have been used to provide estimates of relatedness (Burke et al. 1991; Piper and Rabenold 1992).

Examples of applications of DNA fingerprinting are provided by Rave and colleagues (1994) for captive Hawaiian Geese, Brock and White (1992) for the Puerto Rican Parrot, Ashworth and Parkin (1992) for Rothchild's Mynah, and Haig and co-workers (1994) who reconstructed the pedigree of all living Guam Rails, a species that is extinct in the wild.

Single-locus fingerprinting resolves polymorphisms associated with different hypervariable loci sequentially and therefore addresses most technical, statistical, and theoretical difficulties associated with multilocus minisatellite DNA finger-printing. Under very stringent conditions, a single-locus minisatellite probe detects variation only at a specific hypervariable minisatellite locus, consequently providing allelic diversity (= number of fragments per locus) and allele frequen-cies (= frequency of individual fragments within the population). Sequential application of several probes provides levels of polymorphism comparable with those from multilocus DNA fingerprinting, while also providing estimates of heterozygosity (Jeffreys et al. 1990, 1991; Burke et al. 1991; Hanotte et al. 1992; Scribner et al. 1994). However, the development of single-locus probes is time-consuming and more expensive than multilocus probes, and hence quick screening with multilocus probes may be the best first step.

Microsatellite DNA (Tautz 1989; Weber and May 1989)

Microsatellites are tandem repeats of di-, tri-, or tetranucleotides that are a result of mutation length changes and are found in many vertebrate species. As individual microsatellite arrays are relatively short, they can be amplified by using PCR primers designed to anneal to the conserved flanking regions of each microsatellite "locus." Subsequently, locus-specific microsatellite variation, or simple sequence length polymorphism can be visualized through electrophoresis of the marked PCR fragments (Burke et al. 1991; Schlötterer et al. 1991; Bruford et al. 1992).

Like minisatellite DNA, microsatellites are inherited in Mendelian fashion, and copy numbers of the repeat unit vary greatly within individuals at different loci and among individuals of a population at the same locus. The level of variability differs among microsatellite regions, offering an opportunity to select microsatellite loci that provide the appropriate level of resolution for a study. Microsatellite DNA fingerprinting is ideally suited for parentage analysis, as well as providing allele frequencies essential for estimates of gene flow and genetic diversity. In addition, this PCR-based method allows analysis of minute amounts of DNA (e.g., from hair or fecal samples), providing unrivaled opportunities to address questions of paternity and population genetics in wild animals (e.g., in wild chimpanzees) (Morin and Woodruff 1992). Thus far, microsatellite DNA fingerprinting has been less recommended for quick genetic assays of endangered or threatened species, as suitable microsatellite primers are still not readily available and are labor intensive in development (Edwards et al. 1991; Moore et al. 1991; Rico et al. 1994). However, the probability of detecting variable microsatellite loci and hence their usefulness for conservation genetics has recently increased significantly through the application of microsatellite-enriched genomic libraries (Armour et al. 1994; Fleischer and Loew 1996).

Mitochondrial DNA (Avise et al. 1987; Moritz et al. 1987; Harrison 1989; Wolstenholme 1992; Simon et al. 1994)

Animal mtDNA is a circular molecule 15–20 kilo bases in length and contains about 37 genes (coding for 22 mitochondrial transfer RNA genes [tRNA], two rRNAs, and 13 proteins) and a control region associated with the molecule's replication and transcription. mtDNA is generally maternally inherited and evolves relatively quickly due to the apparent lack of repair mechanisms for mutations during replication (Wilson et al. 1985; Avise et al. 1987; Martin et al. 1992a). Analysis of mtDNA restriction fragments detects sufficient polymorphism to resolve genetic distances among conspecific populations and closely related species (Tarr and Fleischer 1993; Avise 1994).

O'Brien and associates (1990) combined mtDNA analysis with data on protein polymorphism to inform management decisions for the Florida Panther. Menotti-Raymond and O'Brien (1993) used them to evaluate the history, status, and management of African Cheetahs. Taberlet and colleagues (1995) used DNA sequencing and RFLP analysis of mtDNA to delineate the contact zones of two divergent lineages of Scandinavian Brown Bears, to define conservation units, and to guide possible translocation efforts.

DNA Sequencing (Maxam and Gilbert 1977, 1980; Sanger et al. 1977)

DNA sequencing provides the greatest resolution of genetic divergence by actually determining the identity and sequence of all bases within a target region (typically 500 base pairs long). Consequently, all genetic differences between samples are detected, instead of only those that result in restriction site changes. The most commonly used manual sequencing method is called Sanger dideoxy sequencing. It is based on in vitro synthesis of radioactively labeled single-stranded DNA that is interrupted at either an A, T, C, or G nucleotide. DNA pieces ending in different nucleotides are run in separate lanes, and their positions are made visible as individual bands through autoradiography. Each band represents a particular nucleotide, and consecutive bands are separated by one base pair, consequently the band pattern reveals the DNA sequence of a molecule. More recently, automated DNA sequencing with fluorescent dye labels has started to replace the more cumbersome manual sequence analysis. Nuclear and mtDNA can serve as templates for DNA sequence comparisons among individuals and taxa, provided sufficient amounts of purified homologous DNA and suitable sequencing primers are available. The source DNAs most commonly used for DNA sequence analysis are discussed below.

Mitochondrial DNA Control Region

This noncoding region contains sequences that control replication and transcription of the mtDNA molecule and is called D-loop region in vertebrates and A+T-rich region in insects (Fauron and Wolstenholme 1980; Aquadro and Greenberg 1983). It is the only mtDNA region that consistently contains noncoding DNA in many taxa, and although evolutionary rates vary within the control region, it is generally a reliable source for hypervariable mtDNA. This high level of variability has made the control region suitable for sequence comparisons within and between populations and closely related species (Wilson et al. 1985; Harrison 1989; Thomas et al. 1990; Tarr 1995; Morales et al. 1997). For example, Morin and co-workers (1992) used D-loop sequence to identify chimpanzee subspecies and thus recognize hybrids in captive populations and identify the origin of illegally traded animals.

Mitochondrial Protein Coding Genes

DNA coding for proteins is based on a triplet code for amino acids. Several triplets code for the same amino acids. The first and second codon positions are highly conserved, whereas the third position is less constrained in terms of nucleotide changes due to the degenerate nature of the amino acid code. Consequently, most base pair substitutions in the third position do not change amino acid transcription and are considered "silent." Base pair substitutions that result in amino acid replacements generally occur at a lower rate than silent substitutions due to structural and functional constraints. Cytochrome b and ATPase 6 have been used extensively in vertebrate studies to resolve relationships at close and intermediate taxonomic levels (Kocher et al. 1989; Meyer et al. 1990). For example, Bowen and colleagues (1992, 1993) have used cytochrome b sequence data to determine the phylogeny of all known marine turtles, essential information in choosing which taxonomic groups to conserve and in designing recovery plans.

Mitochondrial transfer RNA genes (tRNA) and mitochondrial ribosomal RNA (rRNA) evolve more slowly than protein coding genes (Wolstenholme and Clary 1985), indicating greater structural and functional constraints (Simon et al. 1994). Although their slow evolutionary rates render them generally unsuitable to answer typical conservation genetics questions by themselves, they are occasionally combined with other molecular markers to resolve phylogenies (Mindell and Honeycutt 1990; Hillis and Dixon 1991; Hoelzel et al. 1993; Kretzmann et al. 1997).

Nuclear DNA Introns (Palumbi and Baker 1994)

Nuclear introns have been introduced as a suitable template for DNA sequencing analysis to complement mtDNA data. Universal PCR primers that anneal to exons of highly conserved nuclear genes are used to amplify across nuclear introns. Noncoding introns generally exhibit high rates of evolutionary change, and sequence analysis can reveal high levels of diversity at these nuclear DNA sites. Unlike mtDNA, nuclear DNA reflects the biparental contribution to population structure at several independently segregating sites, and like mtDNA sequencing, the resolution power of nuclear introns sequences is high. Consequently, nuclear introns are expected to find their greatest application in resolving genetic population structure and taxonomic relationships. Using actin intron alleles, Palumbi and Baker (1994) corroborated large-scale movements of Humpback Whales based on mtDNA analysis. They revealed that sequence analysis of mtDNA and nuclear introns predicted different amounts of gene flow between Hawaiian and Californian Humpback Whales, a result consistent with female philopatry and male-biased migration between populations.

In general, nuclear and mtDNA evolve and are transmitted differently, and because mtDNA is haploid and maternally inherited, its effective population size is four times smaller than that of nuclear markers. Consequently, small populations may show genetic diversity at nuclear DNA loci but have no mtDNA variation (Avise 1994, 1995). Consequently, ambiguities in phylogenetic relationships and the genetic structure of natural populations are best resolved by using a combination of nuclear and mtDNA markers (Moritz 1994b; Avise 1995). The value of this approach has been demonstrated in studies revealing hybridization in the history of the endangered Red Wolf (Wayne and Jenks 1991; Roy et al. 1994) and the Florida Panther (O'Brien et al. 1990), as well as in determining the phylogeographic history and infraspecific taxonomy of Leopards (Miththapala et al. 1995).

"Ups and Downs" of Genetic Diversity

Generating Diversity

The amount of genetic diversity present at any point in time in an individual or population is the result of opposing forces that have affected allele frequencies in the past. For example, mutation events, recombination, and immigration are important evolutionary forces that introduce additional alleles and polymorphic sites into a population, whereas selection and inbreeding, as well as random genetic drift, generally homogenize gene pools (for further details, see Hartl and Clark 1989; Loeschcke et al. 1994).

Only mutations can generate novel alleles in entirely monomorphic populations. However, mutations occur rarely and are often deleterious or neutral and therefore cannot be relied on as a major source of genetic variation for short-term genetic management. Immigration offers mixing of gene pools and rapid infusion of new genes into genetically homogeneous populations. The fragmentation of natural habitats, however, decreases natural rates of migration and dispersal and results in population subdivision and eventual isolation.

Conservation biologists can increase gene flow by either encouraging successful dispersal among subpopulations through dispersal corridors or by translocating new genetically distinct breeders to isolated populations. In addition, wild populations can be supplemented with captive-bred individuals through reintroduction programs (for further discussions, see Gibbs 1991; Olney et al. 1994). Numerous threatened or endangered populations have benefited from these management strategies, and consequently, dispersal corridors (e.g., natural areas associated with rivers) and translocation have become an integral part of reserve design (Beier 1993; Dunning et al. 1995; Madsen et al. 1996; Beier and Noss 1998).

Nevertheless, there is a cost to helping gene flow along through human-made corridors or translocations (Simberloff and Cox 1987). Although such population management tools allow the influx of new genetic material into isolated populations, they might also create new problems, such as facilitation of the spread of pathogens and parasites and hybridization (Woodford and Rossiter 1994; Cunningham 1996; Stockwell et al. 1996). Green and Rothstein (1998), for example, reported that translocations of the endangered Black-faced Impalas to private farms have increased the threat of hybridization with resident Common Impalas in Namibia. They agreed with Robinson and associates (1991) that population manipulations (translocations, introductions) should only be carried out after careful genetic, taxonomic, and ecological considerations.

Artificial insemination and embryo transfer are alternative methods of introducing genes into a population. Inseminating females or implanting them with embryos circumvents the problem of mate choice and does not depend on the establishment and mating success of immigrant males. These methods are, however, highly intrusive, cumbersome, and expensive and hence are only practical for captive populations and only recommended in cases in which species survival depends on imminent increase in genetic diversity. Furthermore, reproductive technology of endangered species is a relatively new and still experimental field. Although the techniques have proved useful in livestock management, much more basic research on the reproductive biology of rare species is needed to guarantee safe application in genetic management efforts (see Ballou and Cooper [1992] and Moore et al. [1992] for discussion of reproductive technology and conservation genetics).

In general, increased habitat fragmentation has subdivided many populations to the point at which individual subpopulations would become extinct unless they are managed as part of the larger highly structured population. The extreme complexity of modeling stochastic and deterministic events in such structured populations has hampered efforts to assess genetic and demographic effects simultaneously (Burgman et al. 1993; Ballou et al. 1995; Ratner et al. 1997). However, population viability models are becoming increasingly more sophisticated and valuable in risk assessment of endangered populations (Boyce 1992; Lacy 1993; Kenny et al. 1995; Mills et al. 1996).

Losing Diversity

Inbreeding can be due either to the mating of related individuals (as in selfing or assortative mating) or it may be because, in finite populations, there is a chance for two identical genes to be sampled together. This chance increases when the size of the population decreases. Consequently, levels of homozygosity in inbreeding populations are higher than predicted under random mating (Falconer and Mackay 1996). (Templeton and Read [1994] discuss different measures of inbreeding and their relationship with genetic diversity.)

Natural selection is a significant evolutionary force that can maintain or erode genetic diversity, depending on the relationship between genotypes and fitness. Directional selection, for example, favors extreme phenotypes and their underlying genotypes, and the entire array of intermediate phenotypes are selected against and eventually lost in the population. Consequently, strong selection for particular traits homogenizes the gene pool, ultimately reducing the adaptive potential of a population. This relationship is of concern in captive propagation programs that breed animals for future reintroduction to the wild; any inadvertent selection to improve management in captivity may reduce survival chances back in the wild. In general, the interactions between phenotypes, genotypes, and fitness are far too complex to predict reliably, and conservation biologists generally refrain from managing captive populations based on specific beneficial traits; instead, they try to maintain overall genetic diversity on which natural selection can act.

Maintaining Genetic Diversity and Effective Population Size

Ideal populations are defined as infinite populations, consisting of sexually reproducing diploid organisms that mate at random and have nonoverlapping generations and whose allele frequencies are not affected by migration, mutation, or selection (Wright 1931). The probability of loss of an allele in ideal populations is equal to its allele frequency.

As real populations almost always violate the assumptions of an ideal population, Sewall Wright (1931) introduced the concept of the effective population size (N_e) to evaluate the evolutionary potential of populations that deviate from the ideal. The effective population number is most commonly estimated by relating the variance in allelic frequency (the "variance effective size") or the rate of inbreeding ("inbreeding effective size") of the real to the ideal population (Wright 1931; Crow and Denniston 1988). Accordingly, the effective population size of a real population equals the size of an ideal population that has the same amount of variance in allele frequencies or the same amount of inbreeding as the actual population (Wright 1931, 1938; Crow and Kimura 1970).

Maximizing the inbreeding effective size maintains heterozygosity within local populations, whereas high variance effective size decreases loss of genetic diversity across local populations and significantly affects allelic diversity (Crow and Kimura 1970; Gliddon and Goudet 1994; Ballou and Lacy 1995; Hedrick et al. 1995). In general, N_e is smaller than the actual population size, and loss of genetic variation increases with decreasing effective population size. Numerous factors, such as unequal sex ratio and fluctuating population size, can significantly affect effective population size (for detailed discussion, see Falconer and Mackay 1996). For example, calculations of effective population size are usually based on discrete generation models, which are not applicable to many real populations.

A realistic diploid model with overlapping generations, however, is very complex as it attempts to estimate allele frequency changes per unit time for males and females that might differ in their age-specific birth and death rates (Felsenstein 1971; Lande and Barrowclough 1987). The effective population size per unit time is maximized if the generation lengths of males and females are equal, and it decreases with shorter maturation time. In other words, species that experience delayed sexual maturation and have long generation times (e.g., large-bodied mammals) have a higher probability of retaining genetic diversity for any given period of time (see Hedrick 1992b for review). These species will, however, lose genetic variation at the same rate as short-lived animals following a bottleneck if time is measured in generations.

It is crucial to realize that high genetic diversity in long-lived species (e.g., One-horned Rhinoceros; Dinerstein and McCracken 1990) does not necessarily mean that previous population crashes did not affect them, but instead it may mean that they have not reproduced much since then. Hence, we should realize the opportunity to maintain the genetic diversity stored through such "walking gene banks" by maximizing their effective population size for future generations.

To determine effective population size when several factors are of importance, Chepko-Sade and colleagues (1987) suggested sequential calculation of N_e for all variables by using an iterative process. They used this approach as a first approx-imation to reach more realistic estimates of the evolutionary potential (N_e) for a variety of well-studied natural populations (e.g., wild horses, Black Bears, Dwarf Mongoose), but its theoretical validity has not been tested. Numerous alternative approaches to estimate N_e have been developed to address incomplete data sets and effects of multiple variables (Frankham 1995a; Rockwell and Barrowclough 1995). In addition to demographic approaches, change in various genetic mea-sures (e.g., allozyme heterozygosity, pedigree inbreeding) has been used to esti-mate effective population size (Avise et al. 1988; Tomlinson et al. 1991; Briscoe et al. 1992). Frankham (1995b) argued that genetic versus demographic methods show comparable results, provided the same variables were used to determine short-term estimates of N_e. In general, he concluded that population fluctuations, variance in family size, and unequal sex ratio affect the ratio of effective to actual population size most significantly and that the effective size for most wild popula-tion is disconcertingly small $(N_e /N = 0.10–0.11)$.

In conclusion, loss of genetic variation in endangered or rare species can be minimized through breeding programs that aim to maximize effective population sizes and reduce inbreeding. Outbreeding is generally considered an appropriate measure to maintain or generate genetic diversity; however, individuals should not be outbred indiscriminately. If most genetic variation is found in a number of different geographic regions, regular genetic exchange between those populations would actually reduce overall genetic diversity and homogenize rather than diver-sify the species' gene pool (= genetic cost of dispersal corridors and transloca-tion). Furthermore, populations that experience varying ecological conditions and have been isolated from each other for an extended period of time might show important local adaptations. Under such conditions, gene flow would enhance genetic variation within populations but might also result in reduced fitness of the less well-adapted outbred offspring (outbreeding depression) (Shields 1982; Tem-pleton et al. 1986; Ballou 1995).

In addition, long-term historical divisions within species may be the sign of ongoing adaptive radiation, and individual populations might have to be treated as separate evolutionary entities. Consequently, recommendations for the genetic management of endangered species should be based on an assessment of the present genetic structure and take the population biology and breeding history of individuals into consideration wherever possible (Vrijenhoek et al. 1985; Vrijenhoek 1994; Ballou et al. 1995; Avise and Hamrick 1996).

Who Is to Embark on the Ark?

To maintain biodiversity and to identify those taxonomic units worthy of our protection, we need clearly defined criteria to determine conservation units. Since the 1973 Endangered Species Act mandated the protection of species, determination of species status has necessarily become of great importance to conservation management (U.S. House of Representatives 1973).

The species concept, however, has been at the center of an ongoing debate among evolutionary biologists, and numerous criteria, such as reproductive isolation and ancestral relationships, have been applied to define a species (O'Brien and Mayr 1991; Crozier 1992; Geist 1992; Rojas 1992). For example, Mayr's (1963, 1969) biological species concept defines species as freely interbreeding populations that are reproductively isolated. Criteria of interbreeding ability, however, are of limited use for clarifying taxonomic relationships of discontinuous populations because reproductive barriers among allopatric populations are difficult to prove (Cracraft 1983; McKitrick and Zink 1988). Similarly, distinguishing species on the basis of morphology suffers, in part, from the fact that many morphological traits are considerably affected by environmental conditions (Geist 1987). Therefore phenotypic differences between populations might reflect temporary local adaptations rather than independent evolutionary histories.

By contrast, phylogenetic analysis based on neutral genetic markers can contribute additional measures of the genetic distinctiveness of taxa and may be more reflective of their evolutionary history (Avise et al. 1987; Hillis 1987; Dizon et al. 1992; Moritz 1994b; Wayne et al. 1994; Avise and Hamrick 1996; but see Cronin 1993). In general, a variety of genetic markers in combination with morphometric analysis is preferable to establish the amount of reproductive isolation and phylogenetic uniqueness of a particular taxonomic group. Accordingly, Avise and Ball (1990) suggest that a suite of phylogenetically concordant characteristics should be used to define a taxonomic group as a population of individuals that can be united by one or more derived traits.

Rojas (1992) has pointed out that conservation based on the above typological approaches to species is problematic for several reasons. If nature reserves are designed to ensure survival of a representative sample of "types" or species, unresolved species status will have a major impact on the number of preserved species and ultimately the level of biodiversity. She points out that "the numbers, however, are unlikely to be the same if we are considering biological species, cladistic species, or evolutionary species" (Rojas 1992). In addition, she notes that conserving species as types ignores the importance of preserving geographic variation within species.

Such infraspecific variation, however, is potentially crucial for the long-term survival of a species. Conserving species as evolutionary units addresses the importance of variation within species to preserve the evolutionary potential of the protected organisms and ultimately minimize the extinction probability across their range (Vrijenhoek 1989; Rojas 1992). Although reserve management based on evolutionary units is less subject to "bean counting" of rare types, clear delimiting of taxonomic groups is still essential to avoid inadvertent hybridization and outbreeding depression.

Amendments to the Endangered Species Act (ESA) extended protection of species to include distinct population segments that interbreed when mature. This broader definition of species protection allowed flexibility to preserve species in only part of their range without resolving their taxonomic status (Pennock and Dimmick 1997). As ESA provided no clear definition for "distinct population segments," criteria for identifying fragile populations were highly varied, including uniqueness based on geographic isolation and morphological and genetic differences (Pennock and Dimmick 1997).

The concept of evolutionary significant units (ESU) was developed, in part, to provide biological guidelines for determining conservation units below the species level (Ryder 1986; Vane-Wright et al. 1991; Waples 1991). This attempt to apply the concept of ESUs to identify "distinct population segments" for conservation has resulted in ongoing debates that focus on the biological merits (Dizon et al. 1992; Moritz 1994a; Vogler and DeSalle 1994; Waples 1998) and the fulfillment of the ESA mission (Rohlf 1994; Pennock and Dimmick 1997; Waples 1998). Rojas's (1992) conclusion that "considering the species problem more critically may result in recognizing the limitations of the taxonomic information used; it may also contribute to the refinement of the concepts and methods" still holds, and we continue to benefit from the interaction between systematists and conservation biologists to determine conservation units.

Conclusions

Which species are most at risk of extinction may depend on many interactive aspects of their biology, such as their ecology (e.g., specialists), mating or migratory behavior (e.g., threatened breeding or wintering grounds), demography (e.g., low birth rate), and population genetics (e.g., small effective population size). Populations that are dwindling in size are generally marked by high mortality or low birth rates and eventually may show changes in gene frequencies due to increased population subdivisions and reduced gene flow. Hence, ecologists make use of alarming demographic trends such as low growth rates and conservation geneticists focus on changes in genetic variation (e.g., loss of heterozygosity) to screen for possible threatened populations. Population viability analyses that estimate extinction risks of populations due to both demographic and genetic stochasticity provide powerful monitoring tools that will help to guide the allocation of scarce resources to taxa most likely in need of management (Burgman et al. 1988; Lacy 1993; Nunney and Campbell 1993; Ballou et al. 1995).

Failure to detect signs of increased extinction risk based on the demographic and genetic records of populations should not be interpreted as a guarantee of survival of a taxon, especially not for healthy populations with a limited range or species of a specialized ecological niche that might become endangered by a single catastrophic event. Hence listing taxa as endangered should be based on demographic and genetic indicators of decline, while taking into consideration vulnerabilities of the species due to its specific biology, as well as abiotic factors such as the local political situation (Avise 1994; Ballou et al. 1995; Avise and Hamrick 1996). In their reevaluation of the IUCN threatened species categories, Mace and Lande (1991) applied a similar logic in their development of quantitative criteria to determine endangerment.

Acknowledgments. I appreciate the comments by J. Ballou, R. Frankham, S. Jarvi, C. McIntosh, E. Paxinos, E. Perry, and C. Tarr on a previous draft of this chapter. In addition, I thank Mark Burgman for his exceedingly helpful editorial changes. Ideas and early drafts for this manuscript were developed while I was a Smithsonian Research Fellow and was supported by a California Fish and Game contract to K. Ralls, D. Williams, and R. C. Fleischer.

Literature Cited

Alberts SC, Ober C (1993) Genetic variability in the major histocompatibility complex: a review of non-pathogen-mediated selective mechanisms. Yearbook of Physical Anthropology 36:71–89

Aquadro CF, Greenberg BD (1983) Human mitochondrial DNA variation and evolution: analysis of nucleotide sequences from seven individuals Genetics 103:287–312

Armour JAL, Neumann R, Gobert S, Jeffreys AJ (1994) Isolation of human simple repeat loci by hybridization selection. Human Molecular Genetics 3:599–605

Ashley M (1999) Molecular conservation genetics: assaying the structure of DNA can help protect endangered species. American Scientist 87:28–35

Ashworth D, Parkin DT (1992) Captive breeding: can genetic fingerprinting help? Symposia of the Zoological Society of London 64:135–149

Avise JC (1994) Molecular markers, natural history and evolution. Chapman and Hall, New York

Avise JC (1995) Mitochondrial DNA polymorphism and a connection between genetics and demography of relevance to conservation. Conservation Biology 9:686–690

Avise JC, Ball RM (1990) Principles of genealogical concordance in species concepts and biological taxonomy. Oxford Survey of Evolutionary Biology 7:45–67

Avise JC, Hamrick JL (1996) Conservation genetics: case histories from nature. Chapman and Hall, New York

Avise JC, Arnold J, Ball RM, Bermingham E, Lamb T, Neigel JE, Reeb CA, Saunders NC (1987) Intraspecific phylogeography: the mitochondrial bridge between population genetics and systematics. Annual Review of Ecological Systematics 18:489–522

Avise JC, Ball RM, Arnold J (1988) Current versus historical population sizes in vertebrate species with high gene flow: a comparison based on mitochondrial DNA lineages and inbreeding theory for neutral mutations. Molecular Biology and Evolution 5:331–344

Avise JC, Haig SM, Haig OA, Ryder OA, Lynch M, Geyer CJ (1995) Descriptive genetic studies: application in population management and conservation biology. In: Ballou JD, Gilpin M, Foose T (eds) Population management for survival and recovery. Columbia University Press, New York, pp 183–244

Ballou JD (1995) Inbreeding and outbreeding depression in captive populations. PhD dissertation, University of Maryland, College Park, MD

Ballou JD, Cooper KA (1992) Genetic management strategies for endangered captive populations: the role of genetic and reproductive technology. Symposia of the Zoological Society of London 64:183–206

Ballou JD, Lacy RC (1995) Identifying genetically important individuals for management of genetic diversity in captive populations. In: Ballou JD, Gilpin M, Foose T (eds) Population management for survival and recovery. Columbia University Press, New York, pp 76–111

Ballou JD, Gilpin M, Foose T (eds) (1995) Population management for survival and recovery analytical methods and strategies in small population conservation. Columbia University Press, New York

Beier P (1993) Determining minimum habitat areas and habitat corridors for Cougars. Conservation Biology 7:94–108

Beier P, Noss R (1998) Do habitat corridors provide connectivity? Conservation Biology 12:1241–1252

Bodmer W (1972) Evolutionary significance of the HL-A system. Nature 237:139–145

Bowen BW, Meylan AB, Ross JP, Limpus CJ, Balazs GH, Avise JC (1992) Global population structure and natural history of the Green Turtle (*Chelonia mydas*) in terms of matriarchal phylogeny. Evolution 46:865–881

Bowen BW, Nelson WS, Avise JC (1993) A molecular phylogeny for marine turtles: trait mapping, rate assessment, and conservation relevance. PNAS 90:5574–5577

Boyce MS (1992) Population viability analysis. Annual Review of Ecology and Systematics 23:481–506

Brewer BA, Lacy RC, Foster ML, Alaks G (1990) Inbreeding depression in insular and central populations of Peromyscus Mice. Journal of Heredity 81:257–266

Briles WE, Stone HA, Cole RK (1977) Marek's disease: effects of B histocompatibility alloalleles in resistant and susceptible chicken lines. Science 195:193–195

Briscoe DA, Malpica JM, Robertson A, Smith GJ, Frankham R, Banks RG, Barker JSF (1992) Rapid loss of genetic variation in large captive populations of Drosophila Flies: implications for genetic management of captive populations. Conservation Biology 6:416–425

Brock MK, White BN (1992) Application of DNA fingerprinting to the recovery program of the endangered Puerto Rican Parrot. Proceedings of the National Academy of Science of the USA 89:11121–11125

Brown JL, Eklund A (1994) Kin recognition and the major histocompatibility complex: an integrative review. The American Naturalist 143:435–461

Bruford MW, Hanotte O, Brookfield JFY, Burke T (1992) Single-locus and multilocus DNA fingerprinting. In: Hoelzel AR (ed) Molecular genetic analysis of populations. IRL Press, Oxford, UK, pp 225–269

Burgman MA, Akçakaya HR, Loew SS (1988) The use of extinction models for species conservation. Biological Conservation 43:9–25

Burgman MA, Ferson S, Akçakaya HR (1993) Risk assessment in conservation biology. Chapman and Hall, London

Burke T, Bruford MW (1987) DNA fingerprinting in birds. Nature 327:149–152

Burke T, Davies NB, Bruford MW, Hatchwell BJ (1989) Parental care and mating behavior of Polyandrous Dunnocks *Prunella modularis* related to paternity by DNA fingerprinting. Nature 338:249–251

Burke T, Hanotte O, Bruford MW, Cairns E (1991) Multilocus and single locus minisatellite analysis in population biological studies. In: Burke T, Dolf G, Jeffreys AJ, Wolff R (eds) DNA fingerprinting approaches and applications. Birkhäuser Verlag, Basel, Switzerland, pp 154–168

Byers DL, Waller DM (1999) Do plant populations purge their genetic load? Effects of population size and mating history on inbreeding depression. Annual Review of Ecology and Systematics 30:479–513

Charlesworth D, Charlesworth B (1987) Inbreeding depression and its evolutionary consequences. Annual Review of Ecological Systematics 18:237–268

Chepko-Sade BD, Shields WM, Berger J, Halpin ZT, Jones WT, Rogers LL, Rood JP, Smith AT (1987) The effect of dispersal and social structure on effective population size. In: Chepko-Sade BD, Halpin ZT (eds) Mammalian dispersal patterns. University of Chicago Press, Chicago, IL, pp 287–321

Clarke B, Kirby DRS (1966) Maintenance of histocompatibility polymorphism. Nature 211:999–1000

Cracraft J (1983) Species concepts and speciation analysis. In: Johnson RF (ed) Current ornithology. Plenum Press, New York, pp 150–187

Cronin MA (1993) Mitochondrial DNA in wildlife taxonomy and conservation biology: cautionary notes. Wildlife Society Bulletin 21:339–348

Crow JF, Denniston C (1988) Inbreeding and variance effective population numbers. Evolution 42:482–495

Crow JF, Kimura M (1970) An introduction to population genetics theory. Harper and Row, New York

Crozier R (1992) Genetic diversity and the agony of choice. Biological Conservation 61:11–15

Cunningham AA (1996) Disease risk of wildlife translocations. Conservation Biology 10:349–353

Davis SK, Strassmann JE, Hughes C, Pletscher LS, Templeton AR (1990) Population structure and kinship in Polistes (Hymenoptera, Vespidae): an analysis using ribosomal DNA and protein electrophoresis. Evolution 44:1242–1253

Dinerstein E, McCracken GF (1990) Endangered Greater One-horned Rhinoceros carry high levels of genetic variation. Conservation Biology 4:417–422

Dizon AE, Lockyer C, Perrin WF, Demaster DP, Sission J (1992) Rethinking the stock concept: a phylogenetic approach. Conservation Biology 6:24–36

Doherty PC, Zinkernagel RM (1975) A biological role for the major histocompatibility antigens. Lancet 1:1406–1409

Dunning JB Jr, Borgella R Jr, Clements K, Meffe GK (1995) Patch isolation, corridor effects, and colonization by a resident sparrow in a managed pine woodland. Conservation Biology 9:542–550

Edwards A, Civitello A, Hammond HA, Caskey CT (1991) DNA typing and genetic mapping with trimeric and tetrameric tandem repeats. American Journal of Human Genetics 49:746–756

Egid K, Brown JL (1989) The major histocompatibility complex and female mating preferences in mice. Animal Behaviour 38:548–550

Ellegren H, Hartman G, Johansson M, Andersson L (1993) Major histocompatibility complex monomorphism and low levels of DNA fingerprinting variability in a reintroduced and rapidly expanding population of beavers. Proceedings of the National Academy of Sciences of the USA 90:8150-8153

Falconer DS, Mackay TFC (1996) Introduction to quantitative genetics, 4th ed. Longman, New York

Fauron CM-R, Wolstenholme DR (1980) Intraspecific diversity of nucleotide sequences within the adenine + thymine-rich region of mitochondrial DNA molecules of *Drosophila mauritiana, Drosophila melanogaster* and *Drosophila simulans.* Nucleic Acids Research 8:5391-5410

Felsenstein J (1971) Inbreeding and variance effective numbers in populations with overlapping generations. Genetics 68:581-597

Fleischer RC, Loew SS (1996) Construction and screening of microsatellite-enriched genomic libraries. In: Ferraris J, Palumbi S (eds) Molecular zoology: strategies and protocols. Wiley-Liss, New York

Fleischer RC, Tarr CL, Pratt TK (1994) Genetic structure and mating system in the palila, an endangered Hawaiian Honeycreeper, as assessed by DNA fingerprinting. Molecular Ecology 3:383-392

Fleischer RC, Fuller G, Ledig DB (1995) Genetic structure of endangered Clapper Rail (*Rallus longirostris*) populations in southern California. Conservation Biology 9:1234-1243

Frankham R (1995a) Conservation genetics. Annual Review of Genetics 29:305-327

Frankham R (1995b) Effective population size/adult population size ratios in wildlife: a review. Genetic Research 66:95-107

Frankham R (1995c) Inbreeding and extinction: a threshold effect. Conservation Biology 9:792-799

Frankham R (1998) Inbreeding and extinction: island populations. Conservation Biology 12:665-675

Fu Y, Mamkoong G, Carlson JE (1998) Comparison of breeding strategies for purging inbreeding depression via simulation. Conservation Biology 12:856-864

Garner KJ, Ryder OA (1992) Some applications of PCR to studies in wild life genetics. Symposia of the Zoological Society of London 64:167-181

Geist V (1987) On speciation in Ice Age mammals, with special reference to cervids and caprids. Canadian Journal of Zoology 65:1067-1084

Geist V (1992) Endangered species and the law. Nature 357:274-276

Gibbs JH (1991) Beyond captive breeding: reintroducing endangered mammals to the wild. Symposia of the Zoological Society of London 62. Clarendon Press, Oxford, UK

Gilbert DA, Lehman N, O'Brien SJ, Wayne RK (1990) Genetic fingerprinting reflects population differentiation in the California Channel Island Fox. Nature 344:764-766

Gilbert DA, Packer C, Pusey AE, Stephens JC, O'Brien SJ (1991) Analytical DNA fingerprinting in lions: parentage, genetic diversity and kinship. Journal of Heredity 82:378-386

Gliddon C, Goudet J (1994) The genetic structure of metapopulations and conservation biology. In: Loeschcke V, Tomiuk J, Jain SK (eds) Conservation genetics. Birkhäuser Verlag, Basel, Switzerland, pp 107-114

Green WCH, Rothstein A (1998) Translocation hybridization, and the endangered Black-faced Impala. Conservation Biology 12:475-480

Haig SM, Ballou JD, Derrickson SR (1990) Management options for preserving genetic diversity: reintroduction of Guam Rails to the wild. Conservation Biology 4:290-300

Haig SM, Ballou JD, Casna NJ (1994) Identification of kin structure among Guam Rail founders: a comparison of pedigree and DNA profiles. Molecular Ecology 3:109-119

Hamrick JL, Godt MJW (1996) Conservation genetics of endemic plant species. In: Avise JC, Hamrick JL (eds) Conservation genetics:case histories from nature. Chapman and Hall, New York, pp 281–304

Hanotte O, Cairns E, Robson T, Double MC, Burke T (1992) Cross-species hybridization of a single-locus minisatellite probe in passerine birds. Molecular Ecology 1:127–130

Harrison RG (1989) Animal mitochondrial DNA as a genetic marker in population and evolutionary biology. Trends in Ecology and Evolution 4:6–11

Hartl DL, Clark AG (1989) Principles of population genetics. 2nd ed. Sinauer, Sunderland, MA

Hedrick PW (1992a) Female choice and variation in the major histocompatibility complex. Genetics 132:575–581

Hedrick PW (1992b) Genetic conservation in captive populations and endangered species. In: Kain S, Botsford LW (eds) Applied population biology. Kluwer Academic Publishers, Dordrecht, The Netherlands, pp 45–68

Hedrick PW (1994) Purging inbreeding depression and the probability of extinction: full-sib mating. Heredity 73:363–372

Hedrick PW, Miller P (1992) Conservation genetics: techniques and fundamentals. Ecological Applications 2:30–46

Hedrick PW, Miller PS (1994) Rare alleles, MHC and captive breeding. In: Loeschcke V, Tomiuk J, Jain SK (eds) Conservation genetics. Birkhäuser Verlag, Basel, Switzerland, pp 187–204

Hedrick PW, Thomson G (1988) Maternal-fetal interactions and the maintenance of HLA polymorphism. Genetics 119:205–212

Hedrick PW, Klitz W, Robinson WP, Kuhner MK, Thomson G (1991) Population genetics of HLA. In: Selander R, Clark A, Whittam T (eds) Evolution at the molecular level. Sinauer Associates, Sunderland, MA, pp 248–271

Hedrick PW, Hedgecock D, Hamelberg S (1995) Effective population size in winter-run Chinoock Salmon. Conservation Biology 9:615–624

Hill AVS, Allsop CEM, Kwiatkowski D, Anstey NM, Twumasi P, Rowe PA, Bennett S, Brewster D, McMichael AJ, Greenwood BM (1991) Common west African HLA antigens are associated with protection from severe malaria. Nature 352:595–600

Hillis DM (1987) Molecular versus morphological approaches to systematics. Annual Review of Ecological Systematics 18:23–42

Hillis DM, Dixon MT (1991) Ribosomal DNA: molecular evolution and phylogenetic inference. Quarterly Review of Biology 66:411–453

Hoelzel AR, Halley J, O'Brien SJ, Campagna C, Arnbom T, LeBoeuf B, Ralls K, Dover GA (1993) Elephant Seal genetic variation and the use of simulation models to investigate historical population bottlenecks. Journal of Heredity 84:443–449

Howard JC (1991) Disease and evolution. Nature 352:565–567

Hughes AL (1991) MHC polymorphism and the design of captive breeding programs. Conservation Biology 5:249–251

Hughes AL, Nei M (1988) Pattern of nucleotide substitution at major histocompatibility complex class I loci reveals overdominant selection. Nature 352:565–567

Janeway CA Jr, Travers P (1994) Immunobiology: the immune system in health and disease. Garland Publishing Inc, New York

Jeffreys AJ, Brookfield JFY, Semenoff R (1985a) Positive identification of an immigration test-case using DNA fingerprints. Nature 317:818–819

Jeffreys AJ, Wilson V, Thein SL (1985b) Individual-specific "fingerprints" of human DNA. Nature 316:76–79

Jeffreys AJ, Neumann R, Wilson V (1990) Repeat unit sequence variation in mini-satellites:a novel source of DNA polymorphism for studying variation and mutation by single molecule analysis. Cell 60:473–485

Jeffreys AJ, MacLeod A, Tamaki K, Neil DL, Monckton DG (1991) Minisatellite repeat coding as a digital approach to DNA typing. Nature 317:818–819

Jimenez JA, Hughes KA, Alaks G, Graham L, Lacy R (1994) An experimental study of inbreeding depression in a natural habitat. Science 265:271–273

Jin L, Chakraborty R (1993) A bias-corrected estimate of heterozygosity for single-probe multilocus DNA fingerprints. Molecular Biology and Evolution 10:1112–1114

Jin L, Chakraborty R (1994) Estimation of genetic distance and coefficient of gene diversity from single-probe multilocus DNA fingerprint data. Molecular Biology and Evolution 11:120–127

Karl SA, Avise JC (1993) PCR-based assays of Mendelian polymorphisms from anonymous single-copy nuclear DNA: techniques and applications for population genetics. Molecular Biology and Evolution 10:342–361

Karl SA, Bowen BW, Avise JC (1992) Global population structure and male-mediated gene flow in the Green Turtle (*Chelonia mydas*): RFLP analysis of anonymous nuclear loci. Genetics 131:163–173

Keane B (1990a) Dispersal and inbreeding avoidance in the White-footed Mouse, *Peromyscus leucopus*. Animal Behaviour 40:143–152

Keane B (1990b) The effect of relatedness on reproductive success and mate choice in the White-footed Mouse, *Peromyscus leucopus*. Animal Behaviour 39:264–273

Keller LF, Arcese P, Smith JNM, Hochachka WM, Stearns SC (1994) Selection against inbred song sparrows during a natural population bottleneck. Nature 372:356–357

Kenny JS, Smith JLD, Starfield AM, McDouglas CW (1995) The long-term effects of tiger poaching on population viability. Conservation Biology 9:1127–1133

Kimura M (1968) Genetic variability maintained in a finite population due to mutational production of neutral and nearly neutral isoalleles. Genetic Research 11:247–269

Kimura M, Ohta T (1971) Theoretical aspects of population genetics. Princeton University Press, Princeton, NJ

Klein J (1986) Natural history of the major histocompatibility complex. Wiley, New York

Klein J, Satta Y, O'h Uigin C (1993) The molecular descent of the major histocompatibility complex. Annual Review of Immunology 11:269–295

Kocher TD, Thomas WK, Meyer A, Edwards SV, Pääbo S, Villablanca FX, Wilson AC (1989) Dynamics of mitochondrial DNA evolution in animals: amplification and sequencing with conserved primers. Proceedings of the National Academy of Science of the USA 86:6196–6200

Koehn RK, Eanes WF (1978) Molecular structure and protein variation within and among subpopulations. Evolutionary Biology 11:39–100

Kretzmann MB, Gilmartin WG, Meyer A, Zegers GP, Fain SR, Taylor BF, Costa DP (1997) Low genetic variability in the Hawaiian Monk Seal. Conservation Biology 11:483–490

Kuhnlein U, Zadworny D, Dawe Y, Fairfull RW, Gavora JS (1990) Assessment of inbreeding by DNA fingerprinting: development of a calibration curve using defined strains of chickens. Genetics 125:161–165

Lacy RC (1993) Vortex: a computer simulation model for population viability analysis. Wildlife Resources 20:45–65

Lacy RC, Ballou JD, Prince F, Starfield A, Thompson EA (1995) Pedigree analysis for population management. In: Ballou JD, Gilpin M, Foose TJ (eds) Population management for survival and recovery. Columbia University Press, New York, pp 57–75

Lamont SJ, Bolin C, Cheville N (1987) Genetic resistance to fowl cholera is linked to the major histocompatibility complex. Immunogenetics 25:284–289

Lande R (1981) The minimum number of genes contributing to quantitative variation between and within populations. Genetics 99:541–543

Lande R (1988) Genetics and demography in biological conservation. Science 241:1455–1460

Lande R (1995) Mutation and conservation. Conservation Biology 9:782–791

Lande R, Barrowclough GR (1987) Effective population size, genetic variation and their use in population management. In: Soulé ME (ed) Viable populations for conservation. Cambridge University Press, Cambridge, UK, pp 87–124

Lerner IM (1954) Genetic homeostasis. Oliver and Boyd, Edinburgh, UK

Levitan DR, Grosberg RK (1993) The analysis of paternity and maternity in the marine hydrozoan Hydractinia symbiolongicarpus using randomly amplified polymorphic DNA (RAPD) markers. Molecular Ecology 2:315–326

Lewis PO, Snow AA (1992) Deterministic paternity exclusion using RAPD markers. Molecular Ecology 1:155–160

Lewontin RC, Hubby JL (1966) A molecular approach to the study of genic heterozygosity in natural populations. II Amount of variation and degree of heterozygosity in natural populations of Drosophila pseudoobscura. Genetics 54:595–609

Li, W-H (1997) Molecular evolution. Sinauer Associates, Sunderland, MA

Loeschcke V, Tomiuk J, Jain SK (1994) Conservation genetics. Birkhäuser Verlag, Basel, Switzerland

Loew SS, Fleischer RC (1996) Construction and screening of microsatellite-enriched genomic libraries. In: Ferraris J, Palumbi S (eds) Molecular zoology: strategies and protocols. Wiley-Liss, New York, pp 456–461

Lynch M (1988) Estimation of relatedness by DNA fingerprinting. Molecular Biology and Evolution 5:584–599

Lynch M (1990) The similarity index and DNA fingerprinting. Molecular Biology and Evolution 7:478–484

Lynch M (1991) Analysis of population genetic structure by DNA fingerprinting. In: Burke T, Dolf G, Jeffreys AJ, Wolff R (eds) DNA fingerprinting approaches and applications. Birkhäuser Verlag, Basel, Switzerland, pp 113–126

Lynch M (1996) A quantitative-genetic perspective on conservation issues. In: Avise JC, Hamrick JL (eds) Conservation genetics: case histories from nature. Chapman and Hall, New York, pp 471–501

Mace GM, Lande R (1991) Assessing extinction threats: toward a reevaluation of IUCN threatened species categories. Conservation Biology 5:148–157

Madsen T, Stille B, Shine R (1996) Inbreeding depression in an isolated population of Adders Vipera berus. Biological Conservation 75:113–118

Manning CJ, Potts WK, Wakeland EK, Dewsbury DA (1992) What's wrong with MHC mate choice experiments? In: Doty RL, Müller-Schwarze DM (eds) Chemical signals in vertebrates 6. Plenum, New York, pp 229–235

Martin AP, Naylor GJP, Palumbi SR (1992a) Rates of mitochondrial DNA evolution in sharks are slow compared with mammals. Nature 357:153–155

Martin RD, Dixson AF, Wickings EJ (eds) (1992b) Paternity in primates: genetic tests and theories. Karger, Basel, Switzerland

Maxam AM, Gilbert W (1977) A new method for sequencing DNA. Proceedings of the National Academy of Science of the USA 74:560–564

Maxam AM, Gilbert W (1980) Sequencing end-labelled DNA with base-specific chemical cleavage. Methods in Enzymology 65:499–559

Mayr E (1963) Populations, species and evolution. Belknap Press, Cambridge, MA

Mayr E (1969) The biological meaning of species. Biological Journal of the Linnean Society 1:311–320

McGuire KL, Duncan WR, Tucker PW (1985) Syrian Hamster DNA shows limited polymorphism at class I-like loci. Immunogenetics 22:257–268

McKitrick MC, Zink RM (1988) Species concepts in ornithology. Condor 90:1–14

Menotti-Raymond M, O'Brien SJ (1993) Dating the genetic bottleneck of the African Cheetah. Proceedings of the National Academy of Science of the USA 90:3172–3176

Meyer A, Kocher TD, Basasibwaki P, Wilson AC (1990) Monophyletic origin of Lake Victoria cichlid fishes suggested by mitochondrial DNA sequences. Nature 347:550–553

Miller PS (1995) Selective breeding programs for rare alleles: examples from the Prezewalski's Horse and California Condor pedigrees. Conservation Biology 9:1262–1273

Miller PS, Glasner J, Hedrick PW (1993) Inbreeding depression and male-mating behavior in Drosophila melanogaster. Genetica 88:29–36

Milligan BG, McMurray CK (1993) Dominant vs codominant genetic markers in the estimation of male mating success. Molecular Ecology 2:275–283

Mills LS, Hayes SG, Baldwin C, Wisdom MJ, Citta J, Mattson DJ, Murphy K (1996) Factors leading to different viability predictions for a Grizzly Bear data set. Conservation Biology 10:863–873

Mindell DP, Honeycutt RL (1990) Ribosomal RNA in vertebrates: evolution and phylogenetic applications. Annual Review of Ecological Systematics 21:541–566

Miththapala S, Seidensticker J, O'Brien S (1995) Phylogeographic subspecies recognition in Leopards (Panthera pardus): Molecular genetic variation. Conservation Biology 10(4):1115–1132

Mitton JB (1993) Theory and data pertinent to the relationship between heterozygosity and fitness. In: Thornhill N (ed) The natural history of inbreeding and outbreeding. University of Chicago Press, Chicago, IL, pp 17–41

Mitton JB, Pierce BA (1980) The distribution of individual heterozygosity in natural populations. Genetics 95:1043–1054

Moore HDM, Holt WV, Mace GM (1992) Biotechnology and the conservation of genetic diversity. Symposia of the Zoological Society of London 64. Clarendon Press, Oxford, UK

Moore SS, Sargeant LL, King TJ, Mattick JS, Georges M, Hetzel DJS (1991) The conservation of dinucleotide microsatellites among mammalian genomes allows the use of heterologous PCR primer pairs in closely related species. Genomics 10:654–660

Morales JC, Andau PA, Supriatna J, Zainuddin ZZ, Melnick DJ (1997) Mitochondrial DNA variability and conservation genetics of the Sumatran Rhinoceros. Conservation Biology 11(2):539–543

Morin PA, Ryder OA (1991) Founder contribution and pedigree inference in a captive breeding colony of Lion-tailed Macaques, using mitochondrial DNA fingerprint analyses. Zoo Biology 10:341–352

Morin PA, Woodruff DS (1992) Paternity exclusion using multiple hypervariable microsatellite loci amplified from nuclear DNA of hair cells. In: Martin RD, Dixson AF, Wickings EJ (eds) Paternity in primates: genetic tests and theories. Karger, Basel, Switzerland, pp 63–81

Morin PA, Moore JJ, Woodruff DS (1992) Identification of chimpanzee subspecies with DNA from hair and allele-specific probes. Proceedings of the Royal Society of London 249:293–297

Moritz C (1994a) Defining "evolutionary significant units" for conservation. Trends in Ecology and Evolution 9:373–375

Moritz C (1994b) Applications of mitochondrial DNA analysis in conservation: a critical review. Molecular Ecology 3:401–411

Moritz C, Dowling TE, Brown WM (1987) Evolution of animal mitochondrial DNA: relevance for population biology and systematics. Annual Review of Ecological Systematics 18:269–292

Nei M, Hughes AL (1991) Polymorphism and evolution of the major histocompatibility complex loci in mammals. In: Selander RK, Clark AG, Whittam TS (eds) Evolution at the molecular level. Sinauer, Sunderland, MA, pp 222–247

Nunney L, Campbell KA (1993) Assessing minimum viable population size: demography meets population genetics. Trends in Ecology and Evolution 8:234–239

O'Brien SJ, Evermann JF (1988) Interactive influence of infectious disease and genetic diversity in natural populations. Trends in Ecology and Evolution 3:254–259

O'Brien SJ, Mayr E (1991) Bureaucratic mischief: recognizing endangered species and subspecies. Science 251:1187–1188

O'Brien SJ, Roelke MC, Markes L, Newman A, Meltzer CA, Colly L, Evermann JF, Bush M, Wildt DC (1985) Genetic basis for species vulnerability in the Cheetah. Science 227:1428–1434

O'Brien SJ, Roelke ME, Yuhki N, Richards KW, Johnson W, Franklin WL, Anderson AE, Bass Jr. OL, Belden RC, Martenson JS (1990) Genetic introgression within the Florida Panther *Felis concolor*. National Geographic Research 6:485–494

Olney PJS, Mace GM, Feistner ATC (eds) (1994) Creative conservation. Chapman and Hall, London

Packer C, Gilbert DA, Pusey AE, O'Brien SJ (1991) A molecular genetic analysis of kinship and cooperation in African lions. Nature 351:562–564

Palumbi SR, Baker CS (1994) Contrasting population structure from nuclear intron sequences and mtDNA of Humpback Whales. Molecular Biology and Evolution 11:426–435

Parham P, Ohta T (1996) Population biology and antigen presentation by MHC class I molecules. Science 272:67–74

Pennock DS, Dimmick WW (1997) Critique of the evolutionary significant unit as a definition for "distinct population segments" under the US Endangered Species Act. Conservation Biology 11:611–619

Petit RJ, Mousadik AE, Pons O (1998) Identifying populations for conservation on the basis of genetic markers. Conservation Biology 12:844–855

Philipp DP, Gross MR (1994) Genetic evidence for cuckoldry in Bluegill *Lepomis macrochirus*. Molecular Ecology 3:563–569

Piper WH, Rabenold PP (1992) Using the fragment-sharing estimates from DNA fingerprinting to determine relatedness in a tropical wren. Molecular Ecology 1:69–78

Potts WK, Wakeland EK (1990) Evolution of diversity at the major histocompatibility complex. Trends in Ecology and Evolution 5:181–187

Potts WK, Manning CJ, Wakeland EK (1991) Mating patterns in seminatural populations of mice influenced by MHC genotype. Nature 352:619–621

Quattro JM, Vrijenhoek RC (1989) Fitness differences among remnant populations of the endangered Sonoran Topminnow. Science 245:976–978

Quinn TW, White BN (1987) Identification of restriction-fragment-length polymorphisms in genomic DNA of the Lesser Snow Goose (*Anser caerulescens caerulescens*). Molecular Biology and Evolution 4:126–143

Rabenold PP, Rabenold KN, Piper WH, Haydock J, Zack SW (1990) Shared paternity revealed by genetic analysis in cooperatively breeding tropical wrens. Nature 348:538–540

Ralls K, Ballou J (1982) Effects of inbreeding on infant mortality in captive primates. International Journal of Primatology 3:491–505

Ralls K, Brugger K, Ballou J (1979) Inbreeding and juvenile mortality in small populations of ungulates. Science 206:1101–1103

Ralls K, Ballou J, Templeton A (1988) Estimates of lethal equivalents and the cost of inbreeding in mammals. Conservation Biology 2:185–193

Ratner S, Lande R, Roper BB (1997) Population viability analysis of spring Chinook Salmon in the South Umpqua River, Oregon. Conservation Biology 11(4):879–889

Rave EH, Fleischer RC, Duval F, Black JM (1994) Genetic analyses through DNA fingerprinting of captive populations of Hawaiian geese. Conservation Biology 8:744–751

Reeve HK, Westneat DF, Noon WA, Sherman PW, Aquadro CF (1990) DNA "fingerprinting" reveals high levels of inbreeding in colonies of the Eusocial Naked Mole-rat. Proceedings of the National Academy of Science of the USA 87:2496–2500

Rico C, Rico I, Hewitt G (1994) An optimized method for isolating and sequencing large (CA/GT)n (*n* > 40) microsatellites from genomic DNA. Molecular Ecology 3:181–182

Riedy MF, Hamilton WJ III, Aquadro CF (1992) Excess of non-parental bands in offspring from known pedigrees assayed using RAPD PCR. Nucleic Acids Research 20:918

Robinson TJ, Morris DJ, Fairall N (1991) Interspecific hybridization in the bovidae: sterility of *Alcelaphus buselaphus x Damaliscus dorcas* F1 progeny. Biological Conservation 58:345–356

Rockwell RF, Barrowclough GF (1995) Effective population size and life time reproductive success. Conservation Biology 9(5):1225–1233

Rohlf DJ (1994) There's something fishy going on here: a critique of the National Marine Fisheries Service's definition of species under the Endangered Species Act. Environmental Law 24:617–671

Rojas M (1992) The species problem and conservation: what are we protecting? Conservation Biology 6:552–569

Roy MS, Geffen E, Smith D, Ostrander E, Wayne RK (1994) Patterns of differentiation and hybridization in North American wolf-like canids revealed by analysis of microsatellite loci. Molecular Biology and Evolution 11:553–570

Ryder OA (1986) Species conservation and systematics: the dilemma of subspecies. Trends in Ecology and Evolution 1:9–10

Sanger F, Nicklen S, Coulson AR (1977) DNA sequencing with chain-terminating inhibitors. Proceedings of the National Academy of Science of the USA 74:5463–5467

Schlötterer C, Amos B, Tautz D (1991) Conservation of polymorphic simple sequence loci in cetacean species. Nature 354:63–65

Scribner KT, Arntzen JW, Burke T (1994) Comparative analyses of intra- and interpopulation genetic diversity in *Bufo bufo*, using allozyme, single-locus microsatellite, minisatellite, and multilocus minisatellite data. Molecular Biology and Evolution 11:737–748

Selander RK, Whittam TS (1983) Protein polymorphism and the genetic structure of populations. In: Nei M, Koehn RK (eds) Evolution of genes and proteins. Sinauer, Sunderland, MA, pp 89–114

Shaffer ML (1981) Minimum population sizes for species conservation. BioScience 31:131–134

256   Sabine S. Loew

Sherwin WB, Murray ND, Graves JAM, Brown PR (1991) Measurement of genetic varia-
    tion in endangered populations: Bandicoots (Marsupialia:Peramelidae) as an example.
    Conservation Biology 5:103–108
Shields WM (1982) Philopatry, inbreeding and the evolution of sex. State University of
    New York Press, Albany, NY
Shields WM (1993) The natural and unnatural history of inbreeding and outbreeding. In:
    Thornhill NW (ed) The natural history of inbreeding and outbreeding: theoretical and
    empirical perspectives. University of Chicago Press, Chicago, IL, pp 143–169
Sibley CG, Ahlquist JE (1981) The phylogeny and relationships of the ratite birds as
    indicated by DNA-DNA hybridization. In: Scudder GGE, Reveal JL (eds) Evolution
    today. Hunt Institute Botanical Document, Pittsburgh, PA, pp 301–335
Simberloff D, Cox J (1987) Consequences and costs of conservation corridors. Conserva-
    tion Biology 1:63–71
Simon C, Frati F, Beckenbach A, Crespi B, Liu H, Flook P (1994) Evolution, weighting,
    and phylogenetic utility of mitochondrial gene sequences and a compilation of conserved
    polymerase chain reaction primers. Entomological Society of America 87:651–701
Slade RW (1992) Limited MHC polymorphism in the Southern Elephant Seal: implications
    for MHC evolution and marine mammal population biology. Proceedings of the Royal
    Society 249:163–171
Snell GD (1968) The H-2 locus of the mouse: observations and speculations concerning its
    comparative genetics and polymorphism. Folia Biologica 14:335–358
Stockwell CA, Mulvey M, Vinyard GL (1996) Translocations and the preservation of
    allelic diversity. Conservation Biology 10:1133–1141
Streilein JW, Gerboth-Darden A, Phillips JT (1984) Primordial MHC function may be best
    served by monomorphism. Immunology Today 5:87–88
Taberlet P, Swenson JE, Sandegren F, Bjarvall F (1995) Localization of a contact zone
    between two highly divergent mitochondrial DNA lineages of the Brown Bear *Ursus
    arctos* in Scandinavia. Conservation Biology 9:1255–1261
Tarr CL (1995) Primers for amplification and determination of mitochondrial control-
    region sequences in oscine passerines. Molecular Ecology 4:527–529
Tarr CL, Fleischer RC (1993) Mitochondrial DNA variation and evolutionary relationships
    in the Amakihi complex. The Auk 110(4):825–831
Tautz D (1989) Hypervariability in mitochondrial DNA in a regional population of the
    Great Tit (*Parus major*). Biochemical Genetics 25:95–110
Templeton AR, Read B (1994) Inbreeding: one word, several meanings, much confusion.
    In: Loeschcke V, Tomiuk J, Jain SK (eds) Conservation genetics. Birkhäuser Verlag,
    Boston, MA, pp 91–105
Templeton AR, Hemmer H, Mace G, Seal US, Shields WM, Woodruff DS (1986) Local
    adaptation, coadaptation, and population boundaries. Zoo Biology 5:115–125
Thomas WK, Paabo S, Villablanca FX, Wilson AC (1990) Spatial and temporal continuity
    of Kangaroo Rat populations shown by sequencing mitochondrial DNA from museum
    specimens. Journal of Molecular Evolution 31:101–112
Thornhill NW (ed) (1993) The natural history of inbreeding and outbreeding: theoretical
    and empirical perspectives. University of Chicago Press, Chicago, IL
Tiwari JL, Terasaki PI (1985) HLA and disease associations. Springer Verlag, New York
Tomlinson C, Mace GM, Black JM, Hewston N (1991) Improving the management of a
    highly inbred species: the case of the White-winged Wood Duck *Cairina scutulata* in
    captivity. Wildfowl 42:123–133
U.S. House of Representatives (1973) Report 412, 93rd Congress, 1st Session

Vane-Wright RI, Humphries CJ, Williams PH (1991) What to protect—systematics and the agony of choice. Biological Conservation 55:235–254

Vogler AP, DeSalle R (1994) Diagnosing units of conservation management. Conservation Biology 8:354–363

Vrijenhoek RC (1989) Population genetics and conservation. In: Western D, Pearl MC (eds) Conservation for the twenty-first century. Oxford University Press, New York, pp 89–98

Vrijenhoek RC (1994) Genetic diversity and fitness in small populations. In: Loeschcke V, Tomiuk J, Jain SK (eds) Conservation genetics. Birkhäuser Verlag, Basel, Switzerland, pp 37–53

Vrijenhoek RC, Leberg PL (1991) Let's not throw the baby out with the bathwater: a comment on management for MHC diversity in captive populations. Conservation Biology 5:252–254

Vrijenhoek RC, Douglas M, Meffe GK (1985) Conservation genetics of endangered fish populations in Arizona. Science 229:400–402

Waples RC (1991) Pacific Salmon, *Oncorhynchus* spp, and the definition of "species" under the Endangered Species Act. In: Nielsen JL (ed) Evolution and the aquatic ecosystem: defining unique units in population conservation. Symposium 17. American Fisheries Society, Bethesda, MD, pp 8–27

Waples RS (1998) Evolutionary significant units, distinct population segments, and the Endangered Species Act: reply to Pennock and Dimmick. Conservation Biology 12:718–721

Wayne RK, Jenks SM (1991) Mitochondrial DNA analysis implying extensive hybridization of the endangered Red Wolf: genetic consequences of population decline and fragmentation. Conservation Biology 6:559–569

Wayne RK, Lehman N, Girman D, Gogan PJP, Gilbert DA, Hansen K, Peterson RO, Seal US, Eisenhawer A, Mech ID, Krumenaker RJ (1991) Conservation genetics of the endangered Isle Royale Gray Wolf. Conservation Biology 5:41–51

Wayne RK, Bruford MW, Girman D, Rebholz WER, Sunnucks P, Taylor AC (1994) Molecular genetics of endangered species. In: Olney PJS, Mace GM, Feistner ATC (eds) Creative conservation: interactive management of wild and captive animals. Chapman and Hall, London, pp 92–117

Weber JL, May PE (1989) Abundant class of human DNA polymorphisms which can be typed using the polymerase chain reaction. American Journal of Human Genetics 44:388–396

Wedekind C, Seebeck T, Bettens F, Paepke AJ (1995) MHC-dependent mate preference in humans. Proceedings of the Royal Society of London 260:245–249

Welsh J, McClelland M (1990) Fingerprinting genomes using PCR with arbitrary primers. Nucleic Acids Research 18:7213–7218

Westneat DF (1990) Genetic parentage in indigo buntings. A study using DNA fingerprinting. Behavioral Ecology and Sociobiology 27:67–76

Williams JGK, Kubelik AR, Livak J, Rafalski JA, Tingey SV (1990) DNA polymorphisms amplified by arbitrary primers are useful as genetic markers. Nucleic Acids Research 18:6531–6535

Williams SM, DeSalle R, Strobeck C (1985) Homogenization of geographical variants at the nontranscribed spacer of rDNA in *Drosophila mercatorum*. Molecular Biology and Evolution 2:338–346

Wilson AC, Cann RL, Carr SM, George M, Gyllensten UB, Helm-Bychowski KM, Higuchi RG, Palumbi SR, Prager EM, Sage RD, Stoneking M (1985) Mitochondrial

DNA and two perspectives on evolutionary genetics. Biological Journal of the Linnean Society 26:375–400

Wolstenholme DR (1992) Animal mitochondrial DNA: structure and evolution. In: Jeon KW, Wolstenholme DR (eds) Mitochondrial genomes. International Review of Cytology 141:173–216

Wolstenholme DR, Clary DO (1985) Sequence evolution of *Drosophila* mitochondrial DNA. Genetics 109:725–744

Woodford MH, Rossiter PB (1994) Disease risks associated with wildlife translocation projects. In: Olney PJS, Mace GM, Feistner ATC (eds) Creative conservation. Chapman and Hall, London, pp 178–198

Wright JM (1977) Evolution and the genetics of populations, vol 3: experimental results and evolutionary deductions. University of Chicago Press, Chicago, IL

Wright S (1931) Evolution in Mendelian populations. Genetics 16:97–159

Wright S (1938) Size of population and breeding structure in relation to evolution. Science 87:430–431

Yamazaki K, Boyse EA, Mike V, Thaler HT, Mathieson BJ, Abbot J, Boyse J, Zayas ZA, Thomas L (1976) Control of mating preferences in mice by genes in the major histocompatibility complex. Journal of Experimental Medicine 144:1324–1335

Yamazaki K, Yamaguchi M, Baranoski L, Bard J, Boyse EA, Thomas L (1979) Recognition among mice. Journal of Experimental Medicine 150:755–760

Yamazaki K, Beauchamp GK, Egorov IK, Bard J, Thomas L, Boyse EA (1983) Sensory distinction between H-2b and H-2bm1 mutant mice. Proceedings of the National Academy of Sciences of the USA 80:5685–5688

15

Modeling Problems in Conservation Genetics Using Laboratory Animals

Richard Frankham

Introduction

The biological diversity of the planet is rapidly being depleted. Loss of habitat, overexploitation, pollution, and introduced species are threatening the survival of an ever-increasing number of species. Consequently, many species require human intervention to save them from extinction or at least to optimize their management. Intervention may take the form of habitat protection, habitat restoration, reduction of exploitation, control of predators and diseases (usually introduced), captive breeding, reintroduction, and translocation.

The critical genetic issues in the management of threatened, rare, and endangered species are

1. Inbreeding depression
2. Loss of genetic variation in small populations
3. Fragmentation of populations and reduction in migration
4. Genetic adaptation to captivity and its effect on reintroduction success
5. Taxonomic uncertainties that confound the objectives of conservation efforts.

For issues 1–4, there is considerable theory that has been used in the genetic management of rare and endangered species. However, this theory is typically simple single locus neutralist theory that ignores natural selection, mutation, and linkage. It is notable that experimental studies have shown significant deviations from the predictions of related theory on the relationship between inbreeding and heterozygosity (Mina et al. 1991; Rumball et al. 1994).

Consequently, it is critical that theory underlying genetic management of endangered species be subject to experimental evaluation. However, endangered species are unsuitable for such research as typically they are present in low numbers and are expensive to maintain. Further, their reproduction is often slow and difficult to manage. Laboratory species provide an essential link between theory and wildlife in the field. They have all the realistic genetic variables that are absent in most theoretical models (mutation, migration, natural selection, and linkage), and their reproduction can be controlled so that theory can be adequately tested in controlled replicated studies. Until recently, this area of conservation genetics has been sadly neglected.

The objectives of this chapter are to define the appropriate role for laboratory studies in conservation genetics, to illustrate this with examples mainly from our laboratory, and to point to other studies that need to be done using laboratory species.

Essential Interplay of Theory and Experimentation

An evolving population is a complex system in which multiple genes in linkage groups are subject to mutation, migration, natural selection, and the random events of meiosis. The behavior of the system depends on the breeding system and environmental conditions. One approach to understanding such a complex system is as follows:

1. Build simple mathematical models, firstly keeping all factors constant, then varying one factor at a time, then two factors, and so on.
2. Experimentally evaluate the predictions of the models.
3. On the basis of the experimental results, refine the theoretical models.
4. Subject the refined models to experimental evaluation (i.e., the process of model development, evaluation/monitoring, and redevelopment should be cyclic and ongoing).

Such approaches have been used in quantitative genetics, population genetics, and animal breeding and should be used in conservation genetics. A relevant example of this interplay between theory and experimentation is provided by investigations into factors affecting the long-term evolution of populations. Following the rediscovery of Mendelian inheritance, the multiple-factor hypothesis was developed to explain quantitative genetic variation and experimentally validated (East 1916). Equations were developed to predict response to selection (Fisher 1918; Wright 1921; Lush 1945). These provided reasonable predictions in the short term (Clayton et al. 1957) but not in the long term (Clayton and Robertson 1957). As a consequence, Robertson (1960) developed a theory of limits in artificial selection, based on the loss of genetic variation in finite population sizes and the effects of selection (chance plus selection). Experimental evaluation of this theory in *Drosophila* (Jones et al. 1968) and mice (Eisen 1975) revealed qualitative agreement with predictions but limited quantitative predictability. This work established that larger populations showed more potential for evolutionary change. Genetic analyses of selection lines established that mutations occurring in the lines contributed to long-term selection response (Hollingdale 1971; Frankham et al. 1978; Frankham 1980, 1982; Yoo 1980). Hill (1982) extended the theory of limits in artificial selection by adding mutation to chance and selection. Predictions of Hill's theory have subsequently been evaluated (Frankham 1983; Lopez-Fanjul and Caballero 1990). This is an object lesson in how recommendations for genetic management of endangered species should develop.

Need for Experimental Evaluations of Theory in Conservation Genetics

Experimental evaluation of conservation genetics theory is essential to determine whether it applies in populations of real organisms. A cautionary tale of theory that unexpectedly fell down in practice is selection index theory for simultaneously selecting several traits in animal breeding. This was predicted to be superior to alternative methods (Hazel and Lush 1942) and was applied in chicken breeding without critical experimental evaluation in laboratory animals. Breeding companies who used selection indices to improve egg production stocks went out of business. It was much later discovered (see Hill 1981) that there are statistical problems in estimating the required parameters when there are many characters and the data set is relatively small. Consequently, it is critical that theory in conservation genetics be subject to controlled experimental evaluation.

Role of Laboratory Species

Endangered species are unsuitable for such research in conservation genetics particularly because it may be unethical to risk the fate of an endangered species on an untested procedure (although this has been done frequently). Consequently, evaluations of theory and investigations of problems in conservation genetics are best done by using laboratory animals and plants. Laboratory species are relatively common but have similar genetic behavior to endangered species, so that the information gathered in laboratory experiments can be extrapolated to answer questions and test hypotheses relevant to the endangered species. I first recognized the lack of laboratory animal studies in conservation genetics while preparing a paper on "messages from animal breeding for conservation genetics" in 1990 (Frankham 1992), and as a consequence, David Briscoe and I began systematic experimental investigations of problems in conservation genetics by using *Drosophila*. Independently, Hedrick and colleagues began such work (Miller and Hedrick 1993). Until then, the only experimental evaluation of conservation genetics theory had been done in the context of animal breeding, population, and quantitative genetics (Frankham 1992).

Drosophila melanogaster is useful to model problems in conservation genetics because of its short generation interval, low cost, the detailed knowledge of its genetics, availability of innumerable stocks to aid genetic analysis, and its extensive use as a model for related problems in animal breeding, population genetics, and quantitative genetics (Frankham 1982; Falconer and Mackay 1996). *Drosophila* has proved to be a reliable model for naturally outbreeding diploid eukaryotes. I am aware of no case in which *Drosophila* experiments have yielded qualitative results at variance with equivalent experiments with other such species (Frankham 1982).

Nevertheless, conclusions from laboratory studies are greatly strengthened when complemented by meta-analyses of wildlife data (of the kind done by Ralls and Ballou 1983). Both a laboratory animal study and a meta-analysis of wildlife data have verified the predicted relationship between population size and allozyme genetic variation (Frankham 1996; Woodworth 1996). Similarly, low ratios of effective-to-adult population size found in large *Drosophila* populations (Briscoe et al. 1992) were found to apply to wildlife following a meta-analysis (Frankham 1995a). If a meta-analysis is not possible, it is desirable to carry out studies in a second laboratory species, as has frequently been done in animal breeding (see Ralls and Meadows 1993).

Modeling Problems in Conservation Genetics Using *Drosophila*

Inbreeding and Extinction

The fundamental assumption underlying the application of genetics within conservation biology is that inbreeding and loss of genetic variation increases the risk of extinction. Although there is ample evidence that inbreeding causes depression in components of reproductive fitness in humans, domestic animals, outbred plants, and wildlife in zoos and in nature (see Wright 1977; Ralls et al. 1988; Frankham 1995b; Falconer and Mackay 1996), there is little direct evidence of its effects on extinction. Further, the only paper on this issue (Soulé 1980) failed to distinguish genetic and environmental causes of extinction. Methods were devised to separate these and applied to data sets from *Drosophila melanogaster, D. virilis,* and *Mus musculus* (Frankham 1995c). Inbreeding clearly increased extinction rates. Notably, there was a threshold relationship in all species, with little extinction until intermediate levels of inbreeding but increasing rates of extinction thereafter. Consequently, populations that are not being closely monitored may give little warning of impending inbreeding crises. Subsequently, Saccheri and colleagues (1998) have found that inbreeding increases extinction risk for butterfly populations in nature.

Effective Population Size

Genetic effects of small population size are predicted to depend on the effective population size (N_e) rather than the census size (Wright 1969; Crow and Kimura 1970; Falconer and Mackay 1996). Thus we have concentrated on factors predicted to affect N_e, for instance, variance in family size, unequal sex ratios, and fluctuations in numbers between generations. Much of the relevant theory is more than 60 years old (Wright 1931) but was untested.

Equalization of Family Size

Equalization of family sizes (EFS) is predicted to double effective population sizes and hence reduce rates of inbreeding, loss of genetic variation, and inbreeding depression. Consequently, EFS is recommended for use in the captive breeding of endangered species. It has the potential to double the size of the world's inadequate captive breeding resources. Borlase and associates (1993) performed the first controlled replicated evaluation of this theory. Ten replicate populations of both EFS and variable family size (VFS) controls were tested in a paired comparison. As predicted, EFS led to greater N_e, slower rates of inbreeding, and greater retention of allozyme genetic variation and reproductive fitness.

However, it was surprising that there was no significant difference between the treatments in quantitative genetic variation for abdominal bristle number (Table 15.1). The expectation is that additive genetic variance will decline in proportion to the increased inbreeding coefficient (abdominal bristle number shows predominantly additive genetic variation). There was not even a trend in the predicted direction, as in five cases genetic variation in EFS exceeded that of VFS, and in five the reverse was true.

On the basis of multilocus theory and computer simulations, Bulmer (1980) predicted that random deviations from Hardy-Weinberg equilibrium and linkage equilibrium will cause substantial random deviations from true genetic variances in finite populations. This suggests that a difference in the predicted direction should be evident when our lines have attained Hardy-Weinberg equilibrium and linkage equilibrium after maintenance for many generations by using a large population size. This prediction was verified (Table 15.1).

TABLE 15.1. Genetic variation (V_{G*}) for abdominal bristle number in equalization of family size (EFS) and variable family size (VFS) treatments at generation (G) 11 and after a further 25 generations at an expanded size.[a]

Treatment	V_{G*} G.11	$\sqrt{V_{G*}}$ G.11	V_{G*} G.36	$\sqrt{V_{G*}}$ G.36
EFS	5.47 ± 0.87	2.27 ± 0.19	7.72 ± 1.24	2.71 ± 0.21
VFS	4.97 ± 0.85	2.12 ± 0.22	5.03 ± 1.68	1.96 ± 0.38[b]
Base population	6.63 ± 1.75	2.46 ± 0.33		
EFS > VFS	5/10		8/10	
Paired t-test		0.45		2.27
Probability[c]		0.33		0.025

[a]Square root transformed data are presented, along with results of tests of significance of the differences between the EFS and VFS treatments on the transformed scale. V_{G*} was estimated as $\sigma^2 p - \sigma^2 d$, where $\sigma^2 p$ is the phenotypic variance for fourth plus fifth abdominal bristle number and $\sigma^2 d$ is the variance of the difference between the two counts on the same individual. This measures the total genetic variation, plus a small component of between-fly environmental effects. It has proved to be a reliable means for measuring changes in quantitative genetic variation (see Reeve and Robertson 1954; Latter 1964; Frankham and Nurthen 1981).

[b]One negative value was set to zero to allow transformation.

[c]One-tailed test, as the expectation is directional.

Lande and Barrowclough (1987) recommended monitoring of quantitative genetic variation as part of conservation management. However, our results urge caution in the use of this measure to monitor genetic variation in small populations. Further, simple single-locus models are inadequate to predict the detailed behavior of quantitative genetic variation in small populations.

Harems

The majority of mammals, plus some birds, reptiles, fish, amphibians, and insects, have unequal sex ratios of breeding individuals (i.e., polygamy). Harems are the most extreme form of polygamy. In addition, many species that do not have harems in nature are maintained in captivity by using harems to avoid aggression between males. Harems are predicted to reduce N_e and so increase the rate of inbreeding and loss of genetic variation. Briton and associates (1994) confirmed these predictions. The message is that captive breeders should, as far as possible, avoid harem breeding structures, especially in those species that do not normally have them.

Fluctuating Population Sizes

Fluctuating population sizes are common in natural populations of animals. This is predicted to reduce N_e, to increase the rate of inbreeding, and to reduce genetic diversity. These predictions have been experimentally validated (Woodworth et al. 1994). Wherever possible, populations should be managed so that there is minimal variation in population size over generations.

Equalizing Founder Representation

Many captive populations have been founded at the last moment when only a few individuals of the threatened species remain. These founders often contribute unequally to future generations, such that the rate of inbreeding and loss of genetic variation are increased. Consequently, it is recommended that such populations be managed to equalize founder representation. Experimental evaluation of this recommendation has demonstrated reduced rates of inbreeding and loss of genetic variation but no benefits in reproductive fitness (Loebel et al. 1992). Surprisingly, changes in founder representation were almost completed in the first generation. Computer simulations subsequently showed that this was to be expected when discrete nonoverlapping generations were used. The mating of individuals connects them, such that the fates of underrepresented and overrepresented founders can become inextricable bound. Similar effects have been seen in computer simulations comparing the effects of managing small pedigreed populations using minimum kinship, as compared with maximum avoidance of inbreeding (Ballou and Lacy 1995). Further theoretical investigations of this issue are warranted.

Immigration

The obvious solution when small populations suffer from inbreeding depression is to introduce immigrants. This is often costly and involves risks of introducing disease or causing behavioral disruption. There is no clear-cut theory to predict the benefits of introducing immigrants into partially inbred populations, and most empirical evidence involves hybridizing all individuals, rather than introducing a few immigrants. Spielman and Frankham (1992) evaluated the benefits of introducing a single immigrant into partially inbred populations ($F = 0.5$). The immigrant resulted in an approximate doubling of reproductive fitness.

Rapid Genetic Adaptation to Captivity

An inevitable consequence of maintaining populations of endangered species in captivity is that natural selection operates to adapt populations genetically to the captive environment. This is likely to reduce the chances of successful reintroduction into the wild. The speed of genetic adaptation can be very rapid if selection is strong and generation interval is short (Frankham and Loebel 1992). Even under benign captive conditions, genetic adaptation results in substantial genetic deterioration (selection of traits that facilitate survival in captivity but are deleterious in the wild) (Woodworth 1996). Frankham and Loebel (1992) presented equations for predicting the rate of genetic adaptation to captivity and the means for minimizing it. These predictions need to be experimentally evaluated.

Optimum Management of Populations of Endangered Species Founded from Few Individuals

When a small number of animals is used to found a captive population, inbreeding is inevitable and will lead to inbreeding depression. Based on theoretical considerations, research with *D. mercatorum,* and results of a program with Speke's Gazelle (Templeton and Read 1983, 1984), it was recommended that such species be managed to adapt them to tolerate inbreeding. The recommended procedure has been widely interpreted as an inbreeding program devised to purge deleterious recessives (e.g., Vrijenhoek 1994), but Templeton (personal communication) attributes it primarily to positive selection for epistatic complexes. It involves an increase in total population size, the use of inbred parents, production of inbred offspring, and equalizing founder representation in both parents and offspring. Although the recommended procedure has been used with endangered species, it has not been subjected to a controlled experimental evaluation. Evaluations of different aspects of this procedure and a purging regime have failed to yield meaningful lowering of inbreeding depression (Loebel et al. 1992; Frankham et al. 1993; Wright et al. 1999). Computer simulation studies (Hedrick 1994) indicated that inbreeding should be effective in purging lethals but may be deleterious for genes of small effect. There is controversy regarding the optimum management of such populations; this can only be resolved by clearly defining the procedures and testing them against alternative management programs.

Research Using Other Species

Butterflies (Brakefield and Saccheri 1994), Flour Beetles (Pray et al. 1994), House Flies (Backus et al. 1995), fish (Leberg 1992), and native mice (Brewer et al. 1990) have begun to be used as laboratory species to model problems in conservation genetics. Lacy and colleagues have used native species of mice in work that bridges the gap between laboratory species and wildlife (Brewer et al. 1990; Jimenez et al. 1994). Vrijenhoek and colleagues (see Vrijenhoek 1994) have made innovative use of a combination of field and laboratory studies in two species of desert fish, one of which was endangered.

Other Questions That Need Addressing in Laboratory Species

1. What are the minimum viable population sizes required to avoid inbreeding depression and to avoid loss of evolutionary potential? Franklin (1980) predicted that N_e's of 50 and 500, respectively, were needed. We are currently evaluating these issues.
2. How can genetic adaptation to captivity be minimized and reintroduction success maximized? Equalizing family sizes, extending the generation interval, and minimizing selection in captivity are predicted to minimize genetic adaptation to captivity and hence maximize reintroduction success (Frankham and Loebel 1992; Frankham 1995d). The predicted benefits of equalizing family size in reducing the rate of genetic adaptation to captivity has recently been verified (Frankham and Margan, unpublished data).
3. Is minimizing kinship the optimum procedure for managing small pedigreed populations of endangered species? Several procedures have been, or still are, recommended for the genetic management of such populations. Ballou and Lacy (1995) predicted that minimum kinship is the optimum procedure to retain genetic diversity and allelic diversity, based on computer simulations. However, it was not necessarily this procedure that minimized inbreeding. This procedure requires urgent testing as it is currently being applied to endangered species. It has recently been evaluated in *Drosophila* (Montgomery et al. 1997).
4. What determines the evolutionary potential of populations—heterozygosity or allelic diversity? Most considerations are based on heterozygosity, but Allendorf (1986), Fuerst and Maruyama (1986), and others have raised the question of whether allelic diversity is the critical issue. This issue can only be resolved experimentally.
5. What is the optimum genetic management strategy for populations of endangered species founded from few individuals? This issue is a matter of great

controversy, as indicated above. Clarification of recommendations, further experimental evaluations, and further theoretical developments are required.

6. Does the accumulation of mildly detrimental mutations pose an extinction threat to endangered species? Lande (1995) suggested that this may be as significant a threat to sexual species as environmental stochasticity and catastrophes. The first evaluation of this issue in a sexually reproducing species suggests that it may be of minor importance in conservation (Gilligan et al. 1997).

7. Is a single large population or several small populations of equivalent total size the best means for genetically managing endangered species in captivity? If there are no extinctions, genetic theory favors the several small populations (Kimura and Crow 1963). Experimental results support these predictions (Margan et al. 1998).

8. Do population size bottlenecks reduce evolutionary potential? This is a controversial and unresolved issue (see Frankham 1995b).

9. Can population viability assessment (PVA) software be experimentally evaluated? The prediction of the risk of extinction in populations by using PVA software represents a major advance in conservation biology. However, it is critical that the predictive powers of such software be tested. Experimental testing using laboratory species provides one means for doing this.

10. What are the roles of inbreeding, demographic stochasticity, and environmental stochasticity in extinction? This is a matter of great controversy. Laboratory experimentation has extended our understanding of the potential role of inbreeding and loss of genetic variation (Frankham 1995b, c). It should be possible to devise experiments to greatly increase our understanding of the causes of extinction. In fact, the first use of *Drosophila* to investigate problems in conservation biology was a study of this problem by Forney and Gilpin (1989).

11. What are the optimum genetic management procedures for selfing species? I am aware of no relevant laboratory studies of selfing species. Such studies are urgently required as the management of these species poses issues that cannot be addressed in outbreeding species. *Arabidopsis thaliana* is an obvious candidate, especially as it is now a model species for genome mapping and sequencing studies in plants, as well as a model plant for developmental biology, genetics, and molecular biology.

Relationship Between Computer Simulation and Modeling in Laboratory Animals

Computer simulation provides an invaluable tool for investigating models too complex to investigate analytically and for checking complex theory. However, its role is intermediate between that of theory and experimentation with living organisms. Computer simulation does not encompass the full complexity of the living organism as many parameters are unknown and some variables may yet be unknown (e.g., transposons and split genes have been recent unexpected discoveries). There is a need for the development of more biologically realistic computer software. For example, it should be possible to develop software to evaluate the effects of inbreeding on reproductive fitness as well as on genetic variation. This would be a multilocus program with linkage, recombination, mutation, and deleterious alleles. In this case, there are ample experimental data to evaluate and refine software.

Limitations of Studies with Laboratory Species

Not all issues in conservation genetics are amenable to studies with laboratory species. The most obvious exception is the resolution of taxonomic uncertainties. Clearly, such work must be done with the endangered species itself. Detailed management of a species is aided by knowledge of its genetic variation and breeding structure, such that the species itself must be investigated. For example, details of the relative impact of inbreeding on different components of fitness are likely to be species specific. Information from closely related species is often a useful guide.

When ecological models are required, the choice of laboratory species is much more difficult and challenging. Global model species such as *Drosophila* do not exist for ecological studies as they do in genetics. The choice of appropriate species and management regimes will depend on the question being asked. This has led to some misunderstanding by ecologists of laboratory animal modeling work in conservation genetics.

The Role of Laboratory Animal Research in Conservation Genetics

Laboratory animal research should assume the role of providing rapid and inexpensive evaluations of new theories and problems in conservation genetics. Genetics is much more amenable to model animal and plant work than is ecology, as the genetics of all outbreeding species is remarkably similar, whether it be humans, Cheetahs, Bison, mice, *Drosophila,* or Maize. Much of what we know about genetics in general was first discovered in peas and *Drosophila*. It is desirable to have more than one laboratory animal model in case there are issues in which one species is different from others. For example, there was concern about the low chromosome number in *D. melanogaster.* Comparison of results from different species has established that this is no impediment to its role as a model in population and quantitative genetics (Frankham 1982). Further, there are problems for which another species may be a more suitable experimental animal. Much work has been done on genotype x environmental interaction by using *Tribolium,* as it is easier to control their diet. In animal breeding, *Drosophila,* mice, *Tribolium,* and Japanese Quail have been the main laboratory species used. Typically, new theories are first evaluated in *Drosophila* and then in mice. If the results are favorable, the theory is then applied to chickens, followed by pigs, sheep, and cattle. The operation of conservation genetics may lead to fewer mistakes and the field will develop in a more appropriate manner if laboratory animal studies become an integral component.

Conclusions

1. The theory that underlies many management practices and recommendations in conservation genetics is very simplistic. Consequently, it is essential that it be subjected to controlled replicated experimental evaluation.
2. Such evaluations are impractical in most wildlife species. Consequently, they are best done in laboratory animals and plants.
3. It is critical that there be an interplay between theory and experimentation in conservation genetics and conservation biology generally.
4. Studies have begun recently with laboratory animals to evaluate a range of issues in conservation genetics.
5. It is desirable that more than one laboratory animal model be used, and/or meta-analyses of wildlife data, to establish the generality of conclusions.
6. A selfing plant laboratory model needs to be developed. *Arabidopsis* is an obvious candidate for this role.
7. A wide range of questions in conservation genetics need to be addressed in laboratory species. Several of these are listed.

Acknowledgments. I am grateful to Annette Lindsay, Margaret Montgomery, Paul Sunnuch, and Lynn Woodworth for comments on the manuscript. Our research is supported by Australian Research Council and Macquarie University research grants.

Literature Cited

Allendorf FW (1986) Genetic drift and the loss of alleles versus heterozygosity. Zoo Biology 5:181–190

Backus VL, Bryant EH, Hughes CR, Meffert LM (1995) Effect of migration on inbreeding followed by selection on low-founder-number populations: implications for captive breeding. Conservation Biology 9:1216–1224

Ballou J, Lacy RC (1995) Identifying genetically important individuals for management of genetic diversity in pedigreed populations. In: Ballou J, Gilpin M, Foose T. (eds) Population management for survival and recovery: analytical methods and strategies in small population conservation. Columbia University Press, New York, pp 76–111

Borlase SC, Loebel DA, Frankham R, Nurthen RK, Briscoe DA, Daggard GE (1993) Modeling problems in conservation genetics using captive *Drosophila* populations: consequences of equalizing family sizes. Conservation Biology 7:122–131

Brakefield PM, Saccheri IJ (1994) Guidelines in conservation genetics and the use of the population cage experiments with butterflies to investigate the effects of genetic drift and inbreeding. In: Loeschcke V, Tomiuk J, Jain SK (eds) Conservation genetics. Birkhäuser, Basel, Switzerland, pp 165–179

Brewer BA, Lacy RC, Foster ML, Alaks G (1990) Inbreeding depression in insular and central populations of *Peromyscus* mice. Journal of Heredity 81:257–266

Briscoe DA, Malpica JM, Robertson A, Smith GJ, Frankham R, Banks RG, Barker JSF (1992) Rapid loss of genetic variation in large captive populations of *Drosophila* flies: implications for the genetic management of captive populations. Conservation Biology 6:416–425

Briton J, Nurthen RK, Briscoe DA, Frankham R (1994) Modelling problems in conservation genetics using captive *Drosophila* populations: consequences of harems. Biological Conservation 69:267–275

Bulmer MG (1980) The mathematical theory of quantitative genetics. Clarendon Press, Oxford, UK

Clayton GA, Robertson A (1957) An experimental check on quantitative genetical theory. II. The long-term effects of selection. Journal of Genetics 55:152–170

Clayton GA, Morris JA, Robertson A (1957) An experimental check on quantitative genetical theory. I. Short-term responses to selection. Journal of Genetics 55:131–151

Crow JF, Kimura M (1970) An introduction to population genetics theory. Harper and Row, New York

East EM (1916) Studies on size inheritance in *Nicotiana*. Genetics 1:164–176

Eisen EJ (1975) Population size and selection intensity effects on long-term selection response in mice. Genetics 79:305–323

Falconer DS, Mackay TFC (1996) Introduction to quantitative genetics, 4th ed. Longman, Harlow, UK

Fisher RA (1918) The correlation between relatives on the supposition of Mendelian inheritance. Transactions of the Royal Society of Edinburgh 52:399–433

Forney KA, Gilpin ME (1989) Spatial structure and population extinction: a study with *Drosophila* flies. Conservation Biology 3:45–51

Frankham R (1980) Origin of genetic variation in selection lines. In: Robertson A (ed) Selection experiments in laboratory and domestic animals. Commonwealth Agricultural Bureaux, Farnham Royal, UK, pp 56–68

Frankham R (1982) Contributions of *Drosophila* research to quantitative genetics and animal breeding. Proceeding of the 2nd World Congress on Genetics Applied to Livestock Production 5:43–56

Frankham R (1983) Origin of genetic variation in selection lines. Proceedings of the Thirty-Second National Breeders' Roundtable, St. Louis, Missouri, pp 1–18

Frankham R (1992) Integrating technologies into animal breeding programmes. In: Moore HDM, Holt WV, Mace GM (eds) Biotechnology and the conservation of genetic diversity. Symposia of the Zoological Society of London 64. Clarendon Press, Oxford, UK, pp 207–221

Frankham R (1995a) Effective population size / adult population size ratios in wildlife: a review. Genetical Research 66:95–107

Frankham R (1995b) Conservation genetics. Annual Review of Genetics 29:305–327

Frankham R (1995c) Inbreeding and extinction. Conservation Biology 9:792–799

Frankham R (1995d) Genetic management of captive populations for reintroduction. In: Serena M (ed) Reintroduction biology of Australian and New Zealand fauna. Surrey Beatty and Sons, Chipping Norton, NSW, Australia pp 31–34

Frankham R (1996) Relationship of genetic variation to population size in wildlife. Conservation Biology 10:1500–1508

Frankham R, Loebel DA (1992) Modeling problems in conservation genetics using captive *Drosophila* populations: rapid genetic adaptation to captivity. Zoo Biology 11:333–342

Frankham R, Nurthen RK (1981) Forging links between population and quantitative genetics. Theoretical and Applied Genetics 59:251–263

Frankham R, Briscoe DA, Nurthen RK (1978) Unequal crossing over at the rRNA locus as a source of quantitative genetic variation. Nature 272:80–81

Frankham R, Smith GJ, Briscoe DA (1993) Effects on heterozygosity and reproductive fitness of inbreeding with and without selection on fitness in *Drosophila melanogaster*. Theoretical and Applied Genetics 86:1023–1027

Franklin IR (1980) Evolutionary change in small populations. In: Soulé ME, Wilcox BA (eds) Conservation biology: an evolutionary-ecological perspective. Sinauer, Sunderland, MA, pp 135–140

Fuerst PA, Maruyama T (1986) Considerations on the conservation of alleles and of genic heterozygosity in small managed populations. Zoo Biology 5:171–179

Gilligan DM, Woodworth LM, Montgomery ME, Briscoe DA, Frankham R (1997) Is mutation accumulation a threat to the survival of endangered populations? Conservation Biology 11:1235–1241

Hazel LN, Lush JL (1942) The efficiency of three methods of selection. Journal of Heredity 33:393–399

Hedrick PW (1994) Purging inbreeding depression and the probability of extinction: full-sib mating. Heredity 73:363–372

Hill WG (1981) Assessment of breeding value in selection programs. Proceeding of the Second Conference of the Australian Association of Animal Breeding and Genetics, University of Melbourne, Australia, pp 227–236

Hill WG (1982) Predictions of response to artificial selection from new mutations. Genetical Research 40:255–278

Hollingdale B (1971) Analyses of some genes from abdominal bristle number selection lines in *Drosophila melanogaster*. Theoretical and Applied Genetics 41:292–301

Jimenez JA, Hughes KA, Alaks G, Graham L, Lacy RC (1994) An experimental study of inbreeding depression in a natural habitat. Science 216:271–273

Jones LP, Frankham R, Barker JSF (1968) The effects of population size and selection intensity in selection for a quantitative character in *Drosophila*. II. Long-term response to selection. Genetical Research 12:249–266

Kimura M, Crow JF (1963) On the maximum avoidance of inbreeding. Genetical Research 4:399–415

Lande R (1995) Mutation and conservation. Conservation Biology 9:782–791

Lande R, Barrowclough GF (1987) Effective population size, genetic variation, and their use in population management. In: Soulé ME (ed) Viable populations for conservation. Cambridge University Press, Cambridge, UK, pp 87–123

Latter BDH (1964) Selection for a threshold character in *Drosophila*. I. An analysis of phenotypic variance on the underlying scale. Genetical Research 5:198–210

Leberg PL (1992) Effects of a population bottleneck on genetic diversity as measured by allozyme electrophoresis. Evolution 46:474–494

Loebel DA, Nurthen RK, Frankham R, Briscoe DA, Craven D (1992) Modeling problems in conservation genetics using captive *Drosophila* populations: consequences of equalizing founder representation. Zoo Biology 11:319–332

Lopez-Fanjul C, Caballero A (1990) The effect of artificial selection on new mutations for a quantitative trait. Proceedings of the 4th World Congress on Genetics Applied to Livestock Production 13:210–218

Lush JL (1945) Animal breeding plans, 3rd ed. Iowa State College Press, Ames, IA

Margan SH, Nurthen RK, Montgomery ME, Woodworth LM, Briscoe DA, Frankham R

(1998) Single large or several small? Population fragmentation in the captive management of endangered species. Zoo Biology 17:467–480

Miller PS, Hedrick PW (1993) Inbreeding and fitness in captive populations: Lessons from *Drosophila*. Zoo Biology 12:333–351

Mina NS, Sheldon BL, Yoo BH, Frankham R (1991) Heterozygosity at protein loci in inbred and outbred lines of chickens. Poultry Science 70: 864–872

Montgomery ME, Ballou JD, Nurthen RK, Briscoe DA, Frankham R (1997) Minimizing kinship in captive breeding programs. Zoo Biology 16:377–389

Pray, LA, Schwartz JM, Goodnight CJ, Stevens L (1994) Environmental dependency of inbreeding depression: implications for conservation. Conservation Biology 8:562–568

Ralls K, Ballou J (1983) Extinction: lessons from zoos. In: Schonewald-Cox CM, Chambers SM, MacBryde B, Thomas WL (eds) Genetics and conservation: a reference for managing wild animal and plant populations. Benjamin/Cummings, Menlo Park, CA, pp 164–184

Ralls K, Meadows R (1993) Breeding like flies. Nature 361:689–690

Ralls K, Ballou J, Templeton A (1988) Estimates of lethal equivalents and the cost of inbreeding in mammals. Conservation Biology 2:185–193

Reeve ECR, Robertson FW (1954) Studies in quantitative inheritance. VI. Sternite chaetae number in *Drosophila:* a metameric quantitative character. Zeitschrift für induktive Abstammungs- und Vererbungslehre 86:269–288

Robertson A (1960) A theory of limits in artificial selection. Proceedings of the Royal Society of London 153B:234–249

Rumball W, Franklin IR, Frankham R, Sheldon BL (1994) Decline in heterozygosity under full sib and double first cousin inbreeding in *Drosophila melanogaster.* Genetics 136:1039–1049

Saccheri I, Kuussaari M, Kankare M, Vikman P, Fortelius W, Hanski I (1998) Inbreeding and extinction in a butterfly metapopulation. Nature 392:491–494

Soulé ME (1980) Thresholds for survival: maintaining fitness and evolutionary potential. In: Soulé ME, Wilcox BA (eds) Conservation biology: an evolutionary-ecological perspective. Sinauer, Sunderland, MA, pp 151–169

Spielman D, Frankham R (1992) Modeling problems in conservation genetics using captive *Drosophila* populations: improvement of reproductive fitness due to immigration of one individual into small partially inbred populations. Zoo Biology 11:343–351

Templeton AR, Read B (1983) The elimination of inbreeding depression in a captive herd of Speke's Gazelle. In: Schonewald-Cox CM, Chambers SM, MacBryde B, Thomas WL (eds) Genetics and conservation: a reference for managing wild animal and plant populations. Benjamin/Cummings, Menlo Park, CA, pp 241–261

Templeton AR, Read B (1984) Factors eliminating inbreeding depression in a captive herd of Speke's Gazelle (*Gazella spekei*). Zoo Biology 3:177–199

Vrijenhoek RC (1994) Genetic diversity and fitness in small populations. In: Loeschcke V, Tomiuk J, Jain SK (eds) Conservation genetics. Birkhäuser, Basel, Switzerland, pp 37–53

Woodworth LM (1996) Population size in captive breeding programs. PhD thesis, Macquarie University, Sydney, Australia

Woodworth LM, Montgomery ME, Nurthen RK, Briscoe DA, Frankham R (1994) Modelling problems in conservation genetics using *Drosophila:* consequences of fluctuating population sizes. Molecular Ecology 3:393–399

Wright JW, Treadwell M, Nurthen RK, Woodworth LM, Montgomery ME, Briscoe DA, Frankham R (in press) Modelling problems in conservation genetics using *Drosophila*: purging is ineffective in reducing genetic load. Biodiversity and Conservation

Wright S (1921) Systems of mating. I. The biometric relations between parent and offspring. Genetics 6:111–123

Wright S (1931) Evolution in Mendelian populations. Genetics 16:97–159

Wright S (1969) Evolution and the genetics of populations, vol 2. The theory of gene frequencies. University of Chicago Press, Chicago, IL

Wright S (1977) Evolution and the genetics of populations, vol 3. Experimental results and evolutionary deductions. University of Chicago Press, Chicago, IL

Yoo BH (1980) Long-term selection for a quantitative character in large replicate populations of *Drosophila melanogaster*. II Lethals and visible mutants with large effects. Genetical Research 35:19–31

16

Theoretical Properties of Extinction by Inbreeding Depression Under Stochastic Environments

Yoshinari Tanaka

Introduction

Reduction of genetic diversity is regarded as one of the important factors for extinction of small populations (Frankel and Soulé 1981; Caughley and Gunn 1996). Loss of genetic variation inhibits adaptive evolution and directly reduces fitness by inbreeding depression, which results from biased mating with close relatives or random mating in small populations. Inbreeding depression may induce rapid extinction of populations. This extinction is the consequence of the positive feedback between declines in population size and increases in inbreeding. This phenomenon has been called an extinction vortex (cf. Gilpin and Soulé 1986).

The correlations between fitness and heterozygosity and the rates of inbreeding depression estimated for natural populations provide indirect evidence of inbreeding depression in nature (Mitton and Grant 1984; Zouros 1987; Barrett and Charlesworth 1991; Mitton 1993; Willis 1993; Ouborg and Treuren 1994; Britten 1996; David et al. 1997; Schierup 1998). Although numerous studies have estimated genetic heterozygosity of endangered species (Loeschcke et al. 1994; Avise and Hamrick 1996), the proper biological interpretation of these estimates and their relevance to the extinction process are not completely clear.

There are two shortcomings. First, few examples of population extinction due to inbreeding depression have been established. A field survey on Glanville Fritillary Butterflies (*Melitaea cinxia*) is one of the very few studies that have reported inbreeding depression as a causal factor of extinction in natural populations (Saccheri et al. 1998). Second, the paucity of investigations on the theoretical properties of the extinction vortex by inbreeding depression makes it impossible to evaluate the genetic risks of extinction based on empirical data.

Previous studies have considered the extinction vortex with genetic and population models, and suggest that rapid extinction due to inbreeding depression is restricted but possible with observed genetic parameters (Tanaka 1997, 1998, 2000. Gene frequencies, the inbreeding coefficient, and the equilibrium population size achieved by the balance among mutation, selection, and drift are locally stable, which suggests that an equilibrial population does not spontaneously go extinct due to inbreeding depression. Rapid extinction by inbreeding depression requires that the equilibrium population be perturbed by external nongenetic factors and that the population size before the perturbation had been large enough to maintain recessive deleterious mutations in the population (Tanaka 1997, 1998). In these studies, the perturbation of the equilibrium population was simulated by removal of individuals at a constant rate every generation (Tanaka 1997, 1998) or by assuming the carrying capacity to be exponentially decreasing in time (Tanaka, 2000). Although rapid extinction by inbreeding depression does not occur in a stable population, it is likely for a declining population.

The suggested sufficient conditions required for the extinction vortex to occur are as follows. The equilibrium population size must be large ($>10^5$) so that recessive deleterious genes are maintained with sufficient frequency. As inbreeding depression cannot alone induce extinction because of the local stability of genetic equilibria, nongenetic demographic disturbances are required to reduce the population size at some higher rate than purging selection can eliminate the deleterious genes (and make inbreeding depression irrelevant). Also, the maximum (mutation-free) intrinsic rate of natural increase must be sufficiently small so that reductions of population size positively feed back through the action of inbreeding depression (Tanaka 1997, 1998). The overdominant genes are unlikely to contribute to the extinction vortex by inbreeding depression because high equilibrium segregation loads must precede the inbreeding vortex (Tanaka 1998).

The fluctuation of population size due to environmental stochasticity interacts with inbreeding depression and reinforces the extinction vortex (Tanaka 2000; cf. van Noordwijk 1994). Real populations are subject to random fluctuations of environmental factors, such as temperature, food availability, and predation pressure. This environmental stochasticity brings about random fluctuations in population size. That an extinction vortex due to inbreeding depression may be reinforced or enhanced by environmental stochasticity is important in evaluating the extinction risk due to genetic factors.

In this chapter, I review the models used and show some analytical results on extinction due to inbreeding depression.

Genetic Model

I assumed the genetic mechanism of inbreeding depression involves recessive deleterious genes. The deleterious genes are assumed to be distributed among n diallelic autosomal loci, which are identical with respect to the per-locus mutation rate, the selection coefficient, and the degree of dominance. The model is a simple extension of a one-locus, two-allele model. Linkage disequilibrium and epistatic interaction between loci are neglected for simplicity. It is assumed that three genotypes, AA, Aa, and aa, have mean fitnesses of 1, 1, and $1 - s$, respectively, where s denotes the selection coefficient. Thus all deleterious genes are assumed to be completely recessive.

The mean fitness of a population decreases due to deleterious genes from the maximum value 1 to $1 - L$, where L is the genetic load (Crow and Kimura 1970; Nei 1987). To evaluate the effects of inbreeding depression on extinction of populations, the genetic load must be expressed as reductions of demographic parameters. I assumed that the Malthusian fitness ($W = e^r$, where r is the intrinsic rate of population increase) decreases linearly with the genetic load. Because the realized population growth rate under density dependence is $e^{r(1-N/K)}$, where N is population size and K is carrying capacity, the population growth rate of the mean fitness of a genetically loaded population is $\lambda = (1 - L)e^{r_{max}(1-N/K)}$ where r_{max} is the maximum intrinsic rate of natural increase of a mutation-free population.

The population is assumed to be at demographic and genetic equilibrium. Previous work has suggested the equilibrium is locally stable under realistic parameter values (Tanaka 1998). External factors (e.g., habitat destruction or hunting) that decrease population size from the equilibrium size are required to trigger the inbreeding vortex. Reduction of the population size by such external factors is called "demographic disturbance." To model demographic disturbances, I assume that the carrying capacity K decreases monotonically with time toward a minimum value K_{min}. An exponential function of time was used for K (i.e., $K_t = (K_0 - K_{min})(1 - k)^t + K_{min}$, where k is the rate of decrease for K, and K_0 is the initial carrying capacity).

At equilibrium, the expected inbreeding coefficient and the mean gene frequency over loci are determined by a balance between mutation, genetic drift, and selection. The genetic load is also at equilibrium. The population growth rate at equilibrium is denoted as $\tilde{\lambda} = (1 - \tilde{L})e^{r_{max}(1-N/K)}$, where $\tilde{\lambda}$ and $\tilde{L}$ denote the equilibrium values of λ and L. Because of inbreeding depression, the population growth rate is reduced. Denote the proportional reduction of λ as $1 - \delta_{inb}$. The population growth rate of any inbred population is $\lambda = \tilde{\lambda}\,\delta_{inb}$, where $\delta_{inb} = \dfrac{1 - L}{1 - \tilde{L}}$.

The effect of the equilibrial genetic load on the population growth rate is ignored in the present analysis to determine the extent to which inbreeding depression generated from demographic disturbances contributes to the process of extinction. The equilibrium population growth rate $\tilde{\lambda}$ is assigned arbitrarily.

The per-locus genetic load at the ith locus l_i is defined as $l_i = 1 - \overline{w}_i/w_{max}$, where $\overline{w}_i$ is the marginal mean fitness of the ith locus and w_{max} is the maximum fitness of a locus, assumed to be unity for all loci. Assuming multiplicative fitness without epistatic interaction between loci, the total genetic load L is calculated as

$$L = 1 - \prod_i (1 - l_i)$$

$$= 1 - \prod_i (\overline{w}_i)/W_{max}$$

where W_{max} is the maximum multilocus fitness, $\prod_i w_{max} = 1$. Then $L = 1 - \prod_i \overline{w}_i$. The marginal mean fitness at the ith locus is $\overline{w}_i = 1 - sq_i^2$, where q_i is the gene frequency of the recessive deleterious gene at the ith locus. The total genetic load is approximately

$$L \cong 1 - s\sum_i q_i^2 = 1 - nsE_i(q_i^2)$$

where E_i denotes expectation over all loci.

Gene frequencies of the deleterious genes change by mutation, selection, and genetic drift. Through the action of the random genetic drift, gene frequencies disperse between loci. For simplicity, the joint dynamics of the gene frequencies are summarized as changes in the mean gene frequency over all contributing loci, $\bar{q} = n^{-1}\sum_i q_i$, and the variance of the gene frequencies. From the approximation for L above, the total genetic load, which expresses the net effect of inbreeding depression, is largely determined by the first two moments of the distribution of gene frequencies.

The per-generation change of gene frequency at a locus by selection is derived from Wright's formula, $\Delta_s q_i = \dfrac{q_i(1 - q_i)}{2}\dfrac{\partial \ln \bar{w}_i}{\partial q_i}$, in which $\dfrac{\partial \ln \bar{w}_i}{\partial q_i} \cong -2sq_i$. If all loci are approximately at linkage equilibria, the change in the mean gene frequency by selection is calculated as

$$\Delta_s \bar{q} = E_i(\Delta_s q_i) = E_i[-sq_i^2(1- q_i)]$$
$$\cong - sE_i(q_i^2)$$

The per-generation change of mean gene frequency by mutation is equivalent to the per-locus per-gamete mutation rate $\Delta_m \bar{q} = \mu$ if the mutation is irreversible. Most deleterious mutations that have large adverse effects on fitness are likely to be irreversible.

By random mating in a finite population, gene frequencies starting from an identical initial frequency tend to disperse between independent sets of samplings of gametes (Crow and Kimura 1970; Falconer 1989). The dispersion of gene frequencies is readily expressed by the variance V_q of gene frequencies, which monotonically increases with generations and the inbreeding coefficient F. If the dispersion is independent, the variance of gene frequencies is expressed as $V_q = F\bar{q}(1 - \bar{q})$ (Crow and Kimura 1970; Falconer 1989). The theory of random dispersion of gene frequencies has been successfully applied to genetic differentiation at a locus between local populations (Crow and Kimura 1970; Nei 1987). If there is no gametic correlation between loci, the random dispersion of gene frequencies holds for different loci within a genome. I used this approximate treatment for describing changes in gene and genotypic frequencies by the genetic drift. The standard theory of inbreeding indicates

$$E_i(q_i^2) = \bar{q}^2 + V_q$$

$$= \bar{q}^2 + F\bar{q}(1 - \bar{q})$$

The inbreeding coefficient changes mostly by inbreeding and partly by selection and mutation. The per-generation change in F is expressed by the recurrence equation

$$F_{t+1} = \{1/(2N) + [1 - 1/(2N)]F_t\}(1 - 2\mu)(1 - s\bar{q})$$

(Tanaka 1997, 1998).

Equilibrium

Without demographic disturbance and environmental stochasticity, the population is maintained at a demographic and genetic equilibrium, in which population size, mean gene frequency, and inbreeding coefficient are constant. These equilibrium values were used as the initial values of F and $\bar{q}$ in the simulations. It was assumed that the population had been at a long-term equilibrium before anthropogenic factors started to continually degrade populations and disrupt the equilibria.

The long-term effective population size of a stochatically fluctuating population is smaller than the census population size. If there is no autocorrelation in the fluctuations, the effective population size N_e is $N - \sigma_N^2/N$, where σ_N^2 is the variance of population size (Crow and Kimura 1970). If the variance is mostly due to environmental fluctuation in the population growth rate, it is equivalent to $\sigma_N^2 = (\sigma_e^2 K^2 + K)/(2r)$, where σ_e^2 is the environmental variance of the intrinsic rate of natural increase r (Iwasa 1998). For numerical evaluation of equilibrium values of inbreeding coefficient and mean gene frequency, the effective population size was used in place of the population size.

The equilibrium mean gene frequency resulting from the balance between mutation and selection ($\Delta_s\bar{q} + \Delta_m\bar{q} = 0$) must satisfy the quadratic equation, $(\bar{F} - 1)\tilde{\bar{q}}^2 - \bar{F}\tilde{\bar{q}} + \mu/s \cong 0$, in which a tilde denotes the equilibrium value of a quantity. It is not possible to find a simple analytical solution of the equilibrium mean gene frequency. The numerical simulations used values of $\tilde{\bar{q}}$ approximated from $\Delta_s\bar{q} + \Delta_m\bar{q} = 0$. The demographic and genetic equilibria are locally stable regardless of the equilibrium population size, mutation rate, selection coefficient, or the number of loci (Tanaka 1998).

Stochasticity

A real population in nature is subject to stochasticity. This chapter is concerned with two main kinds of stochasticity, environmental stochasticity and genetic stochasticity (cf. Burgman et al. 1993; Caughley and Gunn 1996). The former includes temporal variation of any environmental factors and results in random fluctuations of the population growth rate. The latter is called random genetic drift because it induces random dispersion of gene frequencies. It results from random samplings of gametes from gene pools in parental generations.

Environmental stochasticity was incorporated by an additional white-noise term representing random fluctuations of the intrinsic rate of natural increase. The recurrence equation for population growth is $N_{t+1} = N_t \exp[r_{max}(1 - N_t/K_t) + \varepsilon_t]\delta_{inb(t)}$, where ε_t is a random normal variate with mean 0 and variance σ_e^2.

Random genetic drift in a multilocus system causes fluctuations in the mean gene frequency of all loci in addition to the dispersion of gene frequencies between loci. The dispersion of gene frequencies between loci was modeled as increases in the variance of gene frequencies and inbreeding coefficient. The effect of genetic drift to the mean gene frequency was incorporated into the model by means of randomly sampling the mean gene frequency in the next generation from normal variates generated from the mean gene frequency after mutation and selection in the present generation. The mean gene frequency in the next generation is $\bar{q}_{t+1} = \bar{q}_t^* + \gamma_t$, where γ_t is a random normal variate with mean 0 and standard deviation $\sqrt{\bar{q}_t^*(1 - \bar{q}_t^*)/(2nN)}$, and $\bar{q}_t^*$ denotes the mean gene frequency after mutation and selection but before reproduction in the tth generation, N is the population size, and n is the number of loci.

Dynamics

The dynamics of the three parameters are summarized by the following joint recurrence equations:

$$\bar{q}_{t+1} = \bar{q}_t - s[\bar{q}^2 + F\bar{q}(1 - \bar{q})] + \mu + \gamma_t$$

$$F_{t+1} = \left[\frac{1}{2N_t} + \left(1 - \frac{1}{2N_t}\right)F_t\right](1 - 2\mu)(1 - s\bar{q}_t)$$

$$N_{t+2} = N_{t+1}\exp\left[r_{max}\left(1 - \frac{N_{t+1}}{K_{t+1}}\right) + \varepsilon_t\right]\delta_{inb(t)}$$

The expression for population size is indexed one generation ahead of the other two variables because the genetic effect on the demographic parameters through survival and reproduction occurs in the next (offspring) generation.

Extinction Vortex of Nonequilibrium Populations

Because inbreeding depression is caused by decreased heterozygosities of recessive deleterious genes, the present analysis focuses on recessive lethal genes rather than weakly deleterious genes with slight recessivity. From experiments using the balancer chromosomes of *Drosophila*, the genomic mutation rate of recessive lethals is approximately 0.03 (Crow and Simmons 1983; Woodruff et al. 1983; Eeken et al. 1987). Throughout the analysis, values of $\mu = 10^{-6}$ and $n = 15000$ were used, which are consistent with the observed genomic mutation rate.

Using the deterministic version of the model [in which v and $var(\gamma)$ are both zero], numerical simulations clarified relationships between extinction due to inbreeding depression and the other population and genetic parameters. We focus on how the extinction vortex is influenced by different rates of demographic disturbance and the equilibrium population size.

With slow rates of demographic disturbance, populations did not become extinct (Fig. 16.1, $k = 0.01$ and 0.02), whereas they did with fast rates (Fig. 16.1,

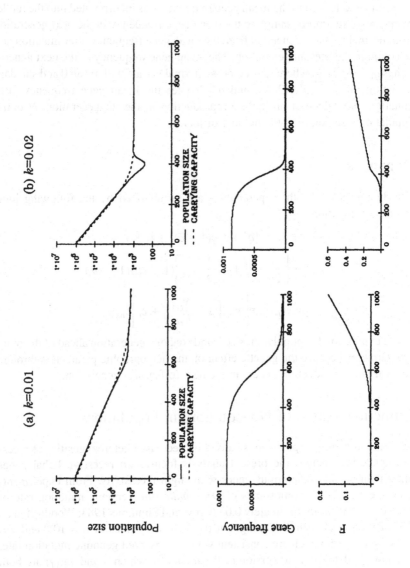

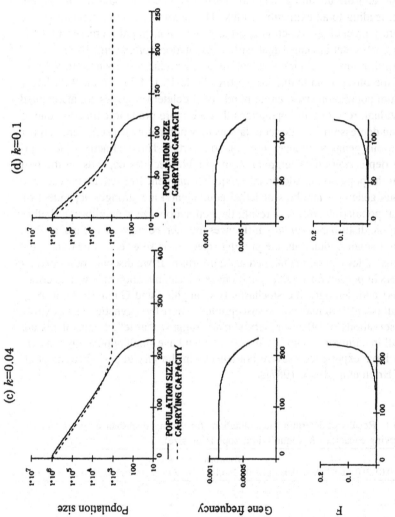

FIGURE 16.1. Predictions with the deterministic model on population trajectories and changes in mean gene frequencies and inbreeding coefficients after the onset of demographic disturbances with four different disturbance rates k [0.01 (a), 0.02 (b), 0.04 (c), and 0.1 (d)]. All other parameter values are common among the simulations ($\mu = 10^{-6}$, $n = 15000$, $\bar{N} = 10^6$, and $s = 1$).

$k = 0.04$ and 0.1) even if other parameter values were unaltered. The discrepancy is a result of differences in the efficiency of the purging selection against the deleterious genes. With slower rates of population decline, the purging selection continued to act long enough to exclude the deleterious genes. With higher rates of demographic disturbances, however, the positive feedback between inbreeding depression and the declining population size exceeded the capacity of purging selection, leading to an extinction vortex. The sharp decrease in population size immediately preceding extinction and the concomitant rapid increase in the inbreeding coefficient are consistent with an extinction vortex (Fig. 16.1).

The equilibrium population size before demographic disturbance strongly influences the onset of an extinction vortex (Table 16.1, Table 16.2). With larger equilibrium population sizes, larger numbers of deleterious genes are maintained per individual, resulting in populations that are more prone to extinction due to inbreeding depression. Smaller populations may not maintain a sufficient number of deleterious genes to induce an extinction vortex. Thus extinction due to inbreeding depression under temporal demographic declines depends on the past history of the population. Additionally, equilibrium gene frequencies of recessive deleterious genes (or numbers of lethal equivalents) are strongly dependent on equilibrial population size. As a result, the population size at equilibrium (before demographic disturbance) greatly influences the inbreeding vortex (Tanaka 1997, 1998). In a small population, the purging selection is so effective that very few deleterious genes can be maintained at equilibrium. If we discount other extinction factors important for small populations (e.g., accumulation of new deleterious mutations) and demographic stochasticity (Caughley and Gunn 1996), a long-term small population may not be susceptible to short-term genetic risk of extinction. Observations of offspring survivorship suggest that lethal equivalents are very small in carnivores such as Cheetahs, which have small population numbers and may have experienced severe bottlenecks in the glacial period (Ralls et al. 1988; O'Brien et al. 1983, 1987).

TABLE 16.1. Results of deterministic simulations for various disturbance rates k and initial carrying capacities K_0 (equilibrium population size).[a]

| K_0 | LE | Disturbance rate k | | | | | | | | | | | | | | | |
		0.01	0.02	0.04	0.06	0.08	0.10	0.12	0.14	0.16	0.18	0.20	0.22	0.24	0.26	0.28	0.30
10^8	299	–	–	+	+	+	+	+	+	+	+	+	+	+	+	+	+
10^{75}	298	–	–	+	+	+	+	+	+	+	+	+	+	+	+	+	+
10^7	292	–	–	+	+	+	+	+	+	+	+	+	+	+	+	+	+
10^{65}	275	–	–	+	+	+	+	+	+	+	+	+	+	+	+	+	+
10^6	213	–	–	–	–	–	–	–	+	+	+	+	+	+	+	+	+
10^{55}	019	–	–	–	–	–	–	–	–	–	–	–	–	–	–	–	–
10^5	005	–	–	–	–	–	–	–	–	–	–	–	–	–	–	–	–
10^{45}	004	–	–	–	–	–	–	–	–	–	–	–	–	–	–	–	–
10^4	003	–	–	–	–	–	–	–	–	–	–	–	–	–	–	–	–

[a]Parameter values are $\mu = 10^{-6}$, $n = 15000$, and $s = 1$. Plus signs denote extinction, and minus signs denote persistence. The LE column gives lethal equivalents at equilibrium.

TABLE 16.2. Results of deterministic simulations for various disturbance rates k and net reproductive rates R_0 at equilibrium.[a]

R_0	Disturbance rate															
	0.01	0.02	0.04	0.06	0.08	0.10	0.12	0.14	0.16	0.18	0.20	0.22	0.24	0.26	0.28	0.30
105	–	–	+	+	+	+	+	+	+	+	+	+	+	+	+	+
110	–	–	+	+	+	+	+	+	+	+	+	+	+	+	+	+
115	–	–	+	+	+	+	+	+	+	+	+	+	+	+	+	+
120	–	–	+	+	+	+	+	+	+	+	+	+	+	+	+	+
125	–	–	–	–	–	–	–	+	+	+	+	+	+	+	+	+
130	–	–	–	–	–	–	–	–	–	–	–	–	–	–	–	–
135	–	–	–	–	–	–	–	–	–	–	–	–	–	–	–	–
140	–	–	–	–	–	–	–	–	–	–	–	–	–	–	–	–

[a]Parameter values are $\mu = 10^{-6}$, $n = 15000$, $\tilde{N} = 10^6$, and $s = 1$. Plus signs denote extinction, and minus signs denote persistence.

Interaction between Inbreeding Depression and Environmental Stochasticity

Simulations using an identical set of parameters, but now with environmental and genetic stochasticity, produce both extinction and persistence (Fig. 16.2). The probability of extinction is influenced by genetic and demographic parameters.

The environmental variance of the population growth rate increases the chance of extinction due to inbreeding depression (Fig. 16.3). Figure 16.3 displays two typical cases with different values for environmental variance but otherwise identical parameters. These parameter values did not induce an extinction by inbreeding vortex in the deterministic model (where $v = 0$ and $var(\gamma) = 0$). To explore the relationship between environmental stochasticity and the chance of an inbreeding vortex, 100 simulations were carried out, each generated from different values of random deviates ε_t and γ_t (Fig. 16.4). The simulations were repeated for different rates of demographic disturbance k. Parameter sets that did not yield extinction in the deterministic simulation sometimes do so when there is environmental variation in population growth rate, even if it is small. When environmental variance is less than 0.05 (the coefficient of variation of r is 1.24), the larger the environmental variance, the more the extinction risk is inflated. Thus inbreeding depression interacts with environmental stochasticity to induce extinction in monotonically decreasing populations. The maximum environmental variance for extinction is surprisingly constant between different rates of demographic disturbance. The reason the curves plotting the proportion of extinctions against environmental variance are modal is that the equilibrium effective population size decreases with large environmental fluctuation in population size so that populations do not maintain deleterious genes.

There is synergistic interaction between inbreeding depression and environmental stochasticity that can induce rapid extinction by inbreeding depression (inbreeding vortex). The simultaneous action of the two factors can induce the

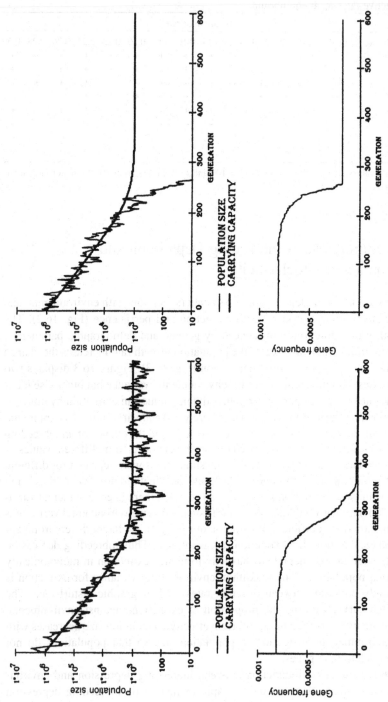

FIGURE 16.2. Examples of population trajectories and gene frequency changes predicted with the stochastic model. All parameter values are equivalent between the two simulations ($\mu = 10^{-6}$, $n = 15000$, $\bar{N} = 10^6$, $k = 0.3$, $v = 0.1$, and $s = 1$).

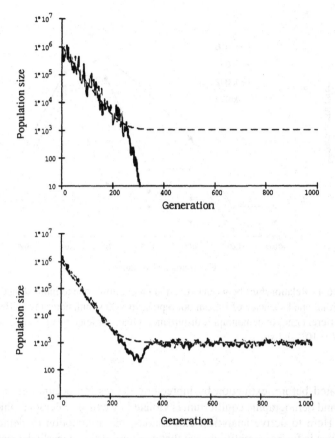

FIGURE 16.3. Example population trajectories showing the influence of environmental stochasticity on extinction by inbreeding depression. The upper graph represents a large environmental variance in the intrinsic rate of natural increase ($v = 0.1$), and the graph below has a small variance ($v = 0.01$). All other parameter values are equivalent between the two simulations ($\mu = 10^{-6}$, $n = 15000$, $\bar{N} = 10^6$, $k = 0.03$, and $s = 1$).

inbreeding vortex even when a single action of either effect does not. The causal mechanism for the synergistic interaction is unclear. Occasional reductions of population size due to environmental fluctuation of population size may facilitate the inbreeding vortex by escaping the effect of purging selection.

Critical Conditions for Rapid Extinction by Inbreeding Depression

A simple analytical expression for the sufficient conditions for a population to become extinct by inbreeding vortex may support the results obtained from numerical simulations.

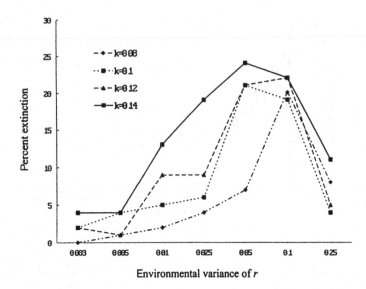

FIGURE 16.4. Relationships between extinction probabilities due to inbreeding depression and environmental variance of r. Each dot represents 100 simulation runs. Different lines denote different rates of demographic disturbance. Other parameter values are $\mu = 10^{-6}$, $n = 15000$, $\bar{N} = 10^6$, and $s = 1$.

As stated before, extinction by inbreeding depression occurs only when the genetic and demographic equilibrium is violated by extrinsic demographic disturbances. Here to derive analytical expressions, the assumption of demographic disturbance is further simplified such that population size linearly decreases for τ generations with rate κ. Then the total decrement of population size is equivalent to $\delta_N = \tau\kappa$. The population size at tth generation is $N(t) = \bar{N}(1 - \kappa t)$. The initial state is assumed to be at genetic and demographic equilibrium. We are interested in the conditions that are sufficient to trigger an extinction vortex during τ generations. If the extinction vortex occurs in that period, the intrinsic rate of population growth must be less than 0 at the τth generation after the onset of demographic disturbance. Provided that $r(\tau) = r_{max} + \ln[1 - L(\tau)]$ and $L(\tau) = 1 - e^{-nsq(\tau)F(\tau)}$, the condition $r(\tau) < 0$ is equivalent to $r_{max} < nsq(\tau)F(\tau)$.

To derive solutions for the dynamics of gene frequency and inbreeding coefficient after the onset of demographic disturbance, I further assumed that the dynamics of $q(t)$ is mostly governed by selection and that of $F(t)$ mostly by inbreeding. Thus $dF(t)/dt = [1 - F(t)]/[2N(t)]$ and $dq(t)/dt = -sq(t)F(t)$. Integrating these with the initial conditions $F(0) = \bar{F}$ and $q(0) = \bar{q}$, the inbreeding coefficient and the gene frequency at the tth generation after the onset of disturbance become $F(t) = 1 - (1 - \bar{F})(1 - \kappa t)$ and $q(t) = \bar{q}\exp\{-st[1 - (1 - F)(1 - \frac{1}{2}\kappa t)]\}$. If we assume a large population at equilibrium, we can further simplify the solutions using $F \ll 1$ and $q \cong \sqrt{\mu/s}$. The sufficient condition for inbreeding vortex to occur within τ generations is

$$\kappa > \frac{s\delta_N^2}{2}\left[\ln\left(\frac{n\delta_N\sqrt{s\mu}}{r_{max}}\right)\right]^{-1}$$

The critical rate of demographic disturbance κ is monotonically increasing with the maximum intrinsic rate of natural increase and the selection coefficient (Fig. 16.5). Higher rates of reduction in population size are needed to trigger the extinction vortex by inbreeding depression when the maximum (mutation-free) intrinsic rate is large. Species with a large intrinsic rate are not prone to extinction by inbreeding depression.

The selection coefficient s does not influence the equilibrium genetic load achieved by mutation and selection. The results in Figure 16.5 indicate, however, that the selection coefficient alters the critical rate of demographic disturbance. Strong selection increases the critical rate and reduces the likelihood of extinction by inbreeding depression. This relationship between κ and s may be explained by the purging selection, which is more effective with larger selection coefficient.

The present results were derived from a standard population genetic model. However, they need empirical checks and further theoretical exploration, perhaps based on individual-based Monte Carlo simulations. In particular, the assumption of linkage equilibrium and the simplified description of stochastic dispersion of gene frequencies among loci may influence the results. Nonetheless, the major results in the present study provide some insights on practical conservation problems. Rapid extinction due to inbreeding depression is likely when a long-term large population, which has not experienced severe bottlenecks in the past, rapidly decreases its abundance due to anthropogenic factors. Environmental fluctuation of population size reinforces the extinction due to inbreeding depression. Evaluation of the minimum viable population for a monotonically declining large popu-

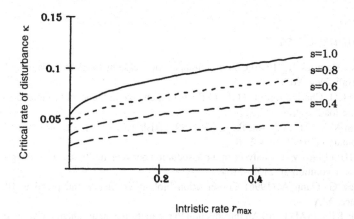

FIGURE 16.5. Dependence of the critical rate of demographic disturbances on maximum (mutation-free) intrinsic rate of natural increase with various selection coefficients. The ultimate loss of population size due to the disturbance is 90% ($\delta_N = 0.9$). Other parameters are $\mu = 10^{-6}$ and $n = 15000$.

lation should take into account the extinction risk aggravated by inbreeding depression.

Conclusions

Throughout the conservation genetics literature, it is frequently stated that inbreeding depression may induce rapid extinction due to positive feedback between inbreeding depression and reduction of population size (i.e., an extinction vortex by inbreeding depression). This chapter has demonstrated that an extinction vortex is likely to occur with realistic parameter values of the genomic mutation rate of lethals or semilethals, the equilibrium population size, the intrinsic rate of natural increase, and the rate of population decline due to nongenetic extrinsic factors. The magnitude of the equilibrium population size, the intrinsic rate of natural increase, and the rate of population decline are especially important in determining whether a population becomes extinct by inbreeding depression.

Simulation models incorporating stochastic fluctuations of population size further indicated that extinction by inbreeding depression is facilitated by environmental fluctuations of population size. The results suggest that there is a positive interaction between genetic and environmental stochasticity that can lead to population extinction by inbreeding depression.

Acknowledgments. I thank Yasushi Harada and Yoh Iwasa for helpful discussions on the topic. I am grateful to the editors, who provided me with the opportunity to contribute to this volume. This work is supported in part by CREST (Core Research for Evolutional Science and Technology) of the Japan Science and Technology Corporation (JST), for which the principal investigator is J. Nakanishi.

Literature Cited

Avise JC, Hamrick JL (1996) Conservation genetics: case histories from nature. Chapman and Hall, New York

Barrett SCH, Charlesworth D (1991) Effects of a change in the level of inbreeding on the genetic load. Nature 352:522–524

Burgman MA, Ferson S, Akçakaya HR (1993) Risk assessment in conservation biology. Chapman and Hall, New York

Britten HB (1996) Meta-analyses of the association between multilocus heterozygosity and fitness. Evolution 50:2158–2164

Caughley G, Gunn A (1996) Conservation biology in theory and practice. Blackwell Science, MA

Crow JF, Kimura M (1970) An introduction to population genetics theory. Harper and Row, New York

Crow JF, Simmons MJ (1983) The mutation load in *Drosophila*. In: Ashburner M, Carson HL, Thompson JN Jr (eds) The genetics and biology of *Drosophila*, vol 3C. Academic Press, New York, pp 1–35

David P, Delay B, Jarne P (1997) Heterozygosity and growth in the marine bivalve *Spisula ovalis:* testing alternative hypotheses. Genetical Research 70:215–223

Eeken JCJ, De Jong AWM, Green MM (1987) The spontaneous mutation rate in *Drosophila simulans.* Mutation Research 192:259–262

Falconer DS (1989) Introduction to quantitative genetics, 3rd ed. Longman Scientific and Technical, New York

Frankel OH, Soulé ME (1981) Conservation and evolution. Cambridge University Press, Cambridge

Gilpin ME, and Soulé ME (1986) Minimum viable populations: the processes of species extinction. In: Soulé ME (ed) Conservation biology: the science of scarcity and diversity. Sinauer Associates, Sunderland, MA, pp 13–34

Iwasa Y (1998) Ecological risk assessment by the use of the probability of species extinction. In: Nakanishi J (org) Proceedings of the 1st International Workshop on Risk Evaluation and Management of Chemicals. Yokohama National University, Yokohama, Japan, pp 42–49

Loeschcke V, Tomiuk J, Jain SK (1994) Conservation genetics. Birkhauser, Basel, Switzerland

Mitton JB (1993) Theory and data pertinent to the relationship between heterozygosity and fitness. In: Thornhill NW (ed) The natural history of inbreeding and outbreeding. The University of Chicago Press, Chicago, IL, pp 17–41

Mitton JB, Grant MC (1984) Associations among protein heterozygosity, growth rate, and developmental homeostasis. Annual Review of Ecology and Systematics 15:479–499

Nei M (1987) Molecular evolutionary genetics. Cambridge University Press, Cambridge, UK

O'Brien SJ, Wildt DE, Goldman D, Merril DR, Bush M (1983) The Cheetah is depauperate in genetic variation. Science 211:459–462

O'Brien SJ, Wildt DE, Bush M, Caro TM, Fitzgibbon C, Aggundey I, Leakey RE (1987) East African Cheetahs: evidence for two population bottlenecks? Proceedings of National Academy and Sciences of the USA 84:508–511

Ouborg NJ, Treuren RV (1994) The significance of genetic erosion in the process of extinction. IV. Inbreeding load and heterosis in relation to population size in the mint *Salvia pratensis.* Evolution 48:996–1008

Ralls K, Ballou J, Templeton A (1988) Estimates of lethal equivalents and the cost of inbreeding in mammals. Conservation Biology 2:185–193

Saccheri I, Kuussaari M, Kankare M, Vikman P, Fortelius W, Hanski I (1998) Inbreeding and extinction in a butterfly metapopulation. Nature 392:491–494

Schierup MH (1998) The effect of enzyme heterozygosity on growth in a strictly outcrossing species, the self-incompatible *Arabis petraea* (Brassicaceae). Hereditas 128:21–31

Tanaka Y (1997) Extinction of populations due to inbreeding depression with demographic disturbances. Researches on Population Ecology 39:57–66

Tanaka Y (1998) Theoretical aspects of extinction by inbreeding depression. Researches on Population Ecology 40:279–286

Tanaka Y (2000) Extinction of populations by inbreeding depression under stochastic environments. Population Ecology (formerly Researches on Population Ecology) 42:(in press)

van Noordwijk AJ (1994) The interaction of inbreeding depression and environmental stochasticity in the risk of extinction of small populations. In: Loeschcke V, Tomiuk J, Jain SK (eds) Conservation genetics. Birkhauser, Basel, Switzerland, pp 131–146

Willis JH (1993) Effects of different levels of inbreeding on fitness components in *Mimulus guttatus*. Heredity 69:562–572

Woodruff RC, Slatko BE, Thompson JN Jr (1983) Factors affecting mutation rates in natural populations. In: Ashburner M, Carson HL, Thompson JN Jr (eds). The genetics and biology of *Drosophila*, vol 3C. Academic Press, New York, pp 37–124

Zouros E (1987) On the relationship between heterozygosity and heterosis: an evolution of evidence from marine mollusks. In: Rattari MC, Scandalios JG, Whitt GS (eds) Isozymes: current topics in biological and medical research, vol 15. Liss, New York, pp 255–270

17
Mathematical Methods for Identifying Representative Reserve Networks

Hugh Possingham, Ian Ball, and Sandy Andelman

Introduction

Many countries have committed to conserving significant amounts of their native biodiversity (McNeely et al. 1990). Biodiversity includes the diversity of ecosystems and the diversity between and within species. Reserve systems dedicated to nature conservation are the cornerstone of most national, regional, and state conservation strategies (Soulé 1991).

In this chapter, we explore the question of which parcels of land, henceforth termed sites, should be selected for reservation. Implicit in our analysis is a limit on the total size of the reserve system. From the perspective of nature conservation alone, one would attempt to have the largest reserve system possible; however, in reality, the extent of any reserve system will be limited by social and economic constraints. Thus, building a reserve network that will conserve biodiversity effectively is not a process of accumulating as much land as possible but, rather, necessitates doing so as efficiently as possible within a constrained area.

We focus on a particular class of reserve design problem in which the goal is to achieve some minimum representation of biodiversity features for the smallest possible cost. In these problems, the objective is to minimize costs, and biodiversity enters as a constraint. For example, governments have agreed to conserve 15% of the pre-European extent of Australia's regional forest types. This sets a constraint, creating an imperative to provide algorithms to minimize opportunity costs to the timber industry or maintenance costs of reserve management, while still satisfying the representation constraint. It is worth noting that there is no guarantee that such a reserve system is adequate—an issue that can only be assessed by using explicit population models (see Chapters 7–13, this volume). To provide some perspective, we briefly review how reserves have been selected in the past before proceeding with specific examples.

How Have We Selected Reserves in the Past?

Most areas now thought of as reserves for "conservation" were not chosen to meet specific biodiversity objectives. Many existing reserves are found in places that are unsuitable for other purposes (e.g., agriculture or urban development) and were chosen for cultural or scenic reasons or as favored holiday destinations (Pressey et al. 1993). Other areas have been selected to protect a few charismatic flagship or umbrella species (Simberloff 1998; Andelman and Fagan, in review) without any guarantee that they will adequately conserve regional biota. Consequently, most existing reserve systems have an excess of high mountains, rocky infertile soil, salt lakes, and inaccessible swamps. Flat, well-drained, fertile land is rarely conserved. Although a few charismatic species such as the Northern Spotted Owl may be well represented within reserves, in general, existing reserve systems provide a poor representation of biodiversity.

What does "classical" conservation biology theory tell us about reserve design? Unfortunately, not much. Early efforts at reserve design were derived from island biogeography theory (MacArthur and Wilson 1963) in which the emphasis was on size, shape, and number of reserves. Many conservation biology texts still begin their discussion of reserve design with diagrams of good and bad reserve design. For example, long thin reserves with a high edge-to-area ratio are considered poor in comparison to compact reserves; well-connected reserve networks are considered better than systems made up of isolated reserves; big reserves are considered better than small reserves; and so on. But these prescriptions offer little explicit guidance for decision makers who are faced with specific choices about how many, which sites, or which configuration to include in a reserve network. The SLOSS debate—should we have a single large reserve or several small reserves—was a product of the island-biogeographic foundation for reserve design theory and ended in the inconclusive answer, "it depends." In particular, close and well-connected patches may be a disadvantage if the arrangement increases correlations among reserves in environmental variation, by inviting disease, exotic species, and/or disturbance events to pass from one patch to another. The disadvantages of such processes may outweigh any advantage to be gained from elevated dispersal rates and increased recolonization probabilities, at least for some species.

For single, well-studied species, it is possible to use population viability analysis to find the best arrangement of habitat patches. It may even be possible to plan a reserve system that will give a particular species an acceptably low extinction probability. This was the aim of intense debate and planning for the Northern Spotted Owl. However, the single species approach avoids the issue that a reserve system designed to be optimal and adequate for a single species is not likely to satisfy the requirements of all species. Rather than blindly trying to maximize the number of species based on the very simple ecological abstractions of island biogeography, the principle of efficiency argues that effective reserve design will seek a reserve system that achieves defined objectives at minimum expense to other land uses, based on empirical information on species distributions.

Minimum Representation Problem

An important criterion for a reserve system is that it represents as much of the available biodiversity as possible (Pressey et al. 1993; Morton et al. 1995). By this criterion, a reserve system should contain at least one example of every vegetation type and/or one population of every species present within the region of interest. Because there are constraints on the amount of land that can be set aside for nature conservation, it would seem prudent to choose a set of sites that achieves comprehensive representation for the minimum cost (Pressey et al. 1993). This is called the minimum representation problem, and if the cost is a linear function of the number of sites in the system, it can be expressed as an integer linear programming problem (Cocks and Baird 1989; Possingham et al. 1993; Underhill 1994; Willis et al. 1996). When only a single occurrence of each species is required and there is a finite number of discrete sites from which to choose, this is termed a set-covering problem.

Consider the following example for the Columbia Plateau, a five-state region in the western United States that has been the focus of several major conservation planning initiatives (e.g., Davis et al., 2000). For a sample of 10 bird species of conservation concern, the objective is to conserve at least one population of every species. The presence or absence of each of the 10 species is known for eight sites, numbered 1–8 in Table 17.1. A "1" in the species by site matrix (Table 17.1) denotes a presence, whereas a "0" denotes an absence. The minimum set problem is to find the smallest number of sites that will represent every species once. In this case, the minimum set reserve system is sites 3 and 5—something that can be

TABLE 17.1. Species by site data for the Columbia Plateau ecoregion.[a,b]

	Site number								
Species	1	2	3	4	5	6	7	8	Species range
Loggerhead Shrike	1	1	1	1	1	1	0	1	7
Western Burrowing Owl	1	1	1	1	0	0	0	1	5
Grasshopper Sparrow	1	1	0	1	1	1	0	0	5
Ferruginous Hawk	1	1	1	0	0	0	1	1	5
Sage Thrasher	1	1	1	1	0	0	1	0	5
Western Sage Grouse	1	0	0	0	1	1	1	0	4
Sage Sparrow	1	0	1	1	0	0	0	0	3
American White Pelican	1	1	1	0	0	0	0	0	3
Bald Eagle	0	1	0	0	1	0	0	0	2
Forster's Tern	0	0	1	0	0	0	0	0	1
Site species richness	8	7	7	5	4	3	3	3	40

[a]Species range is the number of sites in which a species is found.
[b]Sources of data: California Natural Diversity Data System; Idaho Conservation Data Center, Idaho Fish and Game; Oregon Natural Heritage Program; Nevada Natural Heritage Program; Northwest Lepidopterist Society; The Nature Conservancy; Utah Natural Heritage Program; Washington Natural Heritage Program; and Washington Department of Fish and Wildlife.

determined by eye. No other set of two sites conserves all species, although there are plenty of three-site reserve systems that conserve all the species. In the language of mathematical programming, the optimal solution is sites 3 and 5, the optimal value is 2.

There are some interesting lessons from this Columbia Plateau example. First, note that the optimal solution, select sites 3 and 5, and its associated value, two sites, can be obtained by inspection. For bigger data sets, it is not hard to imagine that a solution by inspection will be difficult to find. Given this problem, early workers in the field devised algorithms for finding the minimum set. These algorithms involve selecting complementary sites sequentially, until the objective of representing all species is attained. The most obvious approach to representing every species is to add sites to the reserve system sequentially by selecting the site that adds the most unprotected species to the set that has already been selected. This algorithm is often called the "greedy" algorithm because it greedily attempts to maximize the rate of progress toward the objective at each step. Assuming that initially there are no sites in the reserve system, this algorithm would first select site 1, because it would protect eight species. Once site 1 has been selected, we see that each of sites 2, 3, and 5 will add a single additional species. Regardless of which of the three is chosen first, two more sites are needed to cover every species. The solution is suboptimal because it has three sites. This simple algorithm fails because the final reserve system is inefficient, when compared against the optimal reserve system.

Notice that one of the species in Table 17.1 is only represented in one site. The only place to find Forster's Tern is in site 3. This means site 3 is essential. So, an alternative approach is to first select any sites that are essential and then select sites that add the most unprotected species to the reserve system. This approach, sometimes termed the rarity approach, targets rare species first and then builds a complementary set from there. In this case, once site 3 is selected, the site that adds the most species to site 3 is site 5, which then completes a reserve system of just two sites.

It is interesting to try to devise different sorts of algorithms to solve these types of problems, and many have been tried (Margules et al. 1988; Rebelo and Siegfried 1992; Nicholls and Margules 1993; Pressey et al. 1997). All the algorithms that choose sites sequentially are, however, inefficient in that they are not guaranteed to find the optimal solution.

Consider a second example from the Columbia Plateau, with an expanded data set containing more species and sites (Table 17.2). Again, the minimum set problem is to find the smallest number of sites that will represent every species. The first point is that now it is no longer possible to see the optimal solution by inspection, as it was in Table 17.1. In this case, the minimum set reserve system is sites 3, 9, and 10. No other set of three sites conserves all species, although there are plenty of four-site reserve systems that conserve all the species.

Some insight into why sequential algorithms are inefficient can be gained by considering the number of possible reserve systems for any problem. For example, in Table 17.1 there are eight sites, each of which could be in or out of the

TABLE 17.2. Expanded Columbia Plateau data set.[a,b]

Species	1	2	3	4	5	6	7	8	9	10	11	12	range	rarity
Pallid Bat	1	0	1	1	0	1	1	1	0	1	1	1	9	0.11
Loggerhead Shrike	1	0	0	1	1	1	1	1	1	0	1	1	9	0.11
Western Burrowing Owl	1	1	1	1	1	0	0	1	1	0	0	1	8	0.13
Grasshopper Sparrow	1	1	0	1	1	1	1	0	1	0	0	0	7	0.14
Ferruginous Hawk	1	0	1	1	0	0	0	0	0	0	1	1	5	0.20
Sage Thrasher	0	0	1	1	0	0	1	1	0	0	1	0	5	0.20
Peregrine Falcon	0	1	0	1	1	1	0	0	0	1	0	0	5	0.20
Black-Throated Sparrow	1	1	0	0	1	0	1	0	1	0	0	0	5	0.20
Western Sage Grouse	1	0	0	0	0	0	1	0	0	1	1	0	4	0.25
Sage Sparrow	1	0	1	1	0	0	0	1	0	0	0	0	4	0.25
American White Pelican	1	1	1	0	0	1	0	0	0	0	0	0	4	0.25
Bald Eagle	1	0	1	0	0	1	0	0	0	0	0	0	3	0.33
Forster's Tern	0	1	0	0	1	0	0	0	1	0	0	0	3	0.33
Black Tern	0	1	0	0	0	1	0	0	0	1	0	0	3	0.33
Long-billed Curlew	0	1	0	0	0	0	0	0	0	1	0	0	2	0.50
Pygmy Rabbit	0	0	1	0	1	0	0	0	0	0	0	0	2	0.50
Northern Goshawk	0	0	0	0	0	0	0	1	1	0	0	0	2	0.50
Columbian Sharp-tailed Grouse	0	1	0	0	0	0	0	0	1	0	0	0	2	0.50
Site richness	10	9	8	8	7	7	6	6	6	6	5	4	82	
Rarity score	2	2.6	2	1.3	1.6	1.5	1	1.3	1.4	1.9	0.9	0.5		

[a]The presence or absence of each of 18 threatened species is known for 12 sites, numbered 1–12. A 1 in the species by site matrix below denotes a presence, and a 0 denotes an absence. Range is the total number of sites in which a species is found, rarity is 1/range, site richness is the number of species in each site, and rarity score is the sum of rarity values for the species in the site.
[b]Data sources are listed in Table 17.1.

reserve system. Hence there are $2^8 = 256$ possible systems, ranging from every site in the system to no sites. Now, imagine a more realistic problem, such as the entire Columbia Plateau ecoregion, with nearly 5,000 sites. The number of possible reserve systems is 2^{5000}, a number so big that the problem is intractable, even for the fastest computers.

Formulating the Minimum Representation Problem

The best reserve design for a small problem, such as the example in Table 17.1, can be obtained by inspection. Before considering different methods for solving the problem in larger data sets, we need to formulate it within a decision-theory framework—in this case, the formalism of mathematical programming.

Let the total number of sites be m and the number of different species (or other attributes such as vegetation types) to be represented be n. The information about whether a species is found in a site is contained in a site-by-species ($m \times n$) matrix A whose elements a_{ij} are

$$a_{ij} = \begin{cases} 1 \text{ if species } j \text{ occurs in site } i \\ 0 \text{ otherwise} \end{cases}$$
$$\text{for } i = 1, \ldots m \text{ and } j = 1, \ldots, n$$

Next, define a control variable that determines whether a site is included in the reserve, as the vector X with dimension m and elements x_i, given by

$$x_i = \begin{cases} 1 \text{ if site } i \text{ is included in the reserve} \\ 0 \text{ otherwise} \end{cases}$$
$$\text{for } i = 1, \ldots, m$$

With these definitions, the minimum representation problem is

minimize $\sum_{i=1}^{m} x_i$ (minimize the number of sites in the reserve system)

subject to $\sum_{i=1}^{m} a_{ij} x_i \geq 1$, for $j = 1, \ldots, n$ (subject to each species being represented at least once)

where $a_{ij}, x_i \in \{0,1\}$

This is the integer linear programming formulation of the set-covering problem. It is NP-complete so that the difficulty of guaranteeing an optimum solution increases exponentially with the number of constraints n (Garey and Johnson 1979).

Solving the Minimum Representation Problem: Traditional Methods

In the context of nature reserve design, the minimum-representation problem has been tackled by several authors, in a number of different ways. Kirkpatrick (1983) was first to define the problem and used a heuristic method to find a solution. Such methods rank each potential reserve site according to a set of criteria and then reserve the highest-ranking site. The remaining sites are ranked again and the process continued until all species are represented. Originally, the criterion used in ranking was the number of species in a site not yet represented in the reserve system. Other criteria (e.g., the location of rare species) or combinations of criteria have since been used (e.g., Rebelo and Siegfried 1992; Nicholls and Margules 1993).

As shown above, this approach guarantees a quick answer, but it does not ensure an optimal one (Possingham et al. 1993; Underhill 1994). Another approach to the problem has been to express it in the form of an integer linear program (ILP) and then to use standard mathematical programming techniques such as the branch-and-bound method to find the optimal solution (Cocks and Baird 1989; Church et al. 1996). Pressey and co-workers (1997) compared a variety of heuristic methods (variations of rarity and greedy algorithms) with the optimal solution found by integer linear programming. The ILP package used a branch-and-bound method and found optimal solutions for a reduced version of the original data set of Pressey and associates. The comparison showed that, although solutions obtained with heuristics can be as much as 20% worse than the optimal solution, if a variety of methods are tried, one can probably guarantee being within 5% of the optimum for problems of moderate size. However, ILP fails when the number of reserves is moderately large (more than about 20 or 30), because the problem becomes too big for a solution to be found in reasonable time. Thus, heuristics are fast and easy to understand but may deliver inefficient solutions, whereas integer linear programming methods are likely to fail on large problems. ILPs also have the disadvantage of producing a single optimal solution, whereas in a conservation context, flexibility may be advantageous. It might be useful to identify not one but a range of possible solutions that could meet a particular reservation goal. In the following section, we describe an alternative approach to some of these problems.

Simulated Annealing Solutions: A Comparison

Simulated annealing is a minimization method based on the process of annealing metals and glass (Metropolis et al. 1953; Kirkpatrick et al. 1983). It begins by generating a completely random reserve system. Next, it iteratively explores trial solutions by making sequential random changes to this system. Either a randomly selected site, not yet included in the reserve system, is selected, or a site already in the system is deleted. At each step, the new solution is compared with the previous solution, and the best one is accepted. The advantage of this approach is that potentially it can avoid getting trapped in local optima. It allows the reserve system to move temporarily through suboptimal solution space and thus increases the number of routes by which the global minimum might be reached. Initially, any change to the system is accepted, whether it increases or decreases the value of the system. As time progresses, the algorithm is more and more choosy about which changes it accepts, rejecting those changes that would increase the value of the system by too large an amount. By the end of a simulated annealing run, only changes that improve (i.e., decrease) the value of the system are accepted. At this point, the system soon reaches a local minimum. To reiterate, the central idea behind simulated annealing is that, by allowing bad changes as well as good, local minima are avoided.

The simulated annealing method works as follows:

1. Set input parameters and the maximum number of iterations.
2. Generate an initial reserve system consisting of sites selected at random, and compute objective function.
3. Randomly select a site to add to or delete from the system.
4. Evaluate the resulting change in the objective function: if $e^{\left(\frac{-change}{acceptance\ level}\right)} <$ random number, then accept the change, otherwise reject it.
5. Decrease the acceptance level, and repeat steps 3–5 for the given number of iterations.

TABLE 17.3. New South Wales comparison.[a,b]

Algorithm	Sites in solution	Run time
Greedy	61	1 min
Rarity based	70	2–3 min
Simulated annealing	54	40 min

[a]Adapted from Ball et al. (in press).
[b]Data set is from the western division of New South Wales, Australia. The optimum value for this problem is known to be 54 sites.

The acceptance level determines what size change will be accepted. Negative changes (decreasing the value of the system) will always be accepted. When the acceptance level approaches 0, then the only acceptable changes are those that reduce the value of the system. The use of the exponential term means that the system spends proportionally little time accepting very bad changes and much more time resolving small differences.

The relative ability of this method is demonstrated in an example using data from the western region of New South Wales, Australia (Ball et al., in press). In this region, there are 1,885 sites and 248 species of conservation concern. The results are displayed in Table 17.3. For the same data set, simple greedy and rarity-based selection algorithms achieved results as low as 57 sites (Pressey et al. 1997). This example demonstrates that simulated annealing will generally do better than the simpler heuristics, although it does so at the cost of a slower running time. An additional advantage of simulated annealing is that it will tend to produce a number of solutions rather than a single solution. When using the simple selection algorithms, it is necessary to use a number of different algorithms to generate alternative solutions.

Spatial Reserve Design

One limitation of the minimum set approach is that it does not account explicitly for the spatial relationships among the sites selected for the reserve system. Without some modification or additional constraints, the final reserve system will almost always be highly fragmented and clearly inappropriate. This is a major problem as there are both ecological and economic reasons why reserves should be spatially contagious with low edge-to-area ratios. Long thin reserves with a high edge-to-area ratio will be vulnerable to weed and pest invasions, as well as to "edge effects," caused by biotic interactions such as predation (Fagan et al., 1999) or abiotic factors such as humidity or wind. This can mean that for edge-sensitive species that can only persist in a small core area, the effective area of the reserve may be substantially reduced. The size of the core decreases rapidly as the edge-to-area ratio increases. From an economic perspective, the cost of management often scales more closely with the boundary length of a reserve than with reserve area. From the perspective of logistics, boundaries need to be maintained, and longer boundaries usually mean more neighbors.

Generally, we would like to cluster sites in a reserve system; however, there can be compelling reasons to avoid clustering. Where catastrophes can impact large areas and cause local extinctions, it may be less risky to conserve each species in at least two or three separate places, rather than clustering those sites. Issues such as boundary length, spatial constraints, and costs make the problem significantly more complex. Mathematically, the problem becomes nonlinear, because the cost of adding a site to the reserve system depends on which other sites are already protected and on the spatial relationships between candidate sites and those already reserved. One formulation of a spatial reserve design problem is shown in the following example for section 342I of the Columbia Plateau ecoregion (Bailey 1994). There are 821 sites (subwatersheds) of different sizes. The elements of biodiversity to be conserved in this problem are 113 species, rare plant communities, and common coarse-scale vegetation types. The rare species and plant community data are in the form of presence-absence data, whereas occurrences of vegetation types are measured in the number of hectares of that type found in each site. Thus, in this problem, there are two types of target representation levels, each for different "types" of biodiversity, expressed as ones and zeros for presence-absence, or as area.

When using greedy and rarity-based algorithms, clustering of reserve sites is often achieved by including an adjacency constraint (Nicholls and Margules 1993). Another approach, used here, is to try to minimize the boundary length of the reserve system. For a given area, a smaller boundary length gives a more compact area. To minimize both boundary length and area in this problem, we introduce a boundary length modifier (BLM). The objective to be minimized is now area + boundary length × BLM. By varying the BLM, the relative importance of compactness and size can be balanced. If the BLM is set to zero, the algorithm will ignore boundary length.

A simple measure of the degree of clustering among reserves in the network is the boundary length of the reserve system divided by the area. This is a measure of length per unit area and is intuitive. A more advanced measure is the ratio of the boundary length of the reserve system to the circumference of a circle with the same area as the reserve. A circle is the most compact shape possible, so this is the ratio of the boundary length to the theoretical minimum. As such, it is a dimensionless measure. The formula for this measure is

$$\text{ratio} = \frac{\text{boundary length}}{2\sqrt{\pi} \times \text{area}}$$

As shown in Figure 17.1, as the boundary length modifier is increased, both the boundary length and boundary length/area measures decrease. This occurs at the expense of area. Table 17.4 indicates that when the largest boundary length modifier is used, two-thirds of the area is reserved, and the boundary length is minimized. The best balance between total area and clustering seems to be achieved, with a BLM between 0.5 and 1. Here, the area is increasing, but the boundary length is decreasing at a greater rate.

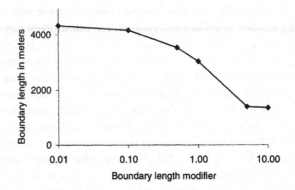

FIGURE 17.1. Boundary length of reserve system (in kilometers) as the boundary length modifier is changed. Note that it is always decreasing. The x-axis has a logarithmic scale, and the zero point is included on the left side.

Figures 17.2 and 17.3 illustrate the results of applying the BLM clustering method. In Figure 17.2, no boundary length multiplier was used, and the resulting reserve system is very fragmented. In Figure 17.3, it is immediately obvious that the reserved sites are well clustered, and there are areas where core sites from Figure 17.2 have been transformed into larger contiguous blocks. The number of outlying sites in the reserve network has been reduced. Those that remain represent the only options for capturing some of the rare species.

The simulated annealing method can be extended beyond the examples here, to incorporate more elaborate spatial requirements for species. For example, if several sites could be affected by the same catastrophe (e.g., a fire or hurricane), we might want to minimize the chances that all reserves for a given species are simultaneously impacted. This can be done by introducing a requirement that species occur in two or three sites, each separated by some minimum distance, while still attempting to minimize total area and boundary length.

TABLE 17.4. A reserve network for one section of the Columbia Plateau, designed with simulated annealing, using a variable boundary length modifier.[a,b]

BLM	Sites	Area	Boundary length	BL/A	BL/Ideal
0	124	1,229,570.51	4,337,428	3.53	11.03
0.1	127	1,242,524.27	4,167,834	3.35	10.55
0.5	160	1,427,321.72	3,519,799	2.47	8.31
1	196	1,771,721.86	3,027,734	1.71	6.42
5	598	5,439,757.03	1,386,868	0.25	1.68
10	643	5,712,270.51	1,344,774	0.24	1.59

[a]Area is in hectares and boundary length in meters. BL/Ideal is the ratio of the boundary of the reserve to the boundary of a circle of the same size.
[b]Data sources are listed in Table 17.1.

FIGURE 17.2. Reserve system selected for section 342I of the Columbia Plateau ecoregion, using simulated annealing, with boundary length modifier set to 0. There are 821 sites (i.e., potential reserves), and the conservation goal is to conserve at least one occurrence of each of 113 species, rare plant communities, and common coarse-scale vegetation types. For species and rare plant communities, occurrences are measured by presence-absence data; for coarse-scale vegetation, occurrences are measured as area in hectares. Data sources are listed in Table 17.1.

Reserve Network Design in an Uncertain World

Despite these kinds of efforts to incorporate spatial considerations into reserve network design, a major limitation of existing approaches is that they remain essentially static (e.g., they are based on a snapshot in time of species incidence or vegetation distribution). They do not deal explicitly with temporal dynamics or uncertainty. Robust and feasible approaches for incorporating temporal dynamics and various types of uncertainty represent one of the greatest challenges to theories and methods for reserve network design.

Data Uncertainty

Although mathematical algorithms for reserve network siting continue to improve, data on the distribution and abundance of biodiversity to feed into these algorithms remain poor. Systematic reserve planning efforts may use explicit

FIGURE 17.3. Reserve system selected for section 342I of the Columbia Plateau, using simulated annealing, with boundary length modifier set to 1. All other details as in legend for Figure 17.2.

mathematical expressions, but they are still opportunistic in that they tend to use whatever data are available. Systematically gathered data on the distribution and abundance of biodiversity are scarce. Available data on species abundance and distribution typically are biased in a number of ways: toward charismatic species, such as mammals, birds and butterflies; toward easily accessible sites; or toward favorite study sites such as field stations or areas close to major universities or museums. For example, in the Columbia Plateau, the density of surveyed sites is inversely correlated with distance from major interstate highways (Andelman and Hansen, unpublished data). In addition, both the amount of data available and the extent of areas surveyed vary widely across different parts of the globe. Sometimes, in the absence of consistent empirical data for a particular region, predicted species or vegetation distributions derived from habitat suitability models or interpretation of satellite imagery are used (e.g., Cocks and Baird 1989). Thus, it becomes critical to understand how sensitive reserve siting algorithms are to variations in data type, quantity, and quality.

Depending on the amount and types of data available, more complex algorithms may not always provide better solutions. For example, simulated annealing approaches can be very sensitive to the choice of input parameters (e.g., Golden and Skiscim 1986; Murray and Church 1996), but these limitations have not been systematically examined in the context of reserve network siting problems. By contrast, the simpler greedy or rarity-based algorithms seem to be relatively robust to the amount of survey effort and to spatial biases in sampling when applied to binary (i.e., presence-absence) data (Andelman and Meir, 2000). Moreover, for any particular set of species or ecosystems, determining the optimal balance of reserve clustering and separation to parameterize a simulated annealing algorithm requires detailed empirical data on species life history and/or the spatial distribution of catastrophes, which often do not exist.

Species/Biological Dynamics

The basic data used with reserve siting algorithms usually consist of presence-absence data (or more often, simply presence data) for particular species and/or vegetation types. The former (and to a lesser extent, the latter) can change. Species appear and disappear from sites independently of whether sites are in, or not in, reserve systems (e.g., Margules et al. 1994), and the frequency with which this happens depends on numerous factors. This means that static approaches may yield reserve networks that may not protect as much biodiversity as one might think. The future of these sorts of problems lies in integrating spatial population modeling with reserve network siting approaches, a significant challenge, both in terms of data acquisition and model complexity (see Andelman et al., 2000, for a recent review).

Landscape Change, Economic and Social Uncertainty

In most cases, it is not possible to both design and implement reserve networks instantly. Even if an optimal set of sites has been identified, it will take decades of negotiation and land purchases to translate that design into a set of reserves on the ground. In the meantime, degrading processes will continue to operate at various spatial scales. If sites are lost before implementation is complete, then efforts to find the optimal reserve system will probably fail. This means, as time passes, not only will the species composition of a site change, the carrying capacity of the site for particular species (i.e., site quality) may also be modified, and some sites may be destroyed entirely. Key sites may be lost before they become available for conservation action, and if these key sites are integral to the overall reserve network goals, then the final system may be poor when measured against the initial goals. Ideally, then, we seek a dynamic theory of reserve design. Such a theory must explicitly consider many stochastic events: the chance of site destruction and degradation, the chance a site becomes available for sale, the dynamics of the financial system providing capital to buy and manage reserves, and the strength of public interest in conservation relative to other land uses. Combining all these factors in a single model would be almost impossible. Nevertheless, incorporating some of these factors in a model is essential. For example, given temporal variation in site availability, for small problem sets, Possingham and colleagues (1993) used stochastic dynamic programming and found it was possible to determine which sites should be acquired, if and when they become available.

Here, we have merely attempted to introduce some of the complexities that will enter into practical reserve design problems, along with some of the mathematical tools for dealing with those complexities. To integrate all these issues in one problem, let alone solve that problem, is difficult. In practice, the best approach will be to build and solve specific problems for specific organizations and localities. For example, when we have some notion of the particular objectives and financial dynamics of an agency or organization, we might attempt to build a dynamic decision support tool. The most common type of problem will be to determine, when an organisation has some existing capital and a variety of sites are available for purchase, which, if any, should be acquired at any given point in time. The problem-solving tool should be tailored to the problem at hand.

Discussion

In this chapter, we have considered the problem of reserve design from the simplest possible formulation—the minimum set, presence-absence problem—to more complex formulations, including explicit spatial constraints. There are many variations and complexities that we have not discussed. For example, in all cases we set biodiversity as a constraint and tried to minimize economic and other costs. An alternative is to set an economic constraint. In this case, if there is fixed capital available for reserve acquisition, or reserves can cover no more than a fixed proportion of the region, the problem is to maximize long-term biodiversity benefits. This problem has rarely been considered.

Regardless of how the problem is formulated or the type of algorithm used, there are additional considerations. In practice, solving the reserve network design problem requires more than just finding the very best solution. Flexibility to explore alternative solutions is one important criterion, because optimality may not be achievable, or its importance may diminish in practical problems (Cocklin 1989a; Andelman et al., 2000). Generally, conservation planners need to be able to evaluate a range of reasonably good solutions (i.e., from an ecological perspective), in the context of other considerations, such as economics or political expediency. Speed of execution also is important, because it facilitates a much greater level of interaction between the planner and the potential solution space. Solutions can be examined and additional constraints added (e.g., the forced inclusion or exclusion of some sites) before running the algorithm again. This can give planners and decision makers a range of good solutions to use in a broader decision-making or negotiation context (Cocklin 1989b; Pressey et al. 1996).

With respect to these criteria, the simple methods are useful. They are quick, which makes it relatively easy for the user to change either the data set (as new information becomes available, or to account for uncertainty) or the selection criteria (to reflect different levels of risk tolerance, or different political or economic considerations), and then quickly reapply the method. For better answers, simulated annealing is an appropriate method. Its use of an objective function with penalties instead of constraints offers both flexibility and efficiency, although we need to know more about how robust it is to variation in types of data or uncertainty. Integer Linear Programming methods, or optimization methods that work in a dynamic framework, such as stochastic dynamic programming, may be harder to justify for potential users, because their computational complexity gives them a "black box" nature.

There are a wide variety of extensions to these problems that are only now being studied. Complex spatial requirements and reserve design under uncertainty are only two such avenues. Other areas of interest include problems in which species or other objectives have different weights (values) and there is a fixed amount of land to be set aside and problems that attempt to simultaneously optimize multiple services (e.g., when a species has one value, say, for ecotourism, when it is in a reserve, and a different value, say, for trophy hunting, when it is outside a reserve). Solutions to these problems—even approximate solutions— will improve the design of effective reserve networks and, ultimately, will contribute to better protection for biodiversity.

Acknowledgments. This work was conducted as part of the Biological Diversity Working Group supported by the National Center for Ecological Analysis and Synthesis, a center funded by NSF (Grant DEB-94-21535), the University of California—Santa Barbara, the California Resources Agency, and the California Environmental Protection Agency.

Literature Cited

Andelman SJ, Meir E (2000) Breadth is better than depth: biodiversity data requirements for adequate reserve networks. Conservation Biology (in press)

Andelman SJ, Fagan W, Davis F, Pressey RL (2000) Tools for conservation planning in an uncertain world. BioScience (in press)

Bailey RG (1994) Ecoregions of the US. USDA Forest Service, Washington, DC

Ball IR, Smith A, Day JR, Pressey RL, Possingham H (in press) Comparison of mathematical algorithms for the design of a reserve system for nature conservation: an application of genetic algorithms and simulated annealing. Journal of Environmental Management

Church RL, Stoms DM, Davis FW (1996) Reserve selection as a maximal covering location problem. Biological Conservation 76:105–112

Cocklin C (1989a) Mathematical programming and resources planning I: the limitations of traditional optimization. Journal of Environmental Management 28:127–141

Cocklin C (1989b) Mathematical programming and resources planning II: new developments in methodology. Journal of Environmental Management 28:143–156

Cocks KD, Baird IA (1989) Using mathematical programming to address the multiple reserve selection problem: an example from the Eyre Peninsula South Australia. Biological Conservation 49:113–130

Davis FW, Stoms DM, Andelman SJ (2000) Systematic reserve selection in the USA: an example from the Columbia Plateau ecoregion. Parks (in press)

Fagan WF, Cantrell RS, Cosner C (1999) How habitat edges change species interactions. American Naturalist 153:165–182

Garey MR, Johnson DS (1979) Computers and intractability: a guide to the theory of NP-completeness. WH Freeman and Company, San Francisco, CA

Golden B, Skiscim C (1986) Using simulated-annealing to solve routing and location problems. Naval Research Logistics Quarterly 33:261–279

Kirkpatrick JB (1983) An iterative method for establishing priorities for selection of nature reserves: an example from Tasmania. Biological Conservation 25:127–134

Kirkpatrick S, Gelatt CD Jr, Vecchi MP (1983) Optimization by simulated annealing. Science 220:671–680

MacArthur RH, Wilson EO (1963) An equilibrium theory of insular zoogeography. Evolution 17:373–387

Margules CR, Nicholls AO, Pressey RL (1988) Selecting networks to maximise biological diversity. Biological Conservation 43:63–76

Margules CR, Nicholls AO, Usher MB (1994) Apparent species turnover probability of extinction and the selection of nature reserves: a case study of the Ingleborough limestone pavements. Conservation Biology 8:398–409

McNeely JA, Miller KR, Reid WV, Mittermeier RA, Werner TB (1990) Conserving the world's biodiversity. IUCN, Gland, Switzerland

Metropolis NA, Rosenbluth M, Rosenbluth A, Teller E (1953) Equation of state calculations by fast computing machines. Journal of Chemical Physics 21:1087–1092

Morton SR, Stafford Smith DM, Friedel MH, Griffen GF, Pickup G (1995) The stewardship of arid Australia: ecology and landscape management. Journal of Environmental Management 43:195–217

Murray AT, Church RL (1996) Applying simulated annealing to location-planning models. Journal of Heuristics 2:31–53

Nicholls AO, Margules CR (1993) An upgraded reserve selection algorithm. Biological Conservation 64:165–169

Possingham HP, Day JR, Goldfinch M, Salzborn F (1993) The mathematics of designing a network of protected areas for conservation. In: Sutton DJ, Pearce CEM, Cousins EA (eds) Decision sciences: tools for today. Proceedings of 12th National ASOR Conference, ASOR, Adelaide, Australia, pp 536–545

Pressey RL, Humphries CJ, Margules CR, Vane-Wright RI, Williams PH (1993) Beyond opportunism: key principles for systematic reserve selection. TREE 8:124–128

Pressey RL, Possingham HP, Margules CR (1996) Optimality in reserve selection algorithms: when does it matter and how much? Biological Conservation 76:259–267

Pressey RL, Possingham PH, Day RJ (1997) Effectiveness of alternative heuristic algorithms for identifying indicative minimum requirements for conservation reserves. Biological Conservation 80:207–219

Rebelo AG, Siegfried WR (1992) Where should nature reserves be located in the cape floristic region South Africa? Models for the spatial configuration of a reserve network aimed at maximizing the protection of floral diversity. Conservation Biology 6:243–252

Simberloff D (1998) Flagships umbrellas and keystones: is single species management passé in the landscape era? Biological Conservation 83:247–257

Soulé ME (1991) Conservation: tactics for a constant crisis. Science 253:744–750

Underhill LG (1994) Optimal and suboptimal reserve selection algorithms. Biological Conservation 70:85–87

Willis CK, Lombard AT, Cowling RM, Heydenrych BJ, Burgers CJ (1996) Reserve systems for limestone endemic flora of the cape lowland fynbos: iterative versus linear programming. Biological Conservation 77:53–62

Index